Tributes
Volume 4

The Way Through Science and Philosophy: Essays in Honour of Stig Andur Pedersen

Tributes Series editor: Dov Gabbay   dov.gabbay@kcl.ac.uk

The Way Through Science and Philosophy:

Essays in Honour of Stig Andur Pedersen

edited by
H. B. Andersen,
F. V. Christansen,
K. F. Jørgensen. and
V. F. Hendricks

ISBN 1-904987-33-8
Published by College Publications
Scientific Director: Dov Gabbay, Vincent F. Hendricks, and John Symons
Managing Director: Jane Spurr
Department of Computer Science
King's College London
Strand, London WC2R 2LS, UK

http://www.collegepublications.co.uk

Cover design by Richard Fraser, www.avalonarts.co.uk
Printed by Lightning Source, Milton Keynes, UK

Dedicated to Stig Andur Pedersen
on the occasion of his 60th birthday

Stig Andur Pedersen, born 18th of January 1943
Professor of Philosophy of Science

Contents

Part I Philosophy, history and applications of mathematics

Preface

This collection of essays is a Festschrift celebrating the 60th birthday of Professor Stig Andur Pedersen. On that occasion we had great pleasure in launching a web site dedicated to Andur,

http://www.andur.ruc.dk

collecting greetings sent from all over the world. At the same time, we initiated the preparation of this Festschrift which includes articles by some of Andur's many colleagues, students and other collaborators over the years.

Roskilde,
April 2006

Henning Boje Andersen
Frederik Voetmann Christiansen
Klaus Frovin Jørgensen
Vincent F. Hendricks

Introduction

Henning Boje Andersen[1], Frederik Voetmann Christiansen[2], Klaus Frovin Jørgensen[3] and Vincent F. Hendricks[3]

[1] Systems Analysis Department, Risø National Laboratory
[2] Department for Medicinal Chemistry, Danish University for the Pharmaceutical Sciences
[3] Section for Philosophy and Science Studies, Roskilde University

This collection of essays spans over a broad scope of topics, ranging from mathematical modelling over realism in philosophy of science to an analysis of apology after medical mistakes. In fact, the range of topics is so broad that the editors realise that few readers will want to try to read all contributions in depth. However, the unusually broad range is nevertheless a fair, but even so still inadequate, reflection of the work of Stig Andur Pedersen, a philosopher and scientist whose range of interests, abilities and production is itself unusually broad. Indeed, while the selection of topics addressed in this collection run a wide gamut of themes, the editors are aware that some areas are under-represented – particularly areas such as logic, ethics and cognitive science. Broadness of knowledge will, as a rule, be gained at the expense of mastery of the subject; but Andur is an exception to this rule. It will be difficult to find another professor who can write and lecture on quantum logic, applications of category theory, simulation of flexible tubes in physiology, utilitarianism and deontology in patient-doctor relations, and the first-person perspective perspective in theories of mind.

In addition to being author, co-author and editor of several influential books ranging from epistemology, philosophy of science, and philosophy of medicine, Andur has had a seminal influence on several research groups and research educational milieus in Denmark and abroad and he has influenced many students and young researchers in their choice and approach to research topics. This Festschrift is therefore a welcome opportunity to acknowledge Andur's prominent role in shaping the Danish philosophical environment in areas such as philosophy and history of science, philosophy of medicine, philosophy of engineering and technology, philosophy of mathematics etc.

The essays that are presented here have been recruited from authors – sometimes joined by co-authors – who have benefitted in various ways from collaboration with Andur, as witnessed by the individual entries in the biographical section at the end of the book. Numerous researchers, scientists and students have benefited from Andur's tutoring, his writings, his organizational talents, and his unfailing generosity in helping colleagues and students.

In their essay below, Frederik Christiansen and Camilla Rump quote Donald Schön who rhetorically asks: "Shall the practitioner stay on the high, hard ground where he can practice rigorously [. . .] or shall he descend to the swamp where he can engage the most important and challenging problems if he is willing to forsake technical rigor?" (Schön, 1983, p. 42). This so-called dilemma 'rigor or relevance' is not a dilemma in Andur's world: He is equally at home in the swampy lowlands and on the high hard grounds – as is witnessed also by Kevin Kelly's humorous essay. Moreover, he has a remarkable gift for seeing which tools from the highlands we might apply in the cultivation of the swamp, when adapted in the right way. A good example of this is Andur's role in the development of the Anesthesia simulators SIMA and SOPHUS where mathematical models are incorporated into artifacts used in the training of anesthesiologists (described in the article by Jesper Larsen *et al.*). Another example is his role in establishing philosophy of medicine and engineering programmes as a part of medical and engineering curricula (touched upon in Søren Holm's article). More recently, his role as director of the Ph.D. research school in Philosophy, History of Science, History of Ideas and Science Studies may be mentioned. These are, however, just a few of initiatives where Andur has had a formative and lasting influence.

The Festschrift is divided into five parts each of which represents a broad area of interest to which Andur has devoted his abilities and energy:

1. Philosophy, history and applications of Mathematics
2. Philosophy and history of science
3. Philosophy of medicine and engineering
4. Metaphysics and epistemology
5. Risk, uncertainty and safety

In the following we will briefly present the essays in the book.

1 Philosophy, history and applications of mathematics

The first part of the Festschrift concerns mathematics, its use, history and philosophy. Symbolic algebra is, together with geometry and analysis, one of the three main branches of this scientific discipline. Jens Høyrup's contribution "Premodern 'algebra': A concise survey of that which was shaped into the technique and discipline we know" is, as indicated by the title, a compact survey of those elements in the history of mathematics that eventually formed and developed into the algebra we know today. Contemporary algebra is normally understood as the general study of structure, relation and quantity. More specifically, it consists of abstract algebra (the study of groups, rings, fields, etc.), linear and universal algebra amongst others. According to Høyrup, algebra as is understood today is an aggregate of a variety of disciplines and practices and cannot be understood as a single discipline. Therefore, algebra did not exist before these different aggregates were merged. Thus, algebra has no history – only the individual constituents of algebra can be said to have a history. Høyrup's article traces the (pre-)history of these constituents, beginning with the great Near Eastern Bronze Age civilizations and ending with the incipient transformation of medieval equation algebra brought about by Cardano and Bombelli.

In "The development of nonlinear programming in post war USA: Origin, motivation, and expansion" Tinne Hoff Kjeldsen provides an analysis of an important aspect in modern history of mathematics. Nonlinear programming is the mathematical theory of finite-dimensional inequality constrained optimization, and it is considered to belong to the applied fields of mathematical programming and operations research. It emerged from joint work of Albert W. Tucker and Harold W. Kuhn. These two scientist were actually never involved in the research for which the discipline is famous – its efficiency with respect to solvability of practical decision problems in "the real world". How did these theoretical mathematicians get involved with this kind of mathematics? Where did nonlinear programming come from and how did it develop into a mathematical theory? Kjeldsen traces the answers to these questions, and provides explanations for the origin and development of nonlinear programming in the context of science support in post war USA.

The concept of function is the topic of Jesper Lützen's "Dirichlet and the flexibility of the new concept of function". It is not unusual to see the modern concept of a function to be attributed to Euler. Lützen has strong reservations about this attribution, however. He observes that one thing is what you say you do, while another is what you *actually* do. The meaning of a concept is not found in its definition – it is found in its use. Following this observation, Lützen points out that in the present context it is true that Euler in fact came close to a formulation of a modern definition of function. On the other hand, Euler never really *used* the properties of the modern function-concept. This prompts the question: 'who introduced the modern use of a function?'. The answer, according to Lützen, is Dirichlet, who used the flexibility of the new concept of function when proving a central theorem that has the convergence of Fourier series as consequence. The central idea to the new concept is that functions need not be globally defined but can be only partially defined. Thus, the (partial) functions can be extended in arbitrary ways. This property, which seems trivial to a contemporary mathematician, was nevertheless not found in Euler's work, but in Dirichlet's. Lützen illustrates Dirichlet's use of the new concept of function by a very elegant and modern proof by Dirichlet. In this way, Lützen's analysis depends crucially on his claim that "a mathematical concept is determined at least as much by its use as by its definition". This is an epistemological claim – pertaining to the epistemology of mathematics. The same claim can be said to be treated in the essay by Klaus Frovin Jørgensen "Kant and the natural numbers". Here the notion of schema introduced by Kant is discussed. Generally Kant argues that any concept is useless, unless it is founded by a schema. The schema of a concept is our cognitive capacity to connect a purely intellectual concept with sensible objects falling under the concept. The schema mediates between the abstract and the concrete. The schema is a rule-governed operation which determines the use of the concept, as the schema determines the extension of the concept. Therefore, different ways of using functions correspond to different *concepts* of functions. Jørgensen provides an interpretation of the so-called transcendental schemata in Kant's theory by showing how *number* is a schema for Kant's quantitative categories – making quantitative judgements possible. Kant is (in)famous for his claim that 'numbers cannot be objects'. Jørgensen shows that his claim is based on his general concept of number being a schema, and schemata are not objects – they make subsumption of objects possible. If one, however, refines Kant's notion of number in such a way that there are i) concepts of particular finite numbers as well as ii) a general notion of numbers, then there is a way out of the problem. Then numbers can be understood as being represented by second-order objects. This can in turn be understood as a proposed solution to the ontological status of numbers which has been an open problem within the philosophy of mathematics.

From the historical and philosophical perspectives on mathematics, we move to the applications of mathematics in Jesper Larsen, Viggo Andreasen, Heine Larsen, Mette Olufsen and Johnny Ottesen's "Cardiovascular modelling at IMFUFA". The essay gives an overview and provides examples of models used in cardiovascular modelling at the Department for Mathematics and Physics (IMFUFA) at Roskilde University. The modelling was initiated by the authors' involvement in the creation of the anaesthesia simulators SOPHUS and SIMA to be used in training of anaesthetists. Stig Andur Pedersen played a key role in the initiation of this interdisciplinary research, and the simulators that eventually grew out of the development efforts are now being used in education. Two models, both developed as part of student projects, are described. The first is an energy-bond graph model for the creation of numerical models of the aorta, and the authors briefly describe the merits of energy-bond graphs for modelling of dynamical physical systems. The other model for simulation of flow in the aorta is based on an analogy between flow in aorta and flow in . . . sewers! It utilises a commercial software package originally created for an urban sewer network, and the basis for the analogy is discussed. In addition, the authors describe the current research in cardiovascular modelling at IMFUFA.

2 Philosophy and history of science

In "Einstein on the history and nature of science" Helge Kragh describes Einstein's attitude to the history and development of science. For Einstein, history of science served primarily practical and teleological purposes. It was a means of thinking about and perspectivizing the current situation in physics. He dealt with historical topics in many publications, most consistently and comprehensively in the semihistorical *The Evolution of Physics* (1938), coauthored by Leopold Infeld. Kragh's essay demonstrates that one may obtain further insight in Einstein's views on the nature of science by considering his historically oriented writings. Kragh describes how the young Einstein was influenced by Ernst Mach's historical writings, but also, and more lastingly, by the ideas of Pierre Duhem, particularly the idea of underdetermination. Further, Kragh argues that there are surprisingly close affinities between the views of Einstein and Popper. The theory of relativity is often considered to be a revolution in science and Einstein an arch-revolutionary. However, according to Kragh, it would be rash to infer from Einstein's frequent use of terms such as "revolution" and "crisis" that he foreshadowed elements of Kuhn's conception of the development of science. In fact, Kragh argues, Einstein's views can in no way be taken as support of Kuhn's ideas of science as a non-progressive series of shifts between incommensurable paradigms. However, Kragh does not rule out entirely the possibility that Kuhn's mature philosophy of science might be more compatible with Einstein's views (although he doubts it).

The reader will be able to pursue this question further by reading Hanne Andersen's essay "How to recognize intruders in your niche". This essay deals with the interpretation of Kuhn's notion of incommensurability and the question of how and when incommensurable theories can be said to compete. The author takes as her starting point Kuhn's rejection of the "referential stability" approach (Sheffler, Putnam), and discusses constructively what should be understood by incommensurability. Kuhn's ideas about incommensurability kept on developing throughout his career, as has been discussed in Hoyningen-Huene's phenomenologically oriented reconstruction of Kuhn's philosophy of science. In the foreword to Hoyningen-Huene's book, Kuhn wrote: "No one, myself included, speaks with as much authority on the nature and development of my ideas" (Hoyningen-Huene, 1993, p. xi). However, with respect to the description of incommensurability Andersen finds Hoyningen-Huene's "double-world" reconstruction of Kuhn's position inconsistent. The double-world interpretation claims that incommensurable theories "target roughly the same object domain, as far as the world-in-itself is concerned, though this 'object domain' isn't graspable in any theory-neutral way". But since the world-in-itself is not epistemically accessible, no such similarity relation can be established. Andersen suggests a different reconstruction of Kuhn's position: That incommensurable theories target roughly the same object domain, as far as *phenomenal* worlds are concerned. But this is only possible in so far as there is an overlap between phenomenal worlds, and Andersen proceeds to unfold what should be understood by such an overlap: When are theories incommensurable, when are they occupying different niches?

From the anti-realist Kuhnian setting we move to discussing varieties of scientific realism in Jan Faye's "Science and reality". Scientific realism is the view that the aim of science is to produce true or approximately true theories about nature. The essay discusses arguments for and against the ontological commitments that scientific theories may entail. Faye argues that scientific realism – in the sense that the semantic content of theories should be understood literally – is not sustainable, but that it is reasonable to maintain realism with respect to entities (e.g. electrons, elements) as has been suggested by Ian Hacking and others. The essay also discusses modern "structural realism" which presents itself as an alternative position claiming to meet some of the important challenges facing scientific realism. However, Faye argues that such a position is neither attractive nor defendable.

The question of scientific realism is also the topic of Kevin Kelly's thought-provoking and humorous "To be or not to be: An ancient Danish dialogue concerning appearance and reality". The essay presents, in the style of the classics, a dialogue between the "gruff viking" Andurtes and the "callow foreign youth" Kevo while strolling on the chalk cliffs on the Danish island Møn. The dialogue concerns the relationship between theory and reality, and how simplicity (Ockham's razor) may be said to be truth-conducive.

3 Philosophy of medicine and engineering

The question of the relationship between theory and truth – and the role of simplicity – is also a central topic in Arne Jakobsen's "Shaping engineering knowledge". But in contrast to both Kelly and Faye who discuss "scientific knowledge", Jakobsen discusses the development of theories specifically made for practical engineering purposes – in this case, the development of theories and norms for concrete constructions. How decisive is truth in the development of such practical engineering methods, and how are claims about validity and utility balanced against each other in the process of developing knowledge and methods used by engineers? In the essay Jakobsen describes the interesting development of plasticity based methods for structural calculations of concrete (a development which has a long history in Denmark), and the advantages of these methods over other methods. Concrete is far from being ideally plastic, but the developed methods have nonetheless steadily gained ground in static calculations of a wide range of concrete constructions, and the methods are claimed to have considerable advantages in their use compared to other principles of construction. Particularly, proponents of these methods stress their simplicity and transparency: using them, engineers are able intuitively to follow how a calculation proceeds.

Frederik Voetmann Christiansen and Camilla Rump's "Nomological Machines in Science and Engineering" presents a similar view of the role of scientific knowledge in engineering. Based on Nancy Cartwright's philosophy of science (particularly her notion of nomological machines) and Kant's description of regulative ideas, Christiansen and Rump argues for a particular antirealist (or perhaps entity realist?) conception of the relationship between theory and technology. The authors find the concept of "nomological machine" as used in the philosophy of science important also with respect to engineering, because it makes it possible to assign a distinct role to scientific knowledge in engineering, without reducing engineering to the application of science. However, science and engineering have different overall goals, and therefore nomological machines play different roles in the respective activities.

Turning from the philosophy of engineering to the philosophy of medicine, Søren Holm reviews the "rise, decline and fall" of the field philosophy of medicine – both as an area of research, and with respect to its role in medical curricula. The first part of "The philosophy of medicine in the late 20th century" traces the philosophy of medicine as a field of academic activity from the late 1960s to the present day, and explores the reasons for the decline from the mid-eighties to the present. Among these are the rise of bioethics and the subsequent crowding out of philosophy of medicine from its niche in medical schools. Another reason is the gradual sub-specialization and independent professionalisation of areas that were central to philosophy of medicine in its early phase (e.g. decision making and research methodology). The second part of the essay considers the developments of central areas of philosophy of medicine: the "concept of disease" debate, the relation between philosophy of science and philosophy of medicine and, finally, the influence of the Kuhnian idea of paradigm shifts in philosophy of medicine. Holm's overview of philosophy of medicine shows an area which in several respects is in decline; nevertheless, the author finishes his review by asking if there is hope for a resurgence of philosophy of medicine, and identifies areas of medicine in need of philosophical analysis.

In "Does modern neuropsychiatry threaten human values?" Raben Rosenberg and Jakob Hohwy argue that cognitive neuropsychiatry is a scientific discipline in which philosophical analysis has an important role to play. Many people worry that modern neuroscience, and in particular neuropsychiatry, is a serious threat to human values and dignity. According to Rosenberg and Hohwy, this worry overlooks the scale of recent progress in neuroscience and neuropsychiatry as well as recent theoretical developments in cognitive neuropsychiatry. Cognitive neuropsychiatry, the extension of cognitive neuroscience into psychiatry, is a new scientific field in which consciousness is not "by-passed", but which shares with neuroscience and neuropsychiatry the idea that the brain is the material substrate for consciousness. The authors argue that this view is compatible with retaining core notions of free will and agency, and is consistent with practical scientific aspirations. On the precise relation between mental and physical states, Rosenberg and Hohwy suggest that only future empirical research gradually can determine whether the former can be reduced to the latter.

4 Metaphysics and Epistemology

In the first essay of this section, "Good news for prisoner A", Jeff Paris analyses the Three Prisoners Puzzle. This puzzle has attracted considerable attention since it was first stated by Martin Gardner in 1961, and solutions to the puzzle may generalize to other areas of probability theory. Previous discussions have generally held that Prisoner A is "an egregious idiot who scarcely deserves even the 2:1 odds of reprieve that he is grudgingly accorded". However, the author jumps to A's defense and proffers a lawyer's argument to show that, at least for a somewhat simplified version of the problem, A can, by the power of thought alone, considerably improve these odds and so perhaps still enjoy at least one last night of carefree slumber.

Epistemic states are the focus of the next essay, "State consciousness – two defective arguments", by Oliver Kauffmann, in which the author examines some arguments for representational theories of state consciousness. A representational theory holds that a conscious state can be exhaustively accounted for in terms of intentionality – what the state is about – and, further, that a mental state must stand in relation to other mental states in order to be conscious. Higher-order representational theories of consciousness seek to explain consciousness in terms of some relation obtaining between a mental state and a higher-order representation – e.g. a thought or belief about it. They also stress that a subject in a conscious state is aware of not only what this state is about but also of being in that state. This feature the author calls "the self-presentational structure of consciousness". Thus higher-order theories suggest that a mental state is conscious whenever the subject has a proper representation of being in that state, where this higher-order representation itself is not conscious. First-order representational theories, on the other hand, deny that any higher-order representation is required and claim we are only aware of what our conscious states represent. First-order representationalists add instead a number of other constraints to be fulfilled for a mental state to be conscious. Kauffmann argues that a well-known theoretical attempt to reconcile the self-presentational character of consciousness with a first-order representational theory fails; and the author further presents a critique of a simple deductive argument for higher-order representationalism. The self-presentational feature of a conscious state might be accomplished by a mental state's representing itself and not by some other state. This suggestion, going back to Brentano, has very recently regained serious attention.

In the next essay, "The logic of temporal beginning", Peter Øhrstrøm examines various notions of temporal beginning. The author focuses on the semantics of statements that state that something is beginning to be or to be the case. The logic of such statements has been studied since antiquity, and in medieval philosophy many logicians dealt with the problems of beginning and ending. The study of the ideas of beginning gives rise to a number of paradoxes and problems. The author argues that the notions of beginning can in fact be presented in a consistent manner using temporal logic. In some cases the beginning may be understood as an instant in time, whereas in other cases we have to make use of durational logic in order to account for the meaning of beginning. Øhrstrøm argues that the temporal logic of A.N. Prior is relevant in all cases if we want a consistent logic dealing with the various notions of temporal beginning. Finally, he suggests that a consistent logic of temporal beginning is an essential background for the discussion of a number of important questions within cosmology, metaphysics and ethics.

Time is also the subject of the next essay, "Becoming through technology", in which Jan-Kyrre Berg Olsen examines conceptions of time and becoming. The author's exposition goes from a critique of deterministic rationality to a discussion of experienced temporality and entropy. The last part of the essay focuses on ways to extend the local temporal viewpoint to a more global point of view, through the use of simple technology such as water clocks, sand glasses, thermometers, or even nature's own technology such as pulse and heartbeat. These are all real world phenomena in which temporal direction is not hidden, the author notes, in contrast to mechanical clocks. Olsen argues that the debate about time refers to essential features of reality itself, and that the nature of time has to do with human cognition; more precisely, with cognition and experience in which true reality is disclosed.

In the final essay of this section, "Interdisciplinary philosophy – or the way Andur looks at it" Vincent F. Hendricks provides an overview of observations made by scientists and philosophers on the implicit but main theme of this book, namely philosophy's engagement with real world problems. Hendricks musters a parade of contemporary philosophers and historic luminaries who share the view that, as it is put by Charles Taylor in one of the representative quotes the author brings, "... philosophy in most aspects is pretty well useless and hopeless unless it's done with other disciplines".

5 Risk, uncertainty and safety

The first essay of this final section is "On imprecise statistical reasoning" by Igor Kozine. Kozine gives an overview and characterization of a recently developed approach to representing and propagating uncertainties in the context of, typically, risk assessment. Imprecise statistical reasoning challenges the central assumption of the Bayesian theory which holds that uncertainty should always be measured by a single (additive) probability measure. Kozine gives a short introduction to the theory of coherent imprecise previsions, recently developed in the form of imprecise statistical reasoning and to which the author has contributed. A hierarchical model developed by the author and the combination rules for aggregating information are also briefly described and discussed. Finally, Kozine describes how, by basing decisions on imprecise statistical values, we may reach results that lead to indecision, and he discusses reasons for holding that indecision is good or bad, respectively.

In the second essay, Martin P. Krayer von Krauss continues the theme of uncertainty in his "The case for in-depth uncertainty analysis in policy relevant science". In addition to possessing the knowledge and know-how required to provide public services, engineers and scientists function as an important source of legitimization for regulatory decisions. According to the liberal philosophy upon which the policy process is based, regulatory authorities should only intervene if development could lead to harmful effects. Their decision to do so should be based on facts, ideally considered within the framework of a rational methodology (e.g. risk assessment) to ensure that the facts are interpreted objectively. This can be quite problematic in situations where scientific knowledge is limited, facts are uncertain, the stakes are high and values are conflicting. The complexity of the regulatory issues scientists and engineers are typically asked to study far surpasses that of typical laboratory problems. The author points out that the umbrella term 'uncertainty' hides important distinctions. The well-established statistical or probabilistic notion of uncertainty leaves out many important aspects of the uncertainty encountered when assessing complex policy problems, such as the uncertainty generated by assumptions and ignorance of cause effect relationships. The precautionary paradigm has emerged in the context of the above realizations. Under this paradigm, the role of scientists and engineers is to participate in the collective effort of producing, evaluating and applying knowledge, considering the interests at stake, and making a necessarily provisional decision. A wide variety of stakeholders must be involved in decision making in order to ensure that a broad spectrum of values is represented. Uncertainty must be highlighted and used as a basis for collective reflection on the appropriateness of policy options. In this context, formal methods are required for experts to make transparent the broad spectrum of uncertainties that typically characterize their assessments.

The following essay, Marlene Dyrløv Madsen discusses the role of apology by doctors or nurses to a patient after a medical mistake leading to harm. This essay, "The nature of apology in the context of healthcare" provides an overview of the complex issues involved in the aftermath of adverse events, affecting patients and staff, and sometimes financial effects on the healthcare provider organization. The author examines the moral arguments for apology as well as the conditions for an apology to work effectively and ethically in healthcare. Different theoretical stances are discussed within the framework of the two ethical positions of utilitarianism and deontology. It is argued that using apology as a means to and end may have a negative effect by destroying the trust relationship between patient and caregiver, and moreover, may distort the essential meaning of apology. Several conditions for apology are suggested: acknowledging the incident as well as its inappropriateness; taking responsibility for the incident; expressing regret; and stating intention to refrain from similar acts in the future. In addition, two pragmatic conditions are proposed: an explanation of why the incident came about and a pledge of making practical amends. It is argued that healthcare organizations and caregivers have a duty to apologize to patients after harm, when justified. A spectrum of acknowledging actions are suggested in terms of when it is justified and necessary to either 'apologize', 'acknowledge' or 'express regret' following harm. The author describes several cases to illustrate the possibility of apologizing in healthcare in the aftermath of harm, and the negative consequences when apologies are not given or are given in the wrong manner.

In the final essay, Bove and Andersen discus theoretical and practical problems associated with identifying and classifying human error. Various taxonomies of human error have been developed for the purpose of analysing incidents and accidents in safety critical domains such as aviation, maritime operations, the military, healthcare etc. Taxonomies are needed for, first, analysing single incidents and accidents and, second, aggregating investigation results from a number of such events in order to identify types of errors and performance shaping factors associated with individual error types. The authors review some current taxonomies of human error and discuss sources of indeterminacy of the conceptualisation and classification of errors. They illustrate their points and by providing a couple of examples of incident reports from Air Traffic Control. Finally, some factors that have a practical influence on the reliability of coding are discussed.

References

Hoyningen-Huene, P. (1993). *Reconstructing Scientific Revolutions – Thomas S. Kuhn's Philosophy of Science*, University of Chicago Press, Chicago.
Schön, D. (1983). *The Reflective Practitioner*, Basic Books, New York.

Philosophy, history and applications of mathematics

Pre-modern "algebra": A concise survey of that which was shaped into the technique and discipline we know

Jens Høyrup

Section for Philosophy and Science Studies, Roskilde University

Since long I owe Andur a schoolbook on the science of medieval Islam. Since that book has not yet been sketched, nor a fortiori written, I dedicate to him at least an article without footnotes, based on work undertaken in part under his benevolent administrative aegis and containing as many footnotes as any dusty pedant could want.

1 Conceptual keys

Contemporary "algebra" is an aggregate of practices, problem types and approaches that have only come to belong together through a historical process. If we wish to describe the early history of algebra we therefore have to make clear which of these practices etc. we discuss.

Elementary algebra today is the practice of solving equations "analytically", and this is the aspect of algebraic thought that is most conspicuous in most mathematical cultures until the outgoing sixteenth century.

An "equation" is the statement that some complex quantity (for instance, the area A of a square) defined in terms of one or more simple quantities (in the example, the side s), or the measure of this complex quantity, equals a certain number or (the measure of) another quantity. "Analysis", as formulated by Viète, is "the assumption of what is searched for as if it were given, and then from the consequences of this to arrive at the truly given" (to assume that s exists, whence $s \times s = A$, $s = \sqrt{A}$).

Since Viète, equations are written in symbols. This was not always the case. It is customary to distinguish (a) "rhetorical" algebra, in which everything is set out in full words; (b) "syncopated" algebra, in which standardized abbreviations or signs are used, but the stenographic expression still represents language; (c) "symbolic algebra", in which the symbolic expression has proper value, and operations are performed directly on this level – as when $\frac{1}{1-x} = 2$ is multiplied by $1 - x$ in order to yield $1 = 2 \cdot (1 - x)$.

In particular with the advent of symbolization, equations can be trained as dealing with abstract number. In all applied equation algebra, however, the abstract numbers intended by the symbols *represent* other kinds of magnitudes – prices, velocities, population densities, etc.

Beyond being a technique for solving equations, algebra is also a theoretical discipline, dealing with the classification of equations, the principles used to solve them, the existence of and relations between solutions, etc. Such concerns are less frequent in pre-modern times. Finally, contemporary algebra encompasses group theory and its kin, which (*inter alia*) grew out of methods developed in traditional theoretical algebra, but which has left the concern with equations behind.

2 Antiquity

Egyptian texts from the early second millennium BCE present us with two basic elements of algebraic thought: a representation in terms of an abstract quantity or "heap"; and the use of a "false position" in an analytical argument, exemplified by the following problem with solution where an unknown quantity is posited falsely but conveniently to be 4 (in paraphrase):

> A heap with its fourth part added produces 15. Assume for convenience that it is 4. Adding its fourth part gives 5. Since we should have 15/5 = 3 times as much, the quantity must instead by $3 \cdot 4 = 12$.

Arguments by false position were also used to solve homogeneous problems of the second degree.

Much more is offered by cuneiform texts from the Old Babylonian period (*c.* 2000–1600 BCE, the mathematical texts being from *c.* 1800–1600).

Firstly, the false position was widely used in first degree problems (as elsewhere until recent centuries). Secondly, a functionally abstract representation by means of measurable segments ("length", "width", "square side") and rectangular and square areas served to treat first- and second-degree problems about quantities of many kinds.

Second-degree problems were solved by means of "naive" cut-and-paste procedures. An example finds two numbers (say, p and q) whose product is 60 and whose difference is 7. The numbers are represented by the length and width of a rectangle with area 60 (see the figure). The excess of length over width is bisected and the outer half moved so as to contain together with the inner half a square

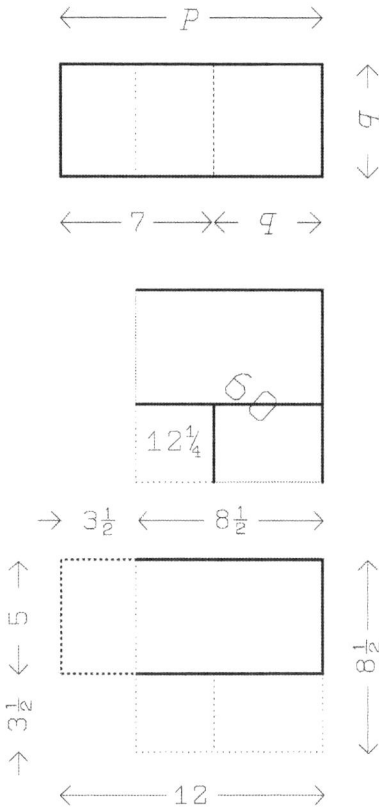

$3\frac{1}{2} \times 3\frac{1}{2}$; adding this small square produces a large square of area $72\frac{1}{4}$ and thus side $\sqrt{72\frac{1}{4}} = 8\frac{1}{2}$. Restoration of the piece that was moved shows that the width of the rectangle must be 5 and its length 12. The procedure is "naive" in the sense that no effort is made to prove that the procedure is correct – this is "seen" immediately.

This problem corresponds to our system $x - y = a$, $xy = b$; the same geometric procedure was used to solve problems about square areas and sides corresponding to $z^2 + m \cdot z = d$ and $z^2 - n \cdot z = e$, and a similar one for the geometric analogues of the system $x + y = a$, $xy = f$ and of the equation $p \cdot z - z^2 = g$. In non-normalized cases corresponding to the equation $r \cdot z^2 + s \cdot z = h$, a change of scale in one dimension was applied, corresponding to the transformation into $rz^2 + s \cdot (rz) = rh$. When linear conditions corresponding to $a \cdot x + b \cdot y = c$ are discussed, explicit terms for coefficients and contributions of the members may turn up. We also encounter two-step procedures corresponding to the change of variable $x' = ax + b$.

By having segments represent areas and volumes, the Babylonian calculators could formulate and solve biquadratic and certain other higher-degree problems. Irreducible third-degree problems were attacked by means of a variant of the false position combined with factorization – which presupposed that an easily factorizable solution was known to exist (all algebraic problems were scribe school problems constructed backwards from the solution, no single problem above the first degree had any practical use).

The description of problems and procedures often employed word signs heavily within the basically syllabic script. It has been claimed that these functioned as algebraic symbols, but since interpretation of the word signs depends on the total text, this cannot be the case (non-mathematical genres, indeed, used word signs just as much). However, certain standard phrases allowed the "nesting" of expressions, achieving part of what modern symbolism does by means of parentheses.

The original inspiration for this naive-geometric algebra appears to have come from "lay", that is, non-scribal practical geometers (surveyors etc.), among whom a small stock of geometric riddles appears to have circulated already in the late third millennium, remaining alive until the late Middle Ages: namely, to find

– one side of a rectangle from the area and the other side;
– the sides from the area and their sum or difference;
– the side of a square from the sum of the area and one or all four sides, or their difference;
– the sides of two squares from the sum of or difference between the areas, and the sum of or difference between the sides;

and a few others. Here, the same analytic naive-geometric procedures were used; but no coefficients appear beyond those occurring "naturally" (*the*, that is, *one* length, *one* width, *one* or *all four* square sides, etc.); nor did the technique serve for representation. In this original context, the technique is thus hardly to be characterized as "algebraic".

The Old Babylonian social system collapsed around 1600 BCE, and the scribe school disappeared together with advanced algebra. The surveyors' tradition survived, however, and inspired a revival of school algebra in Babylonia after 500 BCE; in the Hellenistic age, its stock of riddles swelled, as can be seen in unmistakeable borrowings in Demotic, Indian and Greek practitioners' mathematics. More important, however, was its influence in Greek theoretical mathematics, visible in particular in Euclid's *Elements* II, written somewhere during the third century BCE (the mathematics of book II is likely, however, to go back to the late fifth century).

Elements II.5–6 correspond to the algebraic identity $\left(\frac{x-y}{2}\right)^2 + xy = \left(\frac{x+y}{2}\right)^2$. Similarly, propositions 1–3 correspond to $p \cdot (q + r + \ldots + t) = pq + pr + \ldots + pt$, 4 to $(x+y)^2 = x^2 + y^2 + 2xy$, proposition 7 to $x^2 + y^2 = 2xy + (x-y)^2$, proposition

8 to $4 \cdot \frac{x+y}{2} \frac{x-y}{2} = x^2 - y^2$, propositions 9 and 10 to $x^2 + y^2 = 2 \cdot \left(\left[\frac{x+y}{2} \right]^2 + \left[\frac{x-y}{2} \right]^2 \right)$. Propositions 5 and 6, moreover, allow finding the sides of a rectangle from the area and, respectively, the sum of or difference between the sides, 8 allows finding the sides of two squares from the difference between their areas and the sum of or difference between the sides, 9 and 10 allow finding them from the sum of the areas and the difference between or sum of the sides.

Such implications of the theorems have been noticed since the Middle Ages. In the 1880s they were summed up by H. G. Zeuthen in the claim that the propositions constituted a *geometric algebra*, and the discovery around 1930 of Babylonian algebra (then interpreted as a numerical, not a geometrical technique) gave rise to Otto Neugebauer's further assumption that they represented a translation of Babylonian results into geometric language. The discovery of the geometric nature of the Babylonian technique and of the continuity of the surveyors' tradition allows a reformulation of this thesis: *Elements* II.1–10 constitute a theoretical investigation of the basis of the age-old technique, of the conditions under which the procedures are justified, and of the identities which underlie the solutions – the diagram on which II.6 is based is indeed identical with the one shown above, with the only difference that Euclid does not move areas around but constructs them and demonstrates their equality.

The presence of the surveyors' tradition in the Greek orbit is confirmed by the appearance of some of its riddles in characteristic phrasing in several manuscripts belonging to the pseudo-Heronian corpus (put together in modern times as *Geometrica*).

The same manuscripts contain problems related in structure to the traditional riddles (to find a right triangle whose perimeter equals its area) but indeterminate and meant to be solved in integer numbers, seemingly via factorization and use of identities corresponding to *Elements* II.

Indeterminate algebra searching for rational solutions constitutes the main body of Diophantos's *Arithmetic* (third century CE?); only book I, consisting of pure-number translations of the surveyors' and other traditional mathematical riddles, is in part determinate. The beginnings are simple (e.g., II.8, to split a square number into the sum of two square numbers), but soon matters become intricate (e.g., V.18, to find three numbers whose sum is a square, so that each added to the cube of their sum is also a square). The formulation is syncopated, making use of specific abbreviations for the unknown and its powers, for subtraction (the sum is made by juxtaposition) and for the square root; occasionally, operations take place at the level of symbols, thus approaching genuine symbolic algebra. Mathematicians of later ages have complained that Diophantos's procedures are opaque and do not reveal his underlying basic ideas.

Diophantos explains that his abbreviations belong to an established tradition within "theoretical arithmetic", which is confirmed by papyrological evidence.

Passages in Plato's writings (Republica 587d, Timaeus 31c–32a) suggest that this tradition goes back to the fifth century BCE and was carried by an environment of practical calculators, "logisticians" (but no evidence points toward a level much above Diophantos's book I combined with certain cubic problems).

The Indian Śulbasūtra's from the mid-first millennium BCE, containing rules for altar constructions fulfilling sophisticated mathematical conditions, contain solutions to non-homogeneous metro-geometric problems of the second degree; whether the solutions were found by any kind of algebraic argument is not clear, however. A few centuries later, on the other hand, it is likely that members of the Jaina community solved linear, quadratic and reducible higher-degree equations (the original texts are lost, but agreements between subjects cultivated in Jaina environments in late pre-Christian times and the contents of Mahāvīra's ninth-century *Gaṇita-sāra-saṅgraha* corroborate the assumption). Other evidence for early Indian algebra is a manuscript from Bakhshālī, probably a copy (with commentary) of a late ancient original, and the *Āryabhaṭīya* (499 CE). Taken together, these sources show:

- that the Near Eastern surveyors' tradition had reached India, both in the early and the Seleucid variant (Mahāvīra presents the respective methods in distinct chapters);
- that intricate second-degree problems were solved currently, but in versions that seem independent of the surveyors' riddles (they mostly deal with magnitudes and their square roots, not with magnitudes and their squares or products);
- that equations could be organized in schemes (combined with abbreviations) in which operations were made algorithmically, meaning that a transition to symbolic algebra (though very different from ours, and not allowing nesting whence less productive) had taken place;
- and that astronomers, for purposes of correlating planetary movements with each other, treated indeterminate linear equations.

Even the Chinese first-century CE *Nine Chapters on Arithmetic* betray some familiarity with the Near Eastern metro-geometric algebraic tradition, whose impact however was modest. Fully autochthonous is the creation of a technique (quite similar to our matrix manipulations, and thus another transition to symbolic algebra) for reducing systems of several linear equations (widely circulating riddles, which Diophantos was to treat with different techniques in *Arithmetica* I).

3 Algebras of the mature Middle Ages

The first surviving presentation of the technique from which our algebra developed and took its name is al-Khwārizmī's early ninth-century *Kitāb fi'l-jabr wa'l-*

muqābalah, "Book on restoration and opposition" – according to al-Khwārizmī's preface a brief introduction to an existing art. In the wake of al-Khwārizmī's work, "restoration" came to designate the addition of a subtracted member on both sides of an equation, and "opposition" subtraction on both sides; originally, "restoration" appears also to have encompassed multiplicative completion, and "opposition" to have designated the formation of an equation – or rather, perhaps, of its reduced form (which indeed would often ask for the subtraction of members occurring on both sides).

Pre-al-Khwārizmīan *al-jabr* consisted of two components, which may not have common origin. The core of *al-jabr* proper were fixed rules allowing the solution of equations dealing with a (monetary) possession (*māl*, becoming *census* in Latin), its square root (*jidhr*, becoming *radix*) and a number of dirhams (a coin); negative numbers not being considered, three rules were needed for the simple cases with two members and three for the "mixed" cases. The style (the square root of property) and certain linguistic clues suggest a connection to Indian algebra, most likely through a common ancestor. In *al-jabr*, however, these monetary riddles had become a general representation for second-degree problems.

The *al-jabr* rules went together with a technique for rhetorical transformation of equations, in which the unknown magnitude was spoken of as "a thing" (Arabic *šay'*, Latin *res*, Italian *cosa*), functioning like our *x*. Leonardo Fibonacci speaks of the technique as *regula recta* (a reference to its analytical nature), and treats it independently of *al-jabr* and with examples that suggest a link to elementary Greek "logisticians' algebra".

What made al-Khwārizmī's work pivotal was his introduction of (geometric) proofs for the *al-jabr* rules. The aim was no doubt to bring the presentation of the discipline in agreement with the already familiar Greek norms; the proofs themselves, however, were cut-and-paste proofs borrowed from the surveyors' tradition, only slightly adapted to Greek style.

This may have been a pedagogical advantage, but was deemed unsatisfactory by Thābit ibn Qurrah, a major translator of Greek texts in the second half of the ninth century and a prominent mathematician on his own account. In a small treatise he supplied new proofs for the rules based on *Elements* II.5–6 without even mentioning his predecessor.

The following major Arabic algebraist was Abū Kāmil (*c*. 850 – *c*. 930). He glued the reference to *Elements* II.5–6 to the naive diagrams in the proofs, but added others that produced the *māl* directly and not the root, showing thus that this quantity could be understood as an unknown in its own right (originally, the *māl* had certainly been *the* unknown – but that was long ago). Much in Abū Kāmil's treatise on algebra repeats and expands what al-Khwārizmī had done, but it goes beyond this model in the use of other monetary units as names for auxiliary variables (probably a borrowing from current practice), in its unconstrained use of irrationals, and in the expanded operation with higher powers of the unknown in biquadratic and other reducible problems. In one section he calculates the sides of the regular pentagon and decagon, elsewhere he investigates indeterminate problems of the first and second degree.

Around 1000, al-Karajī produced more striking innovations. His handbook presenting practitioners with "the sufficient about reckoning" (*al-Kāfī*) suggests (*inter alia* through its pre-al-Khwārizmīan use of "restoration" and "opposition") that his starting point was the "low", not the "scientific" al-Khwārizmī-Abū-Kāmil tradition. His major works (*al-Fakhrī*, *al-Badī*) demonstrate familiarity with this tradition as well as with the newly translated Diophantos, but go further by systematizing the treatment of reducible higher-degree equations; by applying the Euclidean theory of irrational magnitudes to number (and expanding it); by formulating an arithmetic of polynomials (including division and root extraction); and, in indeterminate analysis, by formulating principles where Diophantos had only given solutions.

All of this was developed further around the mid-twelfth century by al-Samaw'al, who also extended the notion of "subtractive" magnitudes into a concept of negatives ("subtractive 2" can only be subtracted; but "$n - (-2)$" is meaningfully interpreted as $n + 2$). In order to represent polynomials, he invented a schematic symbolization similar to what was used in Indian algebra.

Already around 1000, al-Bīrūnī and other astronomers had formulated the finding of the chord of a trisected angle (the kind of problem which the Greeks had solved by intersecting conic sections) as a cubic equation, solving it however by numerical, not by algebraic methods. In the context of a full classification of equations until the third degree (14 of which are irreducible cubics), al-Khayyāmī (*c.* 1100) made the reverse step and solved cubic equations by means of intersecting conic sections, identifying also the cases that were not solvable (in positive numbers) and some of those that have several solutions. Certain solutions of this kind had already been obtained by al-Khāzin (d. *c.* 965) and others, as al-Khayyāmī relates.

Developments of a different kind occurred in the Maghreb in the twelfth to fifteenth century, carried by a teacher-student network dense enough to be regarded as a "school" (and indeed organized as a teaching *system* and linked to mosque and madrasah teaching). Its algebra, as evidenced by its neglect of geometrical proofs, was basically in pre-al-Khwārizmīan style. Its essential innovation with regard to the "low" fundament was the development of abbreviations for both unknowns and their powers and for operations; seemingly, this systematic syncopation inspired parallel developments in Italian algebra, ultimately leading to the development of modern symbolic algebra.

In India, astronomers from Brahmagupta (598–*c.* 665) to Bhaskara II (1114– *c.* 1185) followed the lead of Āryabhata I, associating expositions of mathematics with astronomical treatises. The solution of indeterminate linear equations remained an important topic, but Brahmagupta also took up the study of the equations $Nx^2 \pm c = y^2$ (Pell equations), and showed how from one solution (found by trial and error) others can be produced. Bhaskara II formulated a general method. To judge from Brahmagupta's exposition, the pretext was sham astronomical computation, and his purpose the display of professional skill.

In China, the level of mathematics declined in the later first millennium. Between 1247 and 1304, however, a number of works introduce a sophisticated polynomial algebra, working with up to four variables and until degree 14, representing polynomials in a positional notation and solving equations by a procedure seemingly inspired by algorithms for root extraction (the "Horner-Ruffini method"): an approximate solution is found, a new equation for the remainder is derived, to which again an approximate solution is found, etc.

4 Latin Europe

Al-Khwārizmī's *Algebra* was translated twice into Latin in the twelfth century, first by Robert of Chester and next by Gherardo da Cremona. The riddles of the surveyors' tradition became available through Gherardo's translation of an Arabic work on mensuration, the *Liber mensurationum*, and to some extent through Plato of Tivoli's translation of Savasorda's *Collection on Mensuration*.

The echo was faint – the curriculum of the schools had no space for algebra. However, Gherardo's translations were used (at times copied *verbatim*) by Fibonacci in his *Liber abbaci* (1202, revised 1228) and *Practica geometrie* (1220) together with much material he had found in Islamic territory, Constantinople and Provence. The chapter of *Liber abbaci* dealing with "algebra et almuchabala" proves the rules for the mixed cases in ways reminding of Thābit's but possibly invented independently (one, "naive", seems inspired by *Elements* II.4 and does not copy al-Khwārizmī)0. The level and contents of problems are comparable to those of Abū Kāmil (who was only translated in the fourteenth century, without generating any response) and (in his use of the Euclidean theory of irrationals) of al-Karajī. Earlier in the work, as mentioned, rhetorical first-degree *thing*-algebra is used under the name *regula recta*.

Fibonacci was linked to the Italian urban patriciate and to the philo-Arabic Hohenstaufen court. The contemporary university mathematician Jordanus de Nemore (probably active in Paris somewhere between 1210 and 1240) responded differently to the challenge of Arabic algebra. Strongly attached to the metatheoretical ideals of Greek mathematics, and acknowledging that algebra dealt with *number*, he wrote a treatise *On given numbers* that was related to his *Elements of Arithmetic* much as Euclid's *Data* were related to the *Elements* of geometry. It consists of theorems of the form "If certain arithmetical combinations of certain numbers [e.g., their difference and product] are given, then the numbers themselves are given". It thus does not teach the technique of solving equations (often it merely reduces a case to another case that is dealt with previously). Jordanus does not mention algebra at all; what he offers is a theoretical investigation of solvability. However, the theorems are illustrated by numerical illustrations, and these are often unmistakeably borrowed from the Arabic tradition. Readers who knew the latter thus got a hint that Jordanus meant to *replace* (Arabic) algebra by something which was theoretically better.

Jordanus's proofs are arithmetical and general, not based on numerical examples. This was possible because he represented numbers by letters (a technique he had probably developed when proving the validity of the algorithms for calculating with Arabic numerals in an earlier work, and which is also used in his *Elements of Arithmetic*). Since the outcome of every operation is designated by a new letter (thus "the quadruple of d" immediately becomes f.), this should not be mistaken for a symbolic algebra; the letters serve the same purpose as the segments used by Euclid in *Elements* VII–IX.

Fibonacci represents numbers in the same way in a few problems and alternative procedures that may have been added in the revision from 1228. In the fourteenth and fifteenth centuries, respectively, Oresme and Peurbach also refer to Jordanus's *Data* and betray to have understood the particular aim of his treatise; but apart from that it had no perceptible influence.

Traditionally it has been assumed that Fibonacci's *Liber abbaci* and *Practica geometrie* provided the foundation for that Italian "abbacus school" which emerged during the thirteenth century (a school where merchant youth was trained for two years in practical arithmetic); particularly it has been believed that abbacus algebra was derived from the *Liber abbaci*. Closer investigation of early abbacus books shows that this cannot be the case. In general, the *Liber abbaci* is instead an early (but overwhelming) exponent of a mathematical environment of abbacus-school type, which however may have been stronger in his times in the Ibero-Provençal area than in Italy (and may even have been shared by the commercial cultures of the western Mediterranean irrespective of religion). Specifically, even though the earliest Italian abbacus books contain material of a kind which can also be found in the *Liber abbaci* and which comes from shared sources, they contain no algebra at all. When eventually taking up the subject, the abbacus school adopted Arabic algebra via different channels.

The first influential abbacus treatise containing algebra was written by a certain Jacopo da Firenze in Montpellier in 1307; apart from the term *censo* for the second power, it shares nothing with the Latin predecessors. Containing no Arabisms, it must draw on a tradition that was already established in the Romance-speaking world – most likely in Provençal-Catalan area, from which however no other evidence prior to the 1330s has been traced.

Jacopo's algebra is in the "low" Arabic style, as evidenced for instance by the absence of geometric proofs and by the way "restoration" and "opposition" are used (the latter appears as "putting equal to"). It contains correct rules (but no problems) for reducible cubics and quartics, and not the slightest trace of syncopation.

Within three to four decades, Jacopo's treatise together with new imports from the same area had spurred a surprising development (detailed verbal agreements demonstrate that Jacopo was in fact an important contributor, though supplementary inspiration seems to have arrived after a couple of decades). Binomials involving square roots were manipulated with great virtuosity, pure-number problems were created as illustrations of the higher-degree rules, and examples and non-valid rules for non-reducible cubics and quartics were produced and transmitted; they proliferated and remained alive throughout the fifteenth century. The reason for persistence of the latter mathematical scandal (as we would tend to view it) is double: the abbacus masters used them to impress their public and the municipal authorities that might employ them; and solutions contained intricate expressions involving roots, whence fallacies were difficult to expose.

Ongoing contacts to the Maghreb area may be responsible for the introduction of syncopation, for the increasing operation with subtractive quantities, and for the first hints of symbolization (fractions containing polynomials in the denominators and subjected to cross-multiplication, schemes for the arithmetic of polynomials); in any case, it seems certain that all of these were ultimately borrowed from the Maghreb. In the fifteenth century, geometrical proofs were gradually taken up – in part borrowed from Fibonacci, in part coming from al-Khwārizmī, in part from the surveyors' tradition, in part independent.

In 1494, algebra went into print, constituting part of Luca Pacioli's *Summa de arithmetica, geometria, proportioni: et proportionalita*. The wording of the rules and many problems are borrowed from the abbacus tradition, but geometrical proofs copied from Fibonacci form the theoretical basis. Moreover, Luca had discovered that the widely circulating solutions to the non-reducible higher-degree problems were false; restricts himself to giving solutions to the reducible cases; and points out that the others had not been solved *so far.*

The Ibero-Provençal abbacus-like tradition survived into the fifteenth and even the sixteenth century, now in interaction with the strong Italian tradition (but less dependent on it than has been assumed), yet until Nicolas Chuquet's *Triparty* (*c.* 1480, the culmination of the Provencal tradition) without leaving traces of any strong interest in algebra. Towards the mid-sixteenth century, Italian abbacus algebra was adopted by the German *Rechenmeister* under the name of *Coß* (from

Italian *cosa*). This was the source of Robert Recorde's treatment of "the coßike practise" in *The Whetstone of Witte* (1557).

5 The transition phase

Solution of the irreducible third- and fourth-degree cases became pivotal in the transition to modern algebra. The first step was made around 1515 by Scipione del Ferro, who discovered how to resolve the case "cube and roots equal to number". He communicated the rule to his pupils, one of whom used it in a public disputation with Niccolò Tartaglia in 1535 (though the solution was no longer fake, it served the same career purpose as before). Tartaglia managed to find the solution and was persuaded to disclose the rule to Gerolamo Cardano (according to his own account under oath of secrecy, according to Cardano's disciple Ludovico Ferrari without such conditions). When Cardano was informed about del Ferro's earlier discovery he felt free to publish (crediting both del Ferro and Tartaglia) in the *Ars magna* (1545).

Cardano not only published *the rule* for the case in question (and for the related case "cube equal to roots and number", which he may also have received from Tartaglia); he also gave geometric proofs, which he claims to have found himself (which is indeed quite plausible, once he knew the rule his training will have allowed him to recognize the way leading to the goal).

The proof for the latter case can be summarized as follows in modern symbolism: The equation is $x^3 = 3px + n$. We represent x^3 by a cube (see the figure), express x as a sum $x = u + v$, and dissect the cube into 5 pieces corresponding to the transformation $x^3 = (u+v)^3 = u^3 + v^3 + 3uv \cdot (u+v) = (u^3 + v^3) + 3uv \cdot x$ (one of the three pieces $uv \cdot x$ is shown separately). If $u^3 + v^3 = n$, $uv = p$ (whence $u^3 \cdot v^3 = p^3$), $x = u + v$ will fulfil $x^3 = 3px + n$.

Now, the problem $r \cdot s = A$, $r + s = b$ was familiar in the abbacus tradition, from *Elements* II.5, and was part of the surveyors' stock since millennia. Its solution is

$$r = \frac{b}{2} + \sqrt{\left(\frac{b}{2}\right)^2 - A} \ , \ t = \frac{b}{2} - \sqrt{\left(\frac{b}{2}\right)^2 - A}. \tag{1}$$

Substituting p^3 for A, n for b, and finding u and v from u^3 and v^3, we get

$$x = u + v = \sqrt[3]{\frac{n}{2} + \sqrt{\left(\frac{n}{2}\right)^2 - p^3}} + \sqrt[3]{\frac{n}{2} - \sqrt{\left(\frac{n}{2}\right)^2 - p^3}}. \tag{2}$$

This might have been nothing but an ingenious but traditional solution to a traditional problem, had it not been accompanied by other novelties. Firstly, *all* the cubic cases are solved (the case $x^3 + px = n$ analogously, the others by transformation into this or the previous one), the necessary transformations being proved by Euclidean geometry.

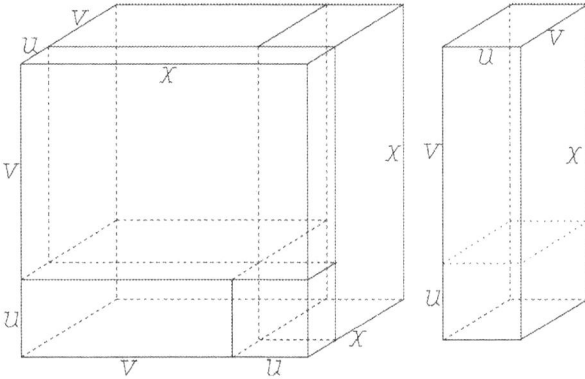

Secondly, Cardano operated without difficulty with negative quantities and solutions (though considering them "fictitious"), which allowed him to clarify the relation between the sets of solutions to related equations and between different solutions to the same problem (and led him to the notion of coinciding solutions). Thirdly, he applied the theory of irrationals in order to find conditions which solutions would have to fulfil.

Also contained in the book is a discovery made by Ferrari: that the complement to be added to a quartic in order to transform it into a biquadratic (and thus resolve it) can be determined by means of a cubic equation.

A curiosity is Cardano's introduction of imaginary and complex solutions – regarded, respectively, as "a second kind of false" solution (the negatives being the first) and as "completely false". He may possibly have been provoked to think about these because they occur in some of the solutions to cubics (e.g., in the above, if $p^3 > (n/2)^2$); but he does not say so, and the example through which he introduces them is the corresponding second-degree problem, $r + t = 10$, $rt = 40$.

In 1572, Bombelli's *L'algebra* was published. Its declared aim was to put into intelligible shape what had so far been written confusedly on the subject – not least by Cardano.

Part of this clarification consisted in the introduction of new symbols. Cardano's style had been purely syncopated; for instance,

$$\sqrt[3]{42 + \sqrt{1700}} + \sqrt[3]{42 - \sqrt{1700}} - 2$$

appears as "R. V. cubica 42. p. R. 1700 p. R. V. cub. 42 m. 1700 m. 2" – "p." representing "più", "m." "meno", "R." "radice", and "V." indicating that the root is taken of two members. Bombelli introduces algebraic parentheses (written $\lfloor \ldots \rfloor$ and used for multiple nesting) and an arithmetical notation for powers, in which $\overset{v}{}$ represents our x^n. Both devices constitute steps toward symbolization. He had been preceded in part by Chuquet in the *Triparty*, but Chuquet's notations were uninfluential.

Bombelli also invented new geometrical constructions "in a plane surface" though "by means of instruments" (not compass and ruler only) that showed the existence of solutions to the cubics even in cases where Cardano's formulas gave them only in the form $\sqrt[3]{a + b\sqrt{-1}} + \sqrt[3]{a - b\sqrt{-1}}$. These solutions also made him take up the study of imaginary numbers (designating them "più di meno" and "meno di meno", respectively $a\sqrt{-1}$ and $-a\sqrt{-1}$) and their arithmetic. He does not refer to Cardano's modest beginnings, and successive work on complex numbers was derived from Bombelli, not Cardano.

With Cardano's and Bombelli's innovations we have reached the threshold to modern algebra. Viète and Descartes belong to the next chapter – of history, not of this volume.

Bibliographic notes

A collective work presenting Babylonian, ancient Egyptian, Greek, Islamic, Latin and later European algebra and algebraic modes of thought is:

> Scholz, E. (ed.) (1990). *Geschichte der Algebra. Eine Einführung*, Lehrbücher und Monographien zur Didaktik der Mathematik, Band 16, B. I. Wissenschaftsverlag, Mannheim etc. A revised English translation has been under way at Oxford University Press for so long that the project must probably be considered dead.

The latest large-scale presentation of ancient Egyptian mathematics, including its quasi-algebraic features is:

> Clagett, M. (1999). *Ancient Egyptian Science. A Source Book*, Vol. III. *Ancient Egyptian Mathematics,* Memoirs of the American Philosophical Society, 232, American Philosophical Society, Philadelphia.

An analysis of Old Babylonian algebra as a naive-geometric technique together with a relatively broad account of the surveyors' tradition is found in:

> Høyrup, J. (2002). *Lengths, Widths, Surfaces: A Portrait of Old Babylonian algebra and its kin*, Studies and Sources in the History of Mathematics and Physical Sciences, Springer, New York.

The surveyors' tradition and its relations to Babylonian, Greek and Islamic mathematics is also presented in:

Høyrup, J. (2001). On a collection of geometrical riddles and their role in the shaping of four to six 'algebras', *Science in Context* **14**: 85–131.

A survey of Islamic algebra from al-Khwīrizmī to al-Samaw'al is:

Anbouba, A. (1978). L'algèbre arabe aux IXe et Xe siècles. Aperçu général, *Journal for the History of Arabic Science* **2**: 66–100.

Jordanus of Nemore's algebra is analyzed in:

Høyrup, J. (1988). Jordanus de Nemore, 13th century mathematical innovator: An essay on intellectual context, achievement, and failure, *Archive for History of Exact Sciences* **38**: 307–363.

Fibonacci's relation to the beginning of the abbacus tradition is the theme of:

Høyrup, J. (2005). Leonardo Fibonacci and *Abbaco* culture: A proposal to invert the roles, *Revue d'Histoire des Mathématiques* **11**: 23–56.

The beginning of abbacus algebra is analyzed in:

Høyrup, J. (2006). Jacopo da Firenze and the beginning of Italian vernacular algebra, *Historia Mathematica* **33**: 4–42.

Various steps in the development of Italian 14th and 15th-century algebra are described in:

Franci, R. and Toti Rigatelli, L. (1985). Towards a history of algebra from Leonardo of Pisa to Luca Pacioli, *Janus* **72**: 17–82.

Commentary: Apart from the Scholz-volume, which is simply the most recent general treatment of the topic (until the twentieth century), most of what is contained in the preceding list presents results and interpretations that go beyond the content of standard histories.

The development of nonlinear programming in post war USA: Origin, motivation, and expansion

Tinne Hoff Kjeldsen

Department of Mathematics and Physics, Roskilde University.

1 Introduction

Nonlinear programming is the mathematical theory of finite dimensional inequality constrained optimisation. In the Mathematics Subject Classification (MSC) system it is classified as a sub-discipline of mathematical programming[1] which itself is placed under the heading of Operations Research and Mathematical Programming as number 90 in MSC, indicating that at least from a mathematics point of view nonlinear programming is tightly connected to operations research (OR). And indeed, nonlinear programming and its main theorem (the Kuhn-Tucker theorem) about necessary conditions for the existence of a solution to a nonlinear programming problem are terms familiar to most people involved with OR, as are the names of the two Princeton mathematicians Albert W. Tucker and Harold W. Kuhn even though they never really practiced what operations research is famous, for namely its ability to solve practical decision problems in many areas of the real-world. As Saul I. Gass phrased it in his paper on Tucker in IFORS' Operational Research Hall of Fame:

> We tend to forget that such successes [in solving real-life decision problems] rest heavily on theoretical mathematical results developed by researchers who had no interest in applications. For over 50 years, especially during the embryonic years of OR, Albert W. Tucker was the pre-eminent example of the theoretical mathematician, teacher, and expositor that graced the field. (Gass, 2004, p. 239)

---

[1] Mathematical programming is an umbrella term for various kinds of finite dimensional inequality constraint optimisation like linear programming, nonlinear programming, dynamical programming, convex programming etc.

The theory of nonlinear programming is one of those theoretical mathematical results developed by theoretical mathematicians with no interest in applications. The notion and with it the research field of the same name entered the world of mathematics in 1950. It happened at the Second Berkeley Symposium in Probability and Statistics where Tucker presented his and Kuhn's joint paper "Nonlinear Programming". The paper was published the following year in a proceedings and is considered to be a classic in the discipline of mathematical programming as well as the founding paper of nonlinear programming. Tucker and Kuhn were especially celebrated for the main theorem – soon to be known as the Kuhn-Tucker theorem – which was conceived as an important result. A lot of significance was (and still is) attached to this theorem – and no wonder – it initiated a whole new research field in applied mathematics.

How did these theoretical mathematicians get involved with this kind of mathematics? Where did nonlinear programming come from and how did it develop into a mathematical theory? The purpose of this paper is to discuss and give answers to these questions as well as to provide explanations for the origin and development of nonlinear programming in post war USA.

After nonlinear programming was established as a mathematical research field it turned out that two other mathematicians, William Karush and Fritz John, had proven the "same" theorem as Kuhn and Tucker already in 1939 and 1948 respectively, and on each ocasion it went almost unnoticed in the mathematical community. This raises two interesting questions: Why were these seemingly identical results perceived so differently in the mathematical communities and why could Kuhn and Tucker's derivation of the result create a new research field in 1950? I have dealt with the first question in the paper "A Contextualized Historical Analysis of the Kuhn-Tucker Theorem in nonlinear programming: The Impact of World War II" (Kjeldsen, 2000a). The second question will be dealt with in this paper together with the questions posed above.

I find the emergence of nonlinear programming to be an interesting event in the history of mathematics, because in order to provide answers with explanatory power to the questions raised mathematical concepts cannot be seen as time, place and context independent. To delimit one's analysis to the mathematical ideas alone is simply not sufficient. A more fruitful approach will be to analyse the historical events from the perspective of – what could be called – mathematical practice, insisting that – despite its universal character – mathematical knowledge is actively produced by mathematicians living at a specific moment of time at a specific place on earth under some historically given circumstances that determine the conditions under which they do their research.[2] In order to explain the emergence of nonlinear

---

[2] For historiographical discussions of these kinds see (Epple, 2004) where also references to studies in the history of mathematics using similar approaches can be found.

programming in 1950 and its development into a mathematical research field we need to view mathematical activities in a broader social context acknowledging that mathematical activities are not only internally driven – mathematics is not a purely autonomous self contained science immune from influences from outside.

To answer the questions I will first introduce the logistics programme of the Office of Naval Research (ONR) within which Kuhn and Tucker's work on non-linear programming took place, followed by a presentation of their mathematical work. Then I will discuss the importance of the duality result and finish with a discussion of the significance of OR for the reception and establishment of nonlinear programming as a new research field in the 1950-60s.

2 The logistics programme of Office of Naval Research

In 1950 Tucker gave his first talk on nonlinear programming, presenting the theorem that later assured him a place in the "Hall of Fame" of OR. Kuhn and Tucker's result did not solve a long outstanding problem in mathematics. On the contrary, they were exploring a new field of applied mathematics that emerged on the basis of work done during and after the war by mathematicians in the US Air Force. The military connection is also present in Kuhn and Tucker's publication of the talk. Here one can read that the work was sponsored by the Office of Naval Research's logistics programme.

The Office of Naval Research was established within the US Navy in 1946 to ensure the continuation of the vitality and thriving of scientific research done during the Second World War. During the first four years of its existence it was the main sponsor for government supported research in the USA. It continued the practise of the war organisation Office of Scientific Research and Development (OSRD) that had been the vehicle for the mobilisation of civilian scientists during the war. Like OSRD, Office of Naval Research supported scientific projects through contracts with scientists working in the universities, projects of which many were proposed by the investigators.[3]

The logistics programme of Office of Naval Research originated in 1948 as a result of the mathematician George B. Dantzig's work with so-called programming planning methods in the US Air Force during and after WW II. An Air Force programme was a huge logistics schedule for Air Force activities. During the war Dantzig had worked on these programmes and taught Air Force staff how to calculate the programmes. The methods they used were slow and inefficient. It took more than 7 months to set up such a programme (Geisler and Wood, 1951, p.

---

[3] For further information on the Office of Naval Research and the mobilisation of civilian scientists in the USA during WW II see (Kjeldsen, 2003a; Schweber, 1988; Owens, 1989; Sapolsky, 1979; Rau, 2000).

191). After the war Dantzig went back to work for the U.S. Air Force Headquarters where he functioned as mathematical advisor. Together with a group of Air Force people he worked on programming planning problems. In October 1947 the Princeton people became aware of this work because Dantzig visited John von Neumann, in von Neumann's capacity as a consultant for the Air Force, to discuss the possibility of solving such an Air Force programme. At this point Dantzig and his group at the Air Force had built a mathematical model for the programming problem, a model they first called "programming in a linear structure" and which soon after became known as a linear programming problem.[4] John von Neumann had just completed the first book on game theory with Oskar Morgenstern and he suggested that Dantzig's programming problem was equivalent to a so-called finite zero-sum two-person game. This connection to game theory provided the linear programming problem with a mathematical foundation in the theory of systems of linear inequalities and the theory of convexity.[5]

With von Neumann as the pivotal point, a summer trial research project financed by Office of Naval Research was set up in the summer of 1948. The Office of Naval Research decided to set up a separate logistics branch with its own research programme. Mina Rees, who was the head of Office of Naval Research's mathematics division, has described the establishment of this programme as she remembered it in 1977:

> ... when, in the late 1940's the staff of our office became aware that some mathematical results obtained by George Dantzig, [...] could be used by the Navy to reduce the burdensome costs of their logistics operations, the possibilities were pointed out to the Deputy Chief of Naval Operations for Logistics. His enthusiasm for the possibilities presented by these results was so great that he called together all those senior officers who had anything to do with logistics, as well as their civilian counterparts, to hear what we always referred to as a "presentation". The outcome of this meeting was the establishment in the Office of Naval Research of a separate Logistics Branch with a separate research program. This has proved to be a most successful activity of the Mathematics Division of Office of Naval Research, both in its usefulness to the Navy, and in its impact on industry and the universities. (Rees, 1977, p. 111)

---

[4] A linear programming problem is – like nonlinear programming – finite dimensional optimisation under inequality constraints but in a linear programming problem all the involved functions are linear functions.

[5] The early history of linear programming can be followed in (Brentjes, 1976b; Dantzig, 1982, 1988, 1991; Dorfman, 1984; Kjeldsen, 2000a,b). For the history of game theory see (Kjeldsen, 2001; Leonard, 1995; Mirowski, 1991; Weintraub, 1992).

3 Tucker's trial summer project

The project started out as a university based trial project in the summer of 1948. Tucker seems to have become involved by a coincidence. He met Dantzig at one of Dantzig's meetings with von Neumann at Princeton, and he showed an interest in the mathematical problems concerning linear programming. Because of that he was contacted by the Office of Naval Research and asked whether he would take on the job as so-called principal investigator for the project. Tucker agreed and he hired two of the department's graduate students – Harold W. Kuhn and David Gale – to work with him over the summer.[6]

After Dantzig's first consultation with von Neumann the latter wrote a note "Discussion of a Maximum Problem" (1947) in which he rewrote a linear programming problem of maximizing a linear form subject to linear inequality constraints into a problem of solving a system of linear inequalities (von Neumann, 1963). This note was circulated privately and together with von Neumann and Morgenstern's book on game theory it furnished the part of departure for the work done by Tucker and his co-workers at Princeton that summer.[7]

Their first results are reported in the paper "Linear programming and the Theory of Games" which they presented the following summer at – what is now regarded as – the first conference on linear programming. The conference was held at the Cowles Commission for Research in Economics in Chicago and the papers were published in the proceedings *Activity Analysis of Production and Allocation* (Koopmans, 1951). That this new area of research was a true child of the military-university cooperation is reflected in the list of participants and their sponsors. The conference itself and the research done by the people in the Cowles Commission, except the work done by C. Hildreth, were supported by the RAND Corporation. Of the remaining twelve contributors two came from RAND, three from the U.S. Department of the Air Force, and seven from different universities. All the university contributions except two were done under contract with Office of Naval Research.

From the organisation of the proceedings and Koopmans introduction to it, it follows that at this time linear programming was mainly perceived as an economic theory and the theory of convexity as a tool "relatively new to economics" (Koopmans, 1951, p. 10). Tucker, Gale, and Kuhn's paper was placed in the third part of the proceedings which had the heading "Mathematical Properties of Convex Sets". In this paper they proved the duality theorem and existence theorems – not for the 'basic' linear programming problem but for a generalized 'matrix' problem that has the ordinary 'basic' problem as a special case. They also gave a new proof of the minimax theorem, which is von Neumann's theorem about existence of optimal

---

[6] See interview in (Albers and Alexanderson, 1985).
[7] Personal communication with Harold W. Kuhn, Princeton, April 23, 1998.

strategies for two-person zero-sum games and showed that such a pair of optimal strategies constitute a solution to the corresponding 'basic' linear programming problem and its dual.

They formulated two dual matrix problems based on the same given matrices A, B and C. In both problems a matrix D, which is either a "minimal" or a "maximal" matrix, is to be found. A matrix D having a certain property is said to be minimal or maximal if no other matrix Δ with the same property is such that $\Delta \leq D$ or $\Delta \geq D$ respectively. Here $\Delta \leq D$ means that each element of the matrix $\Delta - D$ is less than or equal to zero and $\Delta - D \neq 0$ (Gale et al., 1951, pp. 318–319).

Problem 1: To find a maximal matrix D with the property that $Cx \geq Dy$ for some $x \geq 0$, $y > 0$ satisfying the inequality condition $Ax \leq By$ (Gale et al., 1951, p. 319)

Problem 2: To find a minimal matrix D with the property that $B'u \leq D'v$ for some $u \geq 0$, $v > 0$ satisfying the inequality condition $A'u \geq C'v$ (Gale et al., 1951, p. 319)

If B is a column vector b, and C a row vector c', D becomes a scalar δ. The two vectors y and v reduce to positive scalars and can be eliminated. The two problems then become:

Problem (1δ): To find a maximal scalar δ with the property that $c'x \geq \delta$ for some $x \geq 0$ satisfying the inequality condition $Ax \leq b$ (Gale et al., 1951, p. 320)

Problem (2δ): To find a minimal scalar δ with the property that $b'u \leq \delta$ for some $u \geq 0$ satisfying the inequality condition $A'u \geq c$ (Gale et al., 1951, p. 320)

Today these two 'scalar' problems are called the primal and the dual linear programming problems. Nowadays they are often formulated without the condition $c'x \geq \delta$. The condition ensures the existence of finite optimal solutions, which is why Tucker, Kuhn, and Gale had this condition.

Their main results in linear programming were the "existence theorem" and the "duality theorem". The first one says that there exists a matrix D solving both problems, if there exist points $x \geq 0$, $u \geq 0$, $y > 0$, $v > 0$ fulfilling the corresponding inequality conditions. Today we would say that an optimal solution exists if there exist feasible points. In the other theorem, the duality theorem, Tucker, Kuhn, and Gale proved that problem 1 has a solution D, if and only if problem 2 has the same matrix D as a solution. Farkas's lemma was their main tool in the proof of these two theorems (Gale et al., 1951, p. 322).[8]

---

[8] For an account of Farkas's lemma see (Brentjes, 1976a; Kjeldsen, 2002). To consult Farkas's own work see (Farkas, 1901).

Here we see the effects of the military-university complex. Kuhn, Gale, and Tucker treated the linear programming problem as a mathematical research field. Instead of just working on the 'basic' linear programming problem they immediately generalized it without any consideration of the applicability of this generalized 'matrix' version of the problem. This approach is typical for basic research in mathematics as it is conducted in academia.

4 The significance of the duality theorem

The duality result is important from a theoretical as well as an applicational point of view. Here we shall not look into the applications but only mention that the duality theorem played a significant role in the later development of iterative algorithms for the solution of the linear programming problem, because the primal problem will approach the optimal solution from one side while the dual problem will do so from the other side. This means that at each step one will get an upper respective lower bound for the optimal solution.

Theoretically the duality theorem is crucial because it related linear programming to game theory more precisely to zero-sum two-person games, and von Neumann's minimax theorem. This connection provided linear programming with a theoretical mathematical kernel, and exposed the field to traditional mathematical research. It also provided linear programming with a foundation in convex analysis, in which the minimax theorem for zero-sum two-person games was well established at the time.[9]

After the conference in Chicago Tucker went to Stanford on leave. Here he continued to work on linear programming, studying its network nature.[10] In an interview Tucker has explained how this network nature of linear programming made him think of the Kirchoff-Maxwell's law for electrical networks (Albers and Alexanderson, 1985, pp. 342–343). He realised that the electrical network problem can be formulated as an optimisation problem on the minimising of heat-loss. The objective function though would not be linear but quadratic. The standard procedure for solving *equality* constrained problems was the Lagrangian multiplier method. Tucker then wrote to Kuhn and Gale asking them if they wanted to continue their work by extending the duality result for linear programming to quadratic programmes (Kuhn, 1976, pp. 12–13). Gale said no, but Kuhn agreed and according to Kuhn the work developed through a correspondence between Stanford and

---

[9] See (Kjeldsen, 2001).

[10] Among the first applications of linear programming was the transportation problem, which among others Koopmans had presented at the first conference in Chicago, see (Koopmans, 1951, p. 222).

Princeton. Thus Tucker's Office of Naval Research project continued until 1972, after which it became supported by the National Science Foundation.

The point of departure for Kuhn and Tucker's further analysis was to adapt the Lagrangian multiplier method to *inequality* constrained problems. They first examined it for linear programming. The result of that was included in the introduction to their nonlinear programming paper. They wrote:

> *Linear programming* deals with problems such as [...] to maximise a linear function $g(x) = \sum_{i=1}^{n} c_i x_i$ of n real variables $x_1, \ldots, x_n$ (forming a vector x) constrained by $m + n$ linear *inequalities*,
>
> $$f_h(x) = b_h - \sum_{i=1}^{n} a_{hi} x_i \geqq 0, \quad x_i \geqq 0, \quad h = 1, \ldots, m; \ i = 1, \ldots, n.$$
>
> This problem can be transformed as follows into an equivalent saddle value (minimax) problem by an adaptation of the calculus method customarily applied to constraining *equations*. Form the Lagrangian function
>
> $$\phi(x, u) = g(x) + \sum u_h f_h(x).$$
>
> Then, a particular vector x^0 maximises $g(x)$ subject to the $m + n$ constraints if, and only if, there is some vector u^0 with nonnegative components, such that
>
> $$\phi(x, u^0) \leqq \phi(x^0, u^0) \leqq \phi(x^0, u) \quad \text{for all nonnegative } x, u.$$
>
> Such a saddle point (x^0, u^0) provides a solution for a related zero sum two person game [...]. The bilinear symmetry of $\phi(x, u)$ in x and u yields the characteristic duality of linear programming. (Kuhn and Tucker, 1950, p. 481)

If x^0 is a solution to a linear programming problem and (x^0, u^0) is the saddle point for the corresponding Lagrangian function, then u^0 will be an optimal solution of the dual programming problem. Thus, for linear programming there was a very nice connection between saddle points for the associated Lagrangian function and the duality theorem. Keeping in mind that Kuhn and Tucker was trying to extend the duality result for linear programming to nonlinear cases, and that the duality result for linear programming was embedded in this saddle point result, it becomes very natural to take the saddle point property of the Lagrangian function as the point of departure. This was, as we shall see in the next section, exactly what Kuhn and Tucker did.

5 The nonlinear programming paper

Kuhn and Tucker's paper is pivoting around two types of problems; the "maximum problem" and the "saddle value problem", which they defined as follows:

> Maximum Problem: To find an x^0 that maximises $g(x)$ constrained by $Fx \geq 0, x \geq 0$. (Kuhn and Tucker, 1950, p. 483)

Here Fx is a m-vector $(f_1(x), \ldots, f_m(x))$ where $f_1(x), \ldots, f_m(x)$ are differentiable functions of x defined for $x \geq 0$.[11] $g(x)$ is a differentiable function of x also defined for $x \geq 0$ (Kuhn and Tucker, 1950, p. 483). For a differentiable function, $\phi(x, u)$, of a n-vector x with components $x_i \geq 0$ and a m-vector u with components $u_h \geq 0$, the saddle value problem was defined as (Kuhn and Tucker, 1950, p. 482):

> Saddle Value Problem: To find nonnegative vectors x^0 and u^0 such that
>
> $$\phi(x, u^0) \leq \phi(x^0, u^0) \leq \phi(x^0, u) \quad \text{for all} \quad x \geq 0, \ u \geq 0.$$

Thus, the maximum problem is a finite dimensional optimisation problem subject to inequality constraints, and the saddle value problem is the problem of finding a saddle point for a differentiable function of $n + m$ variables.

Kuhn and Tucker connected the two problems in "Theorem 1" and "Theorem 3", which are among the main results in their paper. In the first theorem they proved that a necessary condition for a point x^0 to be a solution to the maximum problem is the existence of a u^0, such that the point (x^0, u^0) satisfy the necessary conditions for being a solution to the saddle value problem for the associated Lagrangian function $\phi(x, u) = g(x) + u'Fx$. In theorem 3 they get the full equivalence between existence of solutions to the maximum problem and the saddle value problem by requiring that the functions g and $f_1, \ldots, f_m$ besides being differentiable also are concave for $x \geq 0$ (Kuhn and Tucker, 1950, p. 486).

Kuhn and Tucker did not succeed in finding a duality for nonlinear programming but they did launch the new research field of nonlinear programming.

6 Fenchel's duality theorem

The theory of convex functions seemed to be a promising tool in the newly emerged field of nonlinear programming. As mentioned above, Kuhn and Tucker did not succeed in expanding their duality theorem for linear programming to the general nonlinear case but they were able to prove that if the involved functions are concave (and differentiable) there will be complete equivalence between the

---

[11] $x \geq 0$ means that all the components x_i of $x = (x_1, \ldots, x_n)$ are nonnegative.

nonlinear programming problem and the saddle value problem for the corresponding Lagrangian function, suggesting the possibility of a duality result for concave functions.

One of the experts on the theory of convexity at that time was Werner Fenchel from the University of Copenhagen in Denmark. He happened to be visiting the USA in the academic year of 1950/51. He ended his visit in Princeton first as a member of the Institute for Advanced Study then as a visiting professor at the University of Princeton (Fuglede, 1989, p. 167). Tucker invited Fenchel to give a series of lectures on convex sets and functions within the logistics project and Fenchel's notes from these lectures were published by the Office of Naval Research in 1953.[12]

It appears from the notes and from letters between Fenchel and Tucker's group about the writing up and publishing of the notes that Fenchel derived his duality result for nonlinear programming during the development of these lectures. Fenchel also reveals that it was his contact with Tucker that turned his attention towards the programming problems, as he writes in the notes:

> The author [Fenchel] wishes to express his gratitude to Professor A. W. Tucker for giving him this opportunity to write this report and for calling his attention to the problems dealt with in the final sections (pp. 105–137). (Fenchel, 1953, Acknowledgement)

The final sections has the title "A Generalized Programming Problem", and here Fenchel proved the first duality result for nonlinear programming – often called Fenchel-duality.[13]

Now, where did Fenchel's duality result come from and what did it look like? In a paper from 1949 Fenchel had introduced the concept *conjugate* convex functions with the purpose of examining the mathematical structure underlying various inequalities – like the Hölder inequality – that appear in analysis. Fenchel showed that to each convex function $f(x_1,...,x_n)$, defined in a convex subset G, of $\mathfrak{R}^n$ and satisfying some conditions of continuity, there corresponds in a unique way a convex subset, Γ, of $\mathfrak{R}^n$ and a convex function $\phi(\xi_1,...,\xi_n)$, defined in Γ and with the same properties as f, such that the inequality

$$x_1\xi_1 + ... + x_n\xi_n \le f(x_1,...,x_n) + \phi(\xi_1,...,\xi_n),$$

---

[12] (Fenchel, 1953, Acknowledgement), (Fenchel, Bidrag til de konvekse funktioners teori, BOX 2, Folder: Manuskripter om konvekse mængder og funktioner. Un-dated but from the years 1953/54. The Fenchel Archive, Department of Mathematics, Copenhagen University, Denmark), (Letter from Tucker to Fenchel, June 11, 1951, BOX 1. The Fenchel Archive, Department of Mathematics, Copenhagen University, Denmark).

[13] See (Kjeldsen, 2003b).

is true for all points $x = (x_1, ..., x_n)$ in G and all points $\xi = (\xi_1, ..., \xi_n)$ in Γ. The correspondence between G, f and Γ, ϕ is symmetric, and Fenchel called the functions f and ϕ for conjugate functions (Fenchel, 1949, pp. 73–75).

Fenchel defined Γ to be the set of all points ξ, for which the function

$$\sum_{i=1}^{n} x_i \xi_i - f(x)$$

is bounded from above in G, and defined

$$\phi(\xi) = \sup_{x \in G} (\sum_{i=1}^{n} x_i \xi_i - f(x)).$$

Then the definition of ϕ ensures that the inequality will be fulfilled. Fenchel proved the theorem by using results about separating hyperplanes.

In his Princeton notes Fenchel argued that in a similar way to each concave function there will always correspond a conjugate function. In order to use this on nonlinear programming problems he considered a closed convex function $f(x)$, defined on a convex set C in $\mathfrak{R}^n$, and a closed concave function $g(x)$, defined on a convex set D. Denoting by $\phi : \Gamma \longrightarrow \mathfrak{R}$ and $\psi : \Delta \longrightarrow \mathfrak{R}$ the conjugates of f and g respectively Fenchel formulated the following two (dual) problems:

PROBLEM I: To find a point x^0 in $C \cap D$, such that $g(x) - f(x)$ as a function in $C \cap D$ has a maximum at x^0.

PROBLEM II: To find a point ξ^0 in $\Gamma \cap \Delta$, such that $\phi(\xi) - \psi(\xi)$ as a function in $\Gamma \cap \Delta$ has a minimum at ξ^0. (Fenchel, 1953, p. 105)

The duality theorem he then derived was that the two problems are connected in the following way:

If the sets $C \cap D$ and $\Gamma \cap \Delta$ are non-empty then $g(x) - f(x)$ is bounded above, $\phi(\xi) - \psi(\xi)$ is bounded below and under some further conditions on the origin in relation to the sets C, D, Γ and Δ, then (Fenchel, 1953, pp. 105–106):
$$\sup_{x \in C \cap D} (g(x) - f(x)) = \inf_{\xi \in \Gamma \cap \Delta} (\phi(\xi) - \psi(\xi)).$$

Fenchel did not explore this further but the lecture notes from his course became a source of inspiration and they had quite an influence on the following development of the theory of convexity in the USA and in the developing of convex programming. Especially R. T. Rockafellar was very much inspired by Fenchel's Princeton lectures. He used the Fenchel duality to build a duality theory for convex programming based on Fenchel's concept of conjugate convex functions.[14] In

---

[14] See e.g. (Rockafellar, 1970).

contrast to Fenchel's duality theorem the later development of the duality theory in nonlinear programming which began in 1959 and continued in the 1960s was motivated by the development of algorithms for the actual solving of quadratic programmes.[15]

7 Expansion: The significance of OR

Tucker's Office of Naval Research programme was just one step in the establishment of nonlinear programming in particular and mathematical programming in general as research disciplines. Another step which played a significant role in securing and further developing those areas in the universities was the establishment of operations research (OR) as a scientific discipline in the USA in the post WW II period.

The story of how OR came into the USA from Britain during WW II and how it – despite resistance from Vannevar Bush, the leader of the scientific mobilisation in the USA – managed to get implemented into the Applied Mathematics Panel late in the war have been told by Eric Rau and Larry Owens in very interesting papers and will not be repeated here (Rau, 2000; Owens, 1989). Instead I will focus on the connection between mathematics and OR to show the importance of OR for the establishment of nonlinear programming in particular and for mathematical programming in general.

The MIT physicist Philip Morse, who had been the leader of one of the most famous OR groups during the war – the ASWORG group[16] – became one of the most influential academics in the shaping of OR as a scientific discipline after the war. Through financial support the Office of Naval Research and the National Research Council played an active role in this move of OR into academia.[17]

In 1948 Morse had the first two OR courses up and running at MIT. They were run by the mathematics department. Other departments followed, offering courses "which include various aspects of operations research, such as linear programming ... " (Morse, 1956, p. 733), which indicates that – at least from Morse's point of view – linear programming was an essential part of the body of OR. But there was no consensus about what this new field called OR was supposed to be. It was a hybrid of different things and techniques and there was a continuously running debate in the Operations Research Society of America's (ORSA's) journal and at ORSA's annual meetings about this question. One of the earliest definitions of OR

---

[15] See e.g. (Wolfe, 1961; Dorn, 1960; Hanson, 1961; Mangasarian, 1962; Stoer, 1963; Mangasarian and Ponstein, 1965).
[16] Antisubmarine Warfare Operations Group.
[17] See (Rees, 1977, p. 111); (Fortun and Schweber, 1993, p. 611).

had been given by Charles Kittels in *Science* in 1947 (Kittel, 1947), a definition that Kimbal and Morse echoed in what is considered to be the first OR textbook:

> Operations research is a scientific method of providing executive departments with a quantitative basis for decisions regarding the operations under their control. (Morse and Kimball, 1951, p. 1).

A definition that doesn't really give much of a hint to the actual content of OR. The discussion among OR people about what OR actually is, or should be, can be followed in the two main journals *Operations Research* and *Operational Research Quarterly* as well as in the proceedings from international OR conferences. The debate was not limited to the definition of OR, but the question whether OR is a science or not was discussed.

Morse tried at an early stage to put an end to the discussions by simply defining OR to be what OR people do:

> We should no longer have trouble explaining the scope and methods of operations research to the layman. We already can say: operations research is the activity carried on by the members of the Operational Research Society; its methods are those reported in our journals. (Morse, 1953, p. 169)

In the beginning OR was heavily loaded with mathematics and Morse was an eager agitator for that. In 1953, in the paper "Trends in Operations Research" published in ORSA's journal, he emphasised game theory as an important future tool in OR that ought to be further developed, and about linear programming he wrote:

> Linear programming is rapidly becoming an important theoretical tool in economics; it deserves equal or greater exploitation in operations research. (Morse, 1953, p. 169)

Two years later, in 1955, he made clear that "[j]ust as with any other field of science, we are finding that we need our own kind of mathematics." (Morse, 1955, p. 383). What he in particular was asking for was basic research in mathematics related to the problems of operations research:

> To obtain complete solutions of these more complicated waiting-line problems will require mathematical abilities of high order. Such solutions will not be achieved in a few months, as casual by products of work on an immediate, practical problem. They probably will only be obtained by using a slower, more fundamental approach, by concentrating on the underlying mathematical relationships, by disregarding, for the time being, the urgencies and extraneous details of specific applications of the theory. This is the usual way that basic theoretical advances are made in science, after all. (Morse, 1955, p. 384)

That Morse regarded basic research in mathematical programming to be especially important for OR becomes clear from his following urgent appeal to a growing generation of OR practitioners:

> But linear programming is only one part of a larger theory of operational programmes, which covers such subjects as dynamic programming, some aspects of search theory and, probably, of game theory. It is hard to visualize, right now, all the mathematical aspects of this broader subject, because they haven't been investigated as yet in any detail. . . .

> It is hard to foresee the considerable usefulness of this general theory in solving many operations problems, particular those concerned with planning. But a great deal of basic research will be needed before the theory will be able to answer our practical needs. Some of the fundamental mathematics has not yet been developed and a great number of the algorithms for solving specific problems have not yet been worked out. Much of this basic work can probably best be done as a long-term study, not subject to the short-term deadlines and crises which occur in the study of immediate, practical problems. It needs a good many man-years of concentrated work. Operations research needs this sort of research. No branch of science can continue to grow unless its underlying theory is continuously being expanded. (Morse, 1955, p. 383)

It follows that Morse regarded mathematics to be the theoretical foundation of OR, and this goes hand in hand with a comment he made in 1948, where he expressed the wish that mathematics students could "be trained in this and related subjects, so that they may contribute to the peace-time applications of this new field [operations research] of applied mathematics . . . " (Morse, 1948, p. 621).

Morse considered OR to be a field of applied mathematics and if we were to take his self-defined definition of OR seriously, a picture emerges that shows, that mathematics and OR were indeed very closely tied in the beginning. Going through the first twenty years of *Operations Research* shows that mathematics and in particular linear programming occupied its share of OR. The very first paper "New Mathematical Methods in Operations Research" in the first number of the journal was devoted to mathematics and every number afterwards have at least one paper on mathematical programming. The papers published in the first 15 years of the journal were classified by Lindsey in 1979. 10 percent of the 1891 papers fell within mathematical programming making it the largest sub discipline of OR. He compared this result with *International Abstracts in Operations Research* during the period 1968–1977, where 17 percent of the papers where about mathematical programming – 12 of them in nonlinear programming. On this basis Lindsey concluded:

One feature common to both of those periods is the large effort in programming. This was proportionately greater after 1967. (Lindsey, 1979,
p. 18)

Also a survey of the first textbooks on OR reveals the importance of linear and
nonlinear programming for OR.[18]

Not everybody, though, held the opinion of Morse regarding the part mathematical programming should take up of OR. Already in 1953 Norman Hitchman
warned against the tendency of making OR synonymous with mathematics:

One main caution seems to stand out glaringly. It concerns the emphasis
which we place on certain specialized fields as to their value in the operations research team. It is not an uncommon observation of today to note
the very great emphasis given to mathematical and physical sciences. Our
new society, ORSA, for example, is playing a pre-eminent part in creating
the impression that mathematics and physics are almost synonymous with
operations research itself. (Hitchman, 1953, p. 242)

So, the opinions about the importance OR should attach to mathematics and
mathematical methods were divided and caused disputes. Some OR people regarded the subjects of mathematical programming as mathematicians excuse for
doing pure mathematics without any reference to application.[19]

The bottom line is that OR provided linear programming with a home where
there was a room for its extensions into nonlinear, dynamical, convex etc. programming. In this context it is no wonder that mathematical programming in general
and linear and nonlinear programming in particular could grow into a discipline of
mathematics in the post war realm in the US.

The critics it faced from some fractions of OR along with the 'bad' reputation
of applied mathematics within the mathematics departments probably was one of
the reasons why the mathematicians active in the field decided to create a home
for themselves. In 1972 the Mathematical Programming Society was founded and
by then mathematical programming had its own summer schools, journals, prices,
textbooks, regularly international conferences, monographs etc. Mathematical programming had become an independent research area.

8 Conclusion

The origin of nonlinear programming as a sub-discipline of mathematics in the
USA in the 1950's can only be understood within the post war context of military

---

[18] See e.g. (Churchman et al., 1957) and (Hillier and Lieberman, 1974).

[19] See (Hitchman, 1953, p. 242) and (Rider, 1992, pp. 231, 234).

supported science in a university culture of academic research. The success of the military collaboration with civilian scientists during the war created a belief that fundamental science was the optimal foundation for warfare. By setting up contract-based projects like Tucker's, where the scientists were not constrained by practical problem solving but were free to explore what ever they found interesting as long as it was just remotely connected to the topics of linear programming and game theory, the soil was prepared for fundamental mathematical research.

More specifically, the duality result for linear programming and its connection to the minimax theorem for two-person zero-sum games was an important source of inspiration for Tucker. This result was interesting from a pure mathematical point of view and it opened up the model of the Air Force logistics problem and made it an interesting research area in itself. In accordance with how basic research proceeds in academia Tucker and his group took the result to the next level and launched the theory of nonlinear programming. The inquires into the foundations of these new mathematical research areas created a framework that embedded the various programming problems in a mixture of pure and applied mathematics that appealed to research mathematicians. The linear programming problem originated directly from concrete, practical logistics problems in the Air Force but the subsequent development into first nonlinear then convex programming, all of it eventually referred to as mathematical programming, did not originate in practical problem solving. This development was not motivated by an urgent need to solve an existing problem here and now, rather it followed the lines of basic research in pure mathematics. It was urged by a quest for understanding and generalization within the realm of abstract mathematics.

As we have seen, also Werner Fenchel was motivated by problems in pure mathematics. It was the contact with Tucker that made him aware of the existence of nonlinear programming and made him realize that his theory for conjugate convex functions could be applied on nonlinear programming to give a duality result. There were no applications outside of mathematics involved in the motivation for his work and there were no considerations at this time about the solvability of these problems, it is purely existence theorems that were presented.

Even though both Kuhn, Tucker, and Fenchel seemingly were motivated by purely theoretical problems and the mathematicians themselves maintain that they felt they were free to do basic research which was governed by the subject itself, the original task was not self-imposed. On the contrary it was posed upon the mathematicians by authorities outside of mathematics. The starting point came from the practical problem of programming planning in the US Air Force. Linear programming originated directly within the military context and its further development into a mathematical research area took place through the interaction between "pure"-research mathematicians in academia and practical problem solving within the US Air Force.

The background for this kind of interaction was the scientific mobilisation during the Second World War with its special contract-system created by the leaders of OSRD. The following post war military financing of science through the Office of Naval Research paved the way for linear programming into academia and exposed it to fundamental research in mathematics with the creation of nonlinear programming as a result. Through its funding the military-university complex served as the mediating link between the Air Force logistics problem and basic research in mathematics, eventually creating together a new mathematical research discipline of mathematical programming.

The formation of the Office of Naval Research logistics programme was one of several interconnected decisive factors in the origin of nonlinear programming. The promised applicability to minimise the costs of logistics, the connection to the military, the flow of military money to basic research in the USA in the wake of WW II, and the fixation of the project in a university setting at the mathematics department at Princeton University are all decisive factors in the explanation of why nonlinear programming emerged in 1950, and they all belong, except for the intellectual challenge associated with generalizations of the duality result, to some extra mathematical realm. Finally linear and nonlinear programming became fundamental tools in operations research – itself a new discipline created by the war and moved into academia through military funding after the war. The significance of OR for the growth and expansion of linear and nonlinear programming into mathematical programming can probably not be overestimated. Even though applied mathematics gained some respect through the war work it continued to be regarded as lower rank, and was to a great extend ignored in mathematics departments.

References

Albers, D. J. and Alexanderson, G. L. (eds) (1985). *Mathematical People, Profiles and Interviews*, Birkhäuser, Boston.

Brentjes, S. (1976a). Bemerkungen zum Beitrag von Julius Farkas zur Theorie der linearen Optimierung, *NTM-Schriftenreihe zur geschicte der Naturwissenschaften, Technik und Medizin* **13**: 21–23.

Brentjes, S. (1976b). Der Beitrag der sowjetischen Wissenschaftler zur Entwicklungen der Theorie der linearen Optimierung, *NTM-Schriftenreihe zur geschicte der Naturwissenschaften, Technik und Medizin* **13**: 105–110.

Churchman, C., Ackoff, R. L. and Arnoff, E. (1957). *Introduction to Operations Research*, John Wiley and Sons, Inc., New York.

Dantzig, G. B. (1982). Reminiscences about the origins of linear programming, *Operations Research Letters* **1**: 43–48.

Dantzig, G. B. (1988). Impact of linear programming on computer development, *OR/MS Today* pp. 12–17.

Dantzig, G. B. (1991). Linear Programming, *in* J. K. Lenstra, A. H. G. R. Kan and A. Schrijver (eds), *History of Mathematical Programming, A Collection of Personal Reminiscences*, Noth-Holland, Amsterdam, pp. 19–31.

Dorfman, R. (1984). The discovery of linear programming, *Annals of the History of Computing* **6**: 283–295.

Dorn, W. S. (1960). Duality in quadratic programming, *Quarterly of Applied Mathematics* **18**: 155–162.

Epple, M. (2004). Knot invariants in vienna and princeton during the 1920s: Epistemic configurations of mathematical research, *Science in Context* **17**: 131–164.

Farkas, J. (1901). Theorie der einfachen Ungleichungen, *Journal für die reine und angewandte Mathematik* **124**: 1–27.

Fenchel, W. (1949). On conjugate convex functions, *Canadian Journal of Mathematics* **1**: 73–77.

Fenchel, W. (1953). *Convex Cones, Sets, and Functions*, Lecture Notes, Department of Mathematics, Princeton University.

Fortun, M. and Schweber, S. S. (1993). Scientists and the legacy of World War II: The case of operations research (OR), *Social Studies of Science* **23**: 595–642.

Fuglede, B. (1989). *Werner Fenchel*, Det Kongelige Danske Videnskabernes Selskab, Oversigt over Selskabets Virksomhed, 1988–1989, Munksgaard, Copenhagen. pp. 163–171.

Gale, D., Kuhn, H. W. and Tucker, A. W. (1951). Linear programming and the theory of games, *in* T. C. Koopmans (ed.), *Activity Analysis of Production and Allocation*, Cowles Commission Monograph, 13, Wiley, New York, pp. 317–329.

Gass, S. I. (2004). IFORS' operational research hall of fame Albert William Tucker, *International Transactions in Operational Research* **11**: 239–242.

Geisler, M. A. and Wood, M. K. (1951). Development of dynamic models for program planning, *in* T. C. Koopmans (ed.), *Activity Analysis of Production and Allocation*, Cowles Commission Monograph, 13, Wiley, New York, pp. 189–215.

Hanson, M. A. (1961). A duality theorem in nonlinear programming with nonlinear constraints, *Australian Journal of Statistics* **3**: 64–72.

Hillier, F. S. and Lieberman, G. J. (1974). *Operations Research.*, Holden-Day, Inc., San Francisco.

Hitchman, N. (1953). What is the mission of operations research?, *Journal of the Operations Research Society of America* **1**: 242–243.

Kittel, C. (1947). The nature and development of operations research, *Science* **105**: 150–153.

Kjeldsen, T. H. (2000a). A contextualized historical analysis of the Kuhn-Tucker theorem in nonlinear programming: The impact of World War II, *Historia Mathematica* **27**: 331–361.

Kjeldsen, T. H. (2000b). The emergence of nonlinear programming: Interactions between practical mathematics and mathematics proper, *The Mathematical Intelligencer* **22**(3): 50–54.

Kjeldsen, T. H. (2001). A history of the minimax theorem: von Neumann's Conception of the minimax theorem – a journey through different mathematical contexts, *Archive for History of Exact Sciences* **56**: 39–68.

Kjeldsen, T. H. (2002). Different motivations and goals in the historical development of the theory of systems of linear inequalities, *Archive for History of Exact Sciences* **56**: 469–538.

Kjeldsen, T. H. (2003a). New Mathematical Disciplines and Research in the Wake of World War II, *in* B. Booss-Bavnbek and J. Høyrup (eds), *Mathematics and War*, Birkhäuser Verlag, Basel, Boston, Berlin, pp. 126–152.

Kjeldsen, T. H. (2003b). Fenchels dualitetssætning, *Matilde, Newsletter, Danish Mathematical Society* **15**: 14–17.

Koopmans, T. C. (ed.) (1951). *Activity Analysis of Production and Allocation*, Cowles Commission Monograph, 13, Wiley, New York.

Kuhn, H. W. and Tucker, A. W. (1950). Nonlinear Programming, *in* J. Neyman (ed.), *Proceedings of the Second Berkeley Symposium on Mathematical Statistics and Probability*, Berkeley, pp. 481–492.

Kuhn, H. W. (1976). Nonlinear programming: A historical view, *SIAM-AMS Proceedings* **9**: 1–26.

Leonard, R. J. (1995). From parlor games to social science: von Neumann, Morgenstern, and the creation of game theory 1928–1944, *The Journal of Economic Literature* **33**: 730–761.

Lindsey, G. R. (1979). Looking back over the Development and Progress of Operational Research, *in* K. B. Haley (ed.), *Operational Research '78*, North-Holland Publishing Company, Amsterdam, pp. 13–31.

Mangasarian, O. L. and Ponstein, J. (1965). Minimax and duality in nonlinear programming, *Journal of Mathematical Analysis and Applications* **11**: 504–518.

Mangasarian, O. L. (1962). Duality in nonlinear programming, *Quarterly of Applid Mathematics* **20**: 300–302.

Mirowski, P. (1991). When Games Grow Deadly Serious: The Military Influence on the Evolution of Game Theory, *in* D. G. Goodwin (ed.), *Economics and National Security. Annual Supplement to Volume 23, History of Political Economy*, Duke University Press, Durham and London, pp. 227–255.

Morse, P. M. and Kimball, G. E. (1951). *Methods of Operations Research*, MIT Press og John Wiley and Sons, Inc., New York.

Morse, P. M. (1948). Mathematical problems in operations research, *Bulletin of the American Mathematical Society* **54**: 602–621.

Morse, P. M. (1953). Trends in operations research, *Operations Research* **1**: 159–165.

Morse, P. M. (1955). Where is the new blood?, *Journal of the Operations Research Society of America* **3**: 383–387.

Morse, P. M. (1956). Training in operations research at the massachusetts institute of technology, *Operations Research* **4**: 733–735.

Owens, L. (1989). Mathematicians at War: Warren Weaver and the Applied Mathematics Panel, 1942–1945, *in* D. E. Rowe and J. McCleary (eds), *The History of Modern Mathematics, vol.II: Institutions and Applications*, Academic Press, Inc., San Diego, pp. 287–305.

Rau, E. (2000). The Adoption of Operations Research in the United States during Worl War II, *in* A. C. Hughes and T. P. Hughes (eds), *Systems, Experts, and Computers*, Dibner Series, J. Z. Buchwald (ed.), MIT Press, Cambridge, MA.

Rees, M. S. (1977). Mathematics and the Government: The Post-War Years as Augury of the Future, *in* D. Tarwater (ed.), *The Bicentennial Tribute to American Mathematics, 1776–1976*, The American Association of America, Buffalo, NY, pp. 101–116.

Rider, R. (1992). Operations Research and Game Theory: Early Connections, *in* E. R. Weintraub (ed.), *Towards a History of Game Theory*, Duke University Press, Durham and London, pp. 225–240.

Rockafellar, R. T. (1970). *Convex Analysis*, Princeton University Press, Princeton, New Jersey.

Sapolsky, H. M. (1979). Academic Science and the Military: The Years Since the Second World War, *in* N. Reingold (ed.), *The Sciences in the American Context: New Perspectives*, Smithsonian Institution Press, Washington, D. C., pp. 379–399.

Schweber, S. S. (1988). The Mutual Embrace of Science and the Military: ONR and the Growth of Physics in the United States after World War II, *in* E. Mendelsohn, M. R. Smith and P. Weingart (eds), *Science, Technology and the Military*, Kluwer Academic Publishers, Dordrecht, The Netherlands, pp. 3–45.

Stoer, J. (1963). Duality in nonlinear programming and the minimax theorem, *Numerische Mathematik* pp. 371–379.

von Neumann, J. (1963). Discussion of a Maximum Problem, *in* A. H. Taub (ed.), *John von Neumann Collected Works*, Vol. 6, Pergamon Press, Oxford, pp. 89–95.

Weintraub, E. R. (ed.) (1992). *Toward a History of Game Theory*, Duke University Press, Durham and London.

Wolfe, P. (1961). A duality theorem for nonlinear programming, *Quarterly of Applied Mathematics* **19**: 239–244.

Dirichlet and the flexibility of the new concept of function

Jesper Lützen

Department for Mathematical Sciences, University of Copenhagen

1 Introduction

Dirichlet's name is often connected with the "modern" concept of function defined as a variable depending in an arbitrary manner on another variable.[1] However, the historian of mathematics Youschkevich (1976) and others have correctly shown that Euler and following him Lagrange, Fourier, Cauchy and other mathematicians formulated similar definitions many decades before Dirichlet. He has therefore concluded, that Euler rather than Dirichlet was the true originator of this concept. In this paper I shall defend Hankel's attribution to Dirichlet, not because I think that priority disputes are particularly interesting, but because I think that dating this concept of function back to Euler's time distorts the history of analysis. I shall argue for Dirichlet's crucial importance in the development of the concept of function by emphasizing the *use* of the concept rather than its definition.

After a brief and general discussion of the definition and use of concepts, I shall argue that mathematicians prior to Dirichlet did not really see the immense difference between the new concept and the old concept of function defined as an analytic expression.[2] Indeed, the new kind of functions were forced upon mathematicians by applications and were at odds with the pre-Cauchyan algebraic and global foundation of analysis that was in harmony with the analytic expressions. Second, I shall point to four different ways in which Dirichlet understood the new concept of function better than his predecessors, if by "better" we mean a way closer to our modern way of understanding the concept: 1. He stated more explicitly than his predecessors that functions need not be given by analytic expressions

---

[1] This attribution can be traced back to Hankel's paper "Untersuchungen über die unedlich oft oscillierenden und unstetigen Funktionen" (Hankel, 1882).

[2] For a more elaborate argument see (Lützen, 2003).

or even by mathematical operations. 2. He defined the first function that was not given by analytic expressions at all. 3. He understood that the generality of the concept necessitated various function classes suitable for various theorems. 4. He understood the local and flexible nature of the new concept of function and put it to constructive use in his proof of convergence of Fourier series. This constructive use of the concept of function which I think I am the first to point out is in my opinion crucial, and shows that with Dirichlet the concept of function entered a new era.

2 Definitions and use of concepts

Mathematical concepts are not defined through definitions alone. This holds true of formalized mathematics, where the whole axiomatic structure is needed in order to fix the meaning of a concept (if one can speak of meaning in formalized systems at all). For example in geometry the definition of a triangle involves points and lines whose meaning is only implicitly fixed through the axioms. It is even more true in informal mathematics where a concept often carries other meanings than those mentioned in the definition of the concept, in some cases even meanings and connotations that partly contradict the ones stated in the definition. For example the 17th and early 18th century definitions of a differential gave only a vague indication of the meaning of the concept. Another example is provided by Cauchy's definition in his *Cours d'Analyse* (1821) of the concept of a limit. Cauchy, like Newton and d'Alembert before him, defined the limit of a *variable* (which is defined as a quantity that can vary) in a way that strikes the modern reader as rather vague. However, as pointed out by Grabiner (1981), when Cauchy sets out to prove complicated theorems about limits he always deals with limits of functions (or series) as we do today, and he characterized the limits by all the εs, δs, Ns, inequality signs and quantifiers that a modern reader would require. Thus Cauchy's *use* of the concept of limit is much closer to modern standards than his definition. It is only by considering Cauchy's proofs and the total structure of his *Cours d'Analyse* that we can really appreciate his importance for the development of the foundations of analysis.

A definition often comes to light as a rationalization that captures for the beginner (some of) the characteristics of a concept which has been shaped by many years of use. Since the use of a concept often predates its definition and mostly characterizes the concept better than its definition a historian cannot rely only on explicitly formulated definitions when he or she wants to understand the earlier meaning and the development of a mathematical concept. He or she must study the concept in action.

In 1984 Rhüthing published a paper containing an interesting list of *some definitions of the concept of function from Joh. Bernoulli to N. Bourbaki*. Such a list is very valuable for a historian but it must not be confused with the history of the concept of function. Only when combined with a thorough study of the *use* of the concept in the period in question, a historical development will begin to take shape.

Indeed, the list of definitions may create the impression that the concept of function was relatively stable in the period from Euler's definition of 1755 through Hankel (1882). However, a study of the *use* of the concept reveals that in spite of the rather similar definitions the concept really underwent great changes during this period and only around Dirichlet's time reached a state close to what we now call the Dirichlet concept of function.

3 Euler's three function concepts

During his long and productive career Euler formulated three different definitions of the concept of function. First in his *Introductio in Analysin Infinitorum* (Euler, 1748a) he defined a function as an analytic expression i.e. as a formula. This concept of function has been named after Euler, and since I shall argue that it is indeed his principal concept I shall follow this tradition. Second, in the same year a discussion with d'Alembert concerning the vibrating string made Euler introduce functions that are expressed by different analytic expressions in different intervals (Euler, 1748b) and later even functions that correspond to an arbitrary hand-drawn curve for which the analytic expression in Euler's words changes from point to point (Euler, 1763). Euler called such functions discontinuous.[3] Third, in his text book on differential calculus (Euler, 1755) he formulated for the first time a definition that to a modern eye looks conspicuously similar to that of Dirichlet:

> Thus when x denotes a variable quantity, then all quantities that depend on x in any manner whatever or are determined by it are called functions of x. (Euler, 1755, p. 4)[4]

It is customary to assume that Euler's reformulation of the concept of function in 1755 was a result of the debate about the vibrating string and was meant to include the E-discontinuous functions as well as the analytic expressions. However, I am not convinced that this is the case. Indeed in the *Institutiones Calculi Differentialis* of 1755 Euler did not contrast his new definition of a function with his earlier one and he uncritically based his differential calculus on the results he had

---

[3] I shall use the term E-discontinuous in order to distinguish it from the modern concept that is more or less due to Cauchy.

[4] Translation from (Rüthing, 1984).

obtained in the *Introductio* without questioning their validity for the new type of functions. Moreover, it is conspicuous that in his papers on the vibrating string and discontinuous functions after 1755 Euler never appealed to his 1755-definition but continued to characterize discontinuous functions by their changing analytic expressions rather than by an arbitrary dependence of y on x.

Be that as it may, Euler definitely extended the concept of function by including functions that were not given by one analytic expression. However these E-discontinuous functions had a rather peculiar place in Euler's analysis. Euler felt forced to introduce them because it was the only way that d'Alembert's solution to the wave equation could handle all initial oscillations of the cord. Thus it was an application that led Euler to extend the concept of function. Later he argued that whenever arbitrary functions turned up in the solution of partial differential equations they were of this more general type; see (Euler, 1763, 1768–70). But the new type of E-discontinuous functions were never harmoniously integrated into Euler's analytic universe, and he continued to reason about the new functions as he had reasoned about analytic expressions.

In particular, Euler never appreciated the local nature of the new functions. To Euler an analytic formula was either valid everywhere (i.e. for all real and complex values of the variables in the formula) or it was not valid at all. Since functions were analytic expressions their properties were therefore the same everywhere. This global aspect of functions was so central to Euler's calculus that he used it also when dealing with E-discontinuous functions. For example he argued (Euler, 1753, §9) that series of the form

$$a_1 \sin x + a_2 \sin 2x + a_3 \sin 3x + \dots$$

could not represent arbitrary functions in the interval $(0, \pi)$ because only functions that are odd and periodic with period 2π can be represented in this way. To a modern reader, who is brought up with Dirichlet's concept of function, this argument makes no sense since a function defined on the interval $(0, \pi)$ cannot be said to be odd or periodic with period 2π. However, to Euler such global aspects were apparently still present in the functions even though we only consider them on an interval.

D'Alembert (1750, §2) immediately called attention to the problems it would cause if E-discontinuous functions were accepted in analysis. In particular he claimed that if the initial oscillation of a string and its periodic and odd prolongation was not expressed by one analytic expression, the problem of the motion of the string could not be solved by the known analysis. In principle Euler agreed that analysis had until then only been developed for analytic expressions, but he maintained that the new application necessitated the inclusion of E-discontinuous functions into analysis. However, he did not realize that such an extension called for a thorough reformulation of analysis.

4 Foundations catches up with the concept of function

Euler's definition of a function as a variable depending in an arbitrary way on an-other variable was repeated by many mathematicians during the next half century. Many of them such as Lagrange (1801, Leçon 1, def. 2) considered the definition as equivalent with the definition of a function as an analytic expression and they all argued with functions as with analytic functions. Euler's global foundation of analysis was developed further to a point where Lagrange could proudly declare that analysis was a part of algebra.[5] Clearly analytic expressions fit this view of analysis well whereas general dependences between variables were strangers in this game.

This changed when Cauchy (1821) declared war upon the generality of algebra and founded a new "rigorous" analysis. He pointed out that algebraic and other identities are usually only valid within given limits and thus dealt a death blow at the idea that functions (even when they are defined by analytic expressions) are defined everywhere and have global properties. His formulation of the basic concepts of analysis (such as continuity) and his proofs of its main results (such as the fundamental theorem of calculus) no longer appealed to analytic expressions and so the new concept of function naturally fitted this new environment. However, Cauchy still did not see much difference between the new general functions and analytic expressions (or analytic functions in the modern sense). For example in the introduction to the *Cours d'Analyse* when he wanted to take exception to Euler's belief in the global nature of functions and their "natural" continuation property, he wrote:

> When the constants or the variables contained in a function become imag-inary (complex), after having been considered real, the notation by which the function is expressed cannot be maintained in calculus except through a new convention designed to fix the sense of this notation under the ex-tended hypothesis. (Cauchy, 1821, p. iv, translation by JL)

This quote makes perfect sense when we think of a function as an analytic expression such as a^x where x is the variable and a is the constant contained in the function. However, if we take Cauchy at his word and consider a function more generally as a variable y depending in an arbitrary way on x it makes little sense to talk about the constants contained in the function. Thus in this quote Cauchy clearly lapsed into speaking of a function as an analytic expression. Moreover, he only late in his life appreciated the great difference between general functions and those analytic functions that he treated in his papers on complex function

---

[5] For a more thorough discussion of the algebraic and global foundation of analysis in the 18th century see (Jahnke, 2002).

theory and which have power series expansions in all points. And although he clearly understood that the theorems of analysis were not applicable in general to all functions, his text books (Cauchy, 1821, 1823), gave the impression that analysis was more or less applicable to continuous functions.

The only mathematician prior to Dirichlet who took the new concept of function seriously and clearly understood its local nature was Fourier. Just as Euler, Fourier needed the new general functions in order to be able to apply analysis to a physical situation, in his case heat conduction. Fourier defined a function as follows:

> In general, the function $f(x)$ represents a succession of values or ordinates each of which is arbitrary. An infinity of values being given to the abscissa x, there are an equal number of ordinates $f(x)$. All have actual numerical values, either positive or negative or null. We do not suppose these ordinates to be subject to a common law; they succeed each other in any manner whatever, and each of them is given as it were a single quantity. (Fourier, 1822, p. 500)[6]

Moreover he understood that the serious adoption of the new concept of function necessitated a new approach to analysis. In particular he pointed out that the definition of the integral as an antiderivative could no longer be used because the derivative no longer existed in general. Instead he interpreted the integral as an area.

5 Dirichlet's definition of a function

Dirichlet's name is associated with the new concept of function primarily because of his definition and use of it in his two papers (1829, 1837) on the convergence of Fourier series. His convergence proof is justly admired as the first rigorous proof of this theorem following many unrigorous arguments due to e.g. Fourier, Poisson and Cauchy.[7] More generally it is the first rigorous proof of a complicated theorem in analysis. Here I shall only deal with those aspects of the papers that relate to the new concept of function.

The first of the papers written in French did not contain a definition of a function at all and the definition in the second paper written in German is rather disappointing at first sight. It reads:

---

[6] Translation from (Rüthing, 1984).

[7] For a discussion of these proofs and the foundations of analysis in the 19th century the reader is referred to (Bottazzini, 1986).

> If every x gives a unique y in such a way that when x runs continuously
> through the interval from a to b then $y = f(x)$ varies little by little, then
> y is called a continuous function of x in this interval. (Dirichlet, 1837, p.
> 135)

Thus Dirichlet did not define a general function but only a continuous function,
and even with a definition that is inferior in clarity to that of Cauchy. The only way
it stands out in comparison to earlier definitions is in requiring that a function be
single valued.[8] However, in the subsequent clarifying remark Dirichlet stressed the
generality of the new concept of function compared with Euler's first concept:

> It is not necessary that y depends on x according to the same law in the
> entire interval. One does not even need to think of a dependence that can
> be expressed through mathematical operations. (Dirichlet, 1837, p. 135)[9]

Thus Dirichlet stressed the arbitrary nature of the dependence of y on x more
clearly than Fourier had done. However, the real difference between Dirichlet and
his predecessors only emerge when we look at his use of the concept.

6 The first pathological function

Though Dirichlet never stated an explicit general definition of a function we may
assume that the above mentioned definition with the continuity requirement removed captures his idea of a general function. Indeed, he freely operated with
discontinuous functions and was the first who defined a (discontinuous) function
without giving it by an analytic expression (or several analytic expressions). He
defined this function as follows:

$$f(x) = \begin{cases} c, & \text{for } x \in \mathbb{Q}, \\ d, & \text{for } x \in \mathbb{R} \backslash \mathbb{Q}, \end{cases}$$

where $c \neq d$. Dirichlet does not seem to have felt the need to give an analytic
expression for this function but contended himself with remarking that "the function thus defined has finite and determinate values for all values of x" (Dirichlet,
1829, p. 132). The aim of this remark seems to be to convince his readers that f
defined in this way really *is* a function. To a modern reader, that seems obvious.
However, though Fourier had stated that each value of $f(x)$ was given as a single value, neither he nor anybody else had taken the liberty to define a function

---

[8] Though Fourier's definition seems to imply single-valuedness he was not consistent in
this respect. Cauchy, on the other hand required that a continuous function be single
valued.

[9] Translation from (Rüthing, 1984).

in such a way before. Dirichlet's function was a novelty in another way. Its graph cannot be drawn. In the discussion about the vibrating string Euler had faced the problem that an arbitrary curve cannot be described by an analytic expression. For this reason he had generalized the concept of a function so that it was as general as the concept of a curve. Dirichlet went a step further: With his interpretation of the concept of a function Dirichlet made this concept even more general than the concept of a curve.

Dirichlet presented his everywhere discontinuous function as an example of a function that could not be integrated, even with Cauchy's new concept of integral (essentially the Riemann integral). For this reason the integrals giving its Fourier coefficients cannot be found, so the function has no Fourier series at all. The function is the first example of a pathological function.[10]

7 The delimitation of function classes

Before Dirichlet there seems to have been a general feeling that there was a natural domain for all of analysis. Euler explicitly formulated the principle that all the rules of analysis were and should be universally valid for all functions (Euler, 1749). Cauchy realized that this principle had to be given up, but as mentioned above his textbooks gave the impression that analysis was generally valid in the domain of continuous functions. He proved that continuous functions were integrable and though he seems to have been aware that continuous functions might not possess derivatives in certain points, he only made the explicitly formulated assumption that they be continuous whenever he differentiated functions. Dirichlet, on the other hand saw that the class of continuous functions was both too large and too small. He wanted to deal with Fourier series of some discontinuous functions and thus had to generalize Cauchy's notion of integral. He showed that many discontinuous functions have an integral but presented his pathological function to show that it would be impossible to integrate all discontinuous functions. He claimed that functions whose points of discontinuity are nowhere dense can be integrated, and thus developed into a Fourier series, but postponed the proof to a later paper that in fact never appeared.

Another class of functions explicitly mentioned by Dirichlet was the class of piecewise continuous piecewise monotonous functions for which he rigorously

---

[10] Or the second such function if we count Cauchy's example e^{1/x^2} of a C^∞ function that is not analytic. (Cauchy, 1823) Such functions show the limitation of general theorems in analysis (or "show up defects in the reasonings of our fathers" as Poincaré (1899) puts it) were cooked up in great numbers in the later part of the 19th century. The appearance of pathological functions is a clear sign that the generality of the new concept of function began to dawn to the mathematical community.

proved that the Fourier series of a function f converges toward the function itself or rather to $\frac{1}{2}(f(x+0)+f(x-0))$ where $f(x\pm0)$ denotes the limit of $f(\xi)$ when ξ tends to x from the right and left. With these examples I wish to argue that Dirichlet realized that with the new concept of function there was no hope of a universal domain of analysis. Each theorem required its own function class that was tailored to the theorem.

In all fairness I must mention that after Dirichlet had proved "in a rigorous way" that Fourier series converge and have the correct sum for all piecewise continuous and piecewise monotonous functions he claimed that he could generalize the result to functions whose points of discontinuity are nowhere dense, i.e. to the functions that he claimed could be integrated. However, as the claim of integrability he relegated the generalization of his convergence result to a later paper. There are good reasons why this paper never appeared. Indeed, du Bois-Reymond later (1876) gave an example of a continuous function whose Fourier series does not converge toward the function. Thus Dirichlet's final delimitation of the functions whose Fourier series converge toward the function turned out to be too large. However, this does not seriously affect my conclusion that Dirichlet was the first mathematician who gave up the dream of a universal domain of analysis and who introduced theorem-generated and proof-generated function classes. After Dirichlet it has gradually become commonplace that an analytic theorem about functions comes with a class of functions for which the the theorem holds true and whose defining properties are explicitly needed in a proof of the theorem.

8 The flexibility of the new concept of function

We have seen that it was applications that forced mathematicians like Euler to generalize the concept of function and we have seen that in so far as functions that were not described by an analytic expression were used at all they were considered as complicating factors that threatened the rigor of mathematical argumentation. Dirichlet was to my knowledge the first mathematician who showed that the greater generality and flexibility of the new concept of function could sometimes make life (or at least proofs) easier without ruining the rigor of the arguments. This flexibility was a result of the local nature of the concept of function. Dirichlet stressed this property of the concept of function in the following remark to his definition of a (continuous) function:

> As long as one has only determined a function on a part of an interval the nature of its continuation to the rest of the interval is left totally arbitrary. (Dirichlet, 1837, p. 136, translation by JL)

He used this property in his proof of the central theorem that has the convergence of Fourier series as a simple consequence:

Theorem: If $f(x)$ is continuous and monotonous on the interval $[a,b] \subseteq \left[0, \frac{\pi}{2}\right]$ then

$$\int_a^b f(x) \frac{\sin nx}{\sin x} dx \tag{1}$$

converges to 0 for n going to infinity except when $a = 0$ in which case it converges to $\frac{\pi}{2} f(0)$.

Dirichlet first proved the theorem when $a = 0$. This proof is justly admired for its ingenuity and its rigor, but I shall not enter into the details. My point relates to the subsequent discussion of the case $a > 0$. Dirichlet first remarked that one could deal with this case through considerations "analogous" to the previous proof. "However", he continued "it is simpler to reduce this new case to those[11] we have considered above" (Dirichlet, 1829, p. 127, translation by JL).[12] He made this reduction by remarking that since the function f is only given on the interval $[a,b]$ where $a > 0$ one can continue it down to zero in an arbitrary way. In particular one can choose to continue it in such a way that it remains continuous and monotonous on the entire interval $[0,b]$. According to the proof he had given for the case $a = 0$ the integrals

$$\int_0^b f(x) \frac{\sin nx}{\sin x} dx \quad \text{and} \quad \int_0^a f(x) \frac{\sin nx}{\sin x} dx$$

where f now designates the extended function both converge toward $\frac{\pi}{2} f(0)$ and by subtracting the second from the first Dirichlet concluded that (1) will converge to zero.

It is my impression that Dirichlet's 18th century predecessors like Euler would have considered this argument as invalid. According to 18th century ways of thinking the function f is globally determined even if we chose to consider it on the interval $[a,b]$ only. Tinkering with its values outside $[a,b]$ in an artificial way will make the argument unconvincing. Indeed, they might argue, how does an argument related to two integrals of the altered function give a result about the original function?

To a modern mathematician who has grown up with Dirichlet's concept of function, such objections seem totally misconceived and irrelevant. We consider Dirichlet's argument to be entirely convincing and rather straightforward. This may also be why the novelty of the argument has not caught the attention of earlier historians. However, Dirichlet's rhetoric strategy indicates that he was aware of the novel nature of his argument. He twice emphasized that a function defined on an interval can be extended in an arbitrary manner outside of the interval: First he stated it after the definition of a (continuous) function and then he repeated it

---

[11] The use of the plural here is a bit confusing. Dirichlet refers to the cases where $a = 0$ and $b \in [0, \pi/2]$.

[12] This argument was repeated in (Dirichlet, 1837, p. 154–5).

when he used it in the proof. Today, this property is considered such an immediate consequence of the concept of function that we do not feel the need to mention it in textbooks on calculus. On the other hand, the unique analytic continuation property of holomorphic functions is considered an almost magical property that is highlighted in every book on complex function theory. Dirichlet's repeated emphasis on our freedom to chose a continuation indicates that he was aware that this went against the practice of his predecessors who had often used arguments similar to the unique analytic continuation property also in cases where Dirichlet and we would think it was not warranted.

One other feature of Dirichlet's presentation of the proof points to his awareness of the novel nature of the argument and the objections it could raise among more traditionally thinking mathematicians: We saw that he formulated the continuation property as follows: when a function f is only defined in an interval $[a, b]$ one can chose a continuation outside the interval in an arbitrary way. However, when he applied the property in the proof of the convergence of Fourier series, the situation was in fact slightly different. Here the function f was in fact defined on the entire interval $\left[0, \frac{\pi}{2}\right]$ and Dirichlet considered a subinterval $[a, b]$ on which f was continuous and monotonous. He then *changed* the function outside of the interval $[a, b]$ in such a way that the new function was continuous and monotonous on the entire interval $[0, b]$. Such a *replacement* of the "true" course of a function with another course in parts of the domain had earlier met with outright rejection from d'Alembert and it had confused Euler. Knowledge of such earlier rejections may have caused Dirichlet to downplay the radical nature of the continuation property. Instead of claiming that one can replace a function with a new one in an interval, he stated that one can arbitrarily continue a function to values for which it was not previously defined. He may have considered this statement as less prone to objections than the replaceability property. Be that as it may, with Dirichlet such constructive usage of the local nature of the new concept of function became acceptable in analysis.

The above mentioned argument can in my opinion be considered a paradigmatic example of the use of the new concept of function. It captures an important aspect of the new concept that is entirely foreign to the old concept. So if we accept that mathematical concepts are constituted by paradigmatic use as well as by explicit definitions one can say that Dirichlet here introduced a new concept of function. In the gradual change of the old concept into a new one Dirichlet's use in the above proof strikes me as the greatest single step in which analytic expressions and their global character was completely discarded and replaced with the arbitrary and locally defined variable dependences. This is the main reason for attributing the new concept to Dirichlet.

About half a century after Dirichlet the concept of function underwent a new change. Indeed to Dirichlet as to Euler a function was a *variable* (depending on another variable). In modern analysis on the other hand, this concept of a variable is not defined in a precise way but is used as a figure of speech only. Instead we define a function as a mapping (i.e. as the dependence itself) and this mapping can in turn be defined as a subset of a Cartesian product. This change from variable to mapping is conceptually important, but precisely because we have kept the old figurative way of talking about functions as variables the change has made less difference in mathematical practice than the change from Euler's (first) concept of function to Dirichlet's concept. This justifies that one sometimes calls the modern concept after Dirichlet as well.

9 Conclusion

I have given four reasons why I consider Dirichlet the true originator of the concept of function defined as a variable depending in an arbitrary way on another variable. The most compelling reason, which to my knowledge has not been emphasized before, is that Dirichlet as the first mathematician saw how to use the flexibility of the new concept constructively in a proof. More specifically he simplified a proof considerably by changing the values of a function on a specific interval. This procedure is no big deal for a modern mathematician, but Dirichlet's way of dealing with it indicates that he was aware of the novel nature of such a step in a proof. It shows that he had thoroughly appreciated the local nature of the new concept of function and knew how to take advantage of it. Earlier mathematicians starting with Euler had formulated function definitions similar to Dirichlet's but no one had used it in ways similar to Dirichlet. In fact his predecessors had often used arguments that were borrowed from the earlier concept of function defined as an analytic expression. This in my opinion shows that the earlier mathematicians had a concept of function that was not identical to Dirichlet's (in fact a concept which was much less general than Dirichlet's concept). This conclusion is based on my conviction that a mathematical concept is determined at least as much by its use as by its definition.

References

Alembert, J. R. d'. (1747). Recherches sur la courbe que forme une corde tentuë mise en vibration, *Mém. Acad.Sci Berlin* **3**: 214–9.

Alembert, J. R. d'. (1750). Addition au mémoire sur la courbe que forme une corde tendüe mise en vibration, *Mém. Acad.Sci Berlin* **6**: 355–66.

Bois-Reymond, P. du. (1876). Untersuchungen über die Konvergenz und Divergenz der Fourierschen Darstellungsformeln, *Abh. bayer. Akad. Wiss.*, II Kl., 12, Pt 2, i-xxiv, 1-102, *Ostwald's Klassiker* **186**.

Bottazzini, U. (1986). *The Higher Calculus*, Springer, New York.

Cauchy, A.-L. (1821). Cours d'analyse de l'École Royale Polytechnique. Première partie, *Analyse algébrique*, Débure frères, Libraires du Roi et la Bibliothèque du Roi, Paris.

Cauchy, A.-L. (1823). *Résumé des leçons données à l'École Royale Polytechnique sur le calcul infinitésimal. Tome premier*, Paris.

Dirichlet, J. P. G. (1829). Sur la convergence des séries trigonomé triques qui servent à représenter une fonction arbitraire entre des limites données, *Journ. reine ang. Math* **4**: 157–69. Page references are to *G. Lejeune Dirichlet's Werke* 1. vol, Georg Reimer, Berlin, 1889, pp. 117–32.

Dirichlet, J. P. G. (1837). Über die Darstellung ganz willkürlicher Funktionen durch sinus- und cosinus-Reihen, *Repert Phys* **1**: 152–74. *Ostwald's Klassiker* **116**: 3-34. Page references are to *Werke* I, 133–60.

Euler, L. (1748a). *Introductio in Analysin Infinitorum*, Vol. 1, Lausanne. The text is also found in *Opera Omnia* (1) 8.

Euler, L. (1748b). Sur la vibration des cordes, *Mém. Acad.Sci Berlin* **4**. The text is also found in *Opera Omnia* (2) 10, 63-77.

Euler, L. (1749). De la controverse entre Mrs. Leibniz et Bernoulli sur les logarithmes des nombres négatifs et imaginaires, *Mém. Acad.Sci Berlin* **5**: 139–79. The text is also found in *Opera Omnia* (1) 17, 195-232.

Euler, L. (1753). Remarques sur les mémoires précédents de M. Bernoulli, *Mém. Acad.Sci Berlin* **9**: 196–222. The text is also found in *Opera Omnia* (2) 10, 233-54.

Euler, L. (1755). *Institutiones Calculi Differentialis*, St. Petersburg. The text also found in *Opera Omnia* (1) 10.

Euler, L. (1763). De usu functionum discontinuarum in analysi, *Novi Comm. Acad. Sci. Petrop.* **11**: 67–102. The text is also found in *Opera Omnia* (1) 23, 74–91.

Euler, L. (1768–70). *Institutiones Calculi Integralis*, Vol. 1–3, St. Petersburg. The text is also found in *Opera Omnia* (1) 11–13.

Fourier, J. (1822). *Théorie Analytique de la Chaleur*, Firmin Dido, Paris. *Oeuvres de Fourier*, 1. vol, Gauthier-Villars, Paris, 1888.

Grabiner, J. V. (1981). *The Origins of Cauchy's Rigorous Calculus*, Dover, Cambridge, MA.

Hankel, H. (1882). Untersuchungen über die unendlich oft oscillirenden und unstetigen Functionen, *Math. Ann.* **20**(1): 63–112. *Ostwald's Klassiker,* **153**, 44-102.

Jahnke, H. N. (2002). The Algebraic Analysis of the 18th Century, *A History of Analysis*, American Mathematical Society, Providence R.I.

Lagrange, J. L. (1801). Leçons sur le calcul des fonctions, Paris. Oeuvres 10.

Lützen, J. (2003). Between Rigor and Applications: Developments in the Concept of Function in Mathematical Analysis, *The Cambridge History of Science vol. 5. The Modern Physical and Mathematical Sciences*, Cambridge, pp. 468–487.

Poincaré, H. (1899). La logique et l'intuition dans la science mathématique et dans l'enseignement, *L'Enseignement Mathématique* **1**: 157–62.

Rüthing, D. (1984). Some Definitions of the Concept of Function from Joh. Bernoulli to N. Bourbaki, *The Mathematical Intelligencer* **6**: 72–7.

Youschkevich, A. P. (1976). The Concept of Function up to the Middle of the 19th Century, *Archive for History of Exact Sciences* **16**: 37–85.

Kant and the natural numbers

Klaus Frovin Jørgensen

Section for Philosophy and Science Studies, Roskilde University

1 Introduction

The ontological status of mathematical objects is perhaps the most important un-solved problem in the philosophy of mathematics. It is thought-provoking that within the philosophy of mathematics there is no agreement on what the mathe-matical objects generally are. It is, moreover, surprising that not even the question of the ontology of the natural numbers is resolved, as these objects arguably are the most fundamental objects in mathematics.

In this paper I present a theory of the natural numbers which is based on the notion of *schema*. In fact the theory is based on a slight generalization of Kant's notion of objects and corresponding schemata. As such the theory is compatible with a kind of constructivism and the result is that numbers certainly do not be-long to some transcendent Platonic realm. Contrary to the Platonic account the following theory presents numbers as a necessary element within *human* cogni-tion. Numbers play a very essential role in our conceptualization of the world. As we shall see, a particular number is not characterized extensionally by objects rep-resenting the number. As we will will be working within a general constructivist epistemology the extension is—as we shall see below—not constant. On the other hand, the properties of numbers are independent of time; thus numbers are deter-mined intensionally. The task of this paper is to state the details of this intensional aspect of the natural numbers. In Kant's theory of schemata this intensional aspect consist of number understood as a schema. Thus we will have to go back and see what Kant understood under this notion.[1]

---

[1] Kant's theory is very interesting in itself. It sheds light upon important topics such as the *objective* and the *subjective* and their interplay; it includes a theory about *mathe-matical reasoning* in general; it initiates a theory about *diagrammatic reasoning*; and it

In a short but unclear chapter entitled Schematism in the *Critique of Pure Reason* Kant outlined what he generally understood under the notion of Schema. Now, the first half of that book has three very important parts: The Deduction, the Schematism and the Principles. All three of them concern the relation between (pure) concepts and sensibility. Roughly it can be said, that the Deduction shows (or is intended to show) that there is harmony between the pure concepts of the understanding and the way appearances are given. This harmony ensures that appearances *can* be cognized under categories. The Schematism shows (or is intended to show) *how* the appearances are subsumed under categories. The Principles show (or are intended to show) what the general *conclusions* to be drawn are.

But the chapter on the Schematism concerns more generally the connection between the intellectual (the conceptual) and the physical (the sensible); it is not only about the schematism of the categories but about schematism of:

1. Empirical concepts,
2. Pure sensible (i.e., geometrical) concepts,
3. Pure concepts of understanding.

According to Kant, the categories can be divided into four different classes: The categories of i) quantity, ii) quality, iii) relation and iv) modality. In this essay I will concentrate mainly on the categories of quantity since the natural numbers are understood as a rule, namely the schema for the categories of quantity.

2 Number as schema

In the chapter on schematism we find the following description of the connection between size, number and schema:

> The pure **schema of magnitude** [*Größe*] (*quantitatis*), however, as a concept of the understanding, is **number,** which is a representation that summarizes [*zusammenfaßt*] the successive addition of one (homogeneous) unit to another. (A142/B182)

It must be admitted that even with regard to some of the most fundamental aspects of his epistemology Kant is not completely clear. Here, however, it seems clear that "magnitude [*Größe*]" is a category—but not which one. In the "Table of Categories" (A80/B106) there are four main divisions of the categories; the first one

---

anticipates the *type-token* distinction found in contemporary epistemology. Thus Kant's theory of schemata is interesting independently of his general theory of knowledge, i.e., his empirical realism and transcendental idealism. For an elaboration on these elements of schematism see my dissertation (Jørgensen, 2005).

being "Of Quantity [*Der Quantität*]". This consists of the three categories *unity,*
plurality and *totality*. Now, "magnitude" could either refer to the collection of the
three categories, or it could be one of them. Let us elaborate a little on this.

Quite generally Kant claims that it is *equivalent* to cognize and to perform
judgments.[2] Therefore "the Clue to the Discovery of all Pure Concepts of the Understanding" (A70/B95) goes via the different ways we form judgments. In respect
to quantity, Kant is completely Aristotelian: There are *singular, particular* and *uni-*
versal judgments (for instance, 'this body has mass', 'some bodies have mass' and
'all bodies have mass'). These different types of judgments lead Kant to the three
categories unity, plurality and totality. The singular judgment corresponds to the
category of unity; the particular judgment to plurality and the universal to totality.[3]
Totally there are 12 categories, and just after stating what "the schema of magnitude" is Kant continues the Schematism by listing the nine remaining schemata—
one after the other (A143–7/B182–7). This can be taken as evidence indicating
that Kant took "the schema of magnitude" to be a schema common to the three
categories unity, plurality and totality. And as Longuenesse notes (1998, p. 254),
the three different categories seem to be involved in the definition of the schema:
unity ("units"); plurality ("successive addition of one (homogeneous) unit to another"); and totality ("a representation that summarizes the successive addition of
one (homogeneous) unit to another"). I think, however, that it is more plausible,
to understand "the schema of magnitude" to be the schema of *totality*.[4] The third
category in all the four divisions is always understood as a "combination of the
first and second in order to bring forth the third" (B111). "Thus **allness** (totality) is
nothing other than plurality considered as a unity" (B111). In *Prolegomena*, when

---

[2] See for instance the important A69/B94:

> We can, however, trace all actions of the understanding back to judgments, so that
> the **understanding** in general can be represented as a **faculty for judging.** For
> according to what has been said above it is a faculty for thinking. Thinking is cog-
> nition through concepts. Concepts, however, as predicates of possible judgments,
> are related to some representation of a still undetermined object.

[3] It is remarkable that it is not even completely clear from Kant's texts precisely what the
correspondence between the Quantity of Judgments and the Table of Categories is in
this case. Just like Paton (1936, II, p. 44) and Longuenesse (1998, p. 249), I understand
Kant in such a way that the (Aristotelian) judgments are given in the traditional order:
Universal–particular–singular, whereas the order of the categories are given as unity–
plurality–universal. Therefore, the order of one of the tables is the reverse of the other.
This interpretation is in contrast with Hartnack (1968, p. 39) and (Tiles, 2004). Tiles has
a refined argument for the opposite of the view expressed here.

[4] Here I am close to Longuenesse (1998, pp. 253–255), although she seems to *identify*
plurality and totality.

listing the categories, Kant terms the first category of Quantity as "Unity (Measure [*das Maß*])" (Ak. 4, p. 303).[5] Therefore, *Unity* is the concept we use when we judge something to be a unit; 'this body—taken as a unit—has mass'; but also 'this unit can be taken as a unit, when we want to count—it can be seen as (giving rise to) a measure'. A determination of a unit 'body' together with a judgment 'here is a plurality of divisible bodies' can form a new judgment, when we think another unit: 'All bodies in *this collection* are divisible'.

> [T]o bring forth the third [pure] concept requires a special act of the understanding, which is not identical with that act performed in the first and second. Thus the concept of a **number** (which belongs to the category of allness) is not always possible wherever the concepts of multitude and of unity are (e.g., in the representation of the infinite) (B111).

Therefore, the category of totality is not reducible to unity and plurality and, moreover, it is "number" which belongs to totality. If we re-read the definition of "the schema of magnitude" in this light we see that it corresponds very well to the act performed, when totality—as a combination of unity and plurality—is performed: A certain unit (measure) has been determined; and we have encountered a plurality of these units. When reflecting *on* this plurality, we form a set out of the homogeneous elements and *enumerate* (summarize) the elements. This enumeration ends with a number, which is the number of elements in the set.[6] Therefore, when we apply "the schema of magnitude" we think unity in a given plurality. Generally "the schema of magnitude" is the ability humans have for determining finite extensions of empirical concepts. For instance, 'there are five fingers on my

---

[5] I use (Ak. *volume*, p. *page*) when I refer to *Kants Gesammelte Werke* in the so-called Akademie version.

[6] But as Kant notes, we cannot reduce the third category to the two foregoing, because sometimes we encounter an infinite plurality of units which do not form a set. In this way, the set-class distinction from set theory is reflected in Kant's epistemology. We can for instance, form a sequence of growing finite (in the sense that the volume is finite) empirical spaces. The space containing my office; the space containing my city; the space containing my country, the earth and so on. All these spaces are possible objects of intuition. But the *collection* of all finite space—the absolute empirical space—which is needed for Newtonian mechanics in the definition of absolute motion is not a possible object of intuition and "absolute space is *in itself* therefore nothing and indeed no object at all" as Kant writes in *Metaphysical Foundation* (Ak. 4, p. 481). I can measure all the finite spaces, but their collection is not measurable. Nevertheless, the concept of "*motion in absolute* (immovable) *space*" "in general natural science is unavoidable" (Ak. 4, p. 558), so "[a]bsolute space is therefore necessary, not as a concept of an actual object, but rather as an idea", i.e., as a concept of *reason* (Ak. 4, p. 560). In set theory any *set* of ordinals is itself measurable by an ordinal, but the class of all ordinals, is not a set, because it is not measurable by any ordinal.

left hand': The unit is 'finger'; the context is 'my left hand'; and there are five of them in total. The rule determining this act of enumerating and counting is the schema called "number".

But how precisely do we operate with this schema? It is a transcendental schema and there are certain important differences between the transcendental schemata and the geometrical schemata. "The schema of a pure concept of the understanding, on the contrary [to empirical and geometrical schemata], is something that can never be brought to an image at all" (A142/B181).[7] Rather than producing images, transcendental schemata provide a "transcendental time-determination" (A138–9/B178–9) of the objects given in experience. As empirical objects are given to us in outer sense (space) and time-determination is a determination of moments according to inner sense,[8] this determination proceeds *mediately* by way of inner images.[9] Thus the schematism of pure concepts—in contrast to geometrical and empirical concepts—is more about *reflection* on images, rather than *construction*. This observation somehow runs counter to Shabel's (2003, p. 109) claim that the diagrammatic reasoning in Euclid—which Kant supplies an epistemological analysis of in terms of geometrical schemata—"provides an interpretive model for the function of a transcendental schema". Nevertheless, it will become clear that Kant's theory of geometrical schemata and his theory of transcendental schemata have many properties in common. The most important one being that schemata provide a foundation and explanation of the use of types and tokens.

In the following I will give my interpretation of the transcendental "**schema of magnitude**". As will be clear this really is an interpretation. Kant does not write much about the transcendental schemata and many of the details are 'left to the reader as an easy exercise'. But this exercise is not an easy one, as Kant certainly is not very clear on the whole issue. This is in contrast with the geometrical schemata about which Kant's theory is much clearer. In consequence of this, I will in the first place not stay close to the text. This is for the sake of giving a coherent (as coherent as it can be, at least) interpretation of the schema of magnitude. After having provided my interpretation I will go back to Kant's text and see whether or

---

[7] In fact, this fundamental difference between the arithmetical schema number and the geometrical schemata leads Kant to reject the possibility axioms for arithmetic. This well-known claim of Kant is expressed on pages A163–4/B205. This view seems rather awkward in a contemporary understanding and I will elaborate on it below.

[8] "Time is nothing other than the form of inner sense, i.e., of the intuition of our self and our inner state. For time cannot be a determination of outer appearances; it belongs neither to a shape or a position, etc., but on the contrary determines the relation of representations in our inner state." (A33/B49-50)

[9] "[T]ime is an *a priori* condition of all appearance in general, and indeed the immediate condition of the inner intuition (of our souls), and thereby also the mediate condition of outer appearances." (A34/B51)

not I am able to supply difficult passages with meaning. I think my interpretation should be more than just compatible with what Kant originally wrote. It should also shed new light on difficult passages. If I succeed in doing this my interpretation will be justified.

Let me return for a moment to the empirical schemata which I will give the following interpretation. Let us view the collection of all the different images representing empirical objects as a constructive and non-monotone open-ended universe. It is not monotonic as our empirical concepts may vary over time and it is constructive in the sense that we produce images in time in accordance with rules. Furthermore, it is open-ended as our collection of empirical concepts is certainly not fixed once and for all. However, given a point i in time we can take the universe of images which are *in principle constructible* according to the collection of empirical concepts we may possess at i. Let us call this snapshot U_i. Any empirical concept partitions U_i in two sets. One of the sets consisting of the images representing the concept, the other set consisting of those which do not represent the concept.[10]

"[M]agnitude (*quantitatis*)" concerns the question: "How big is something?" (A163/B204). An example could be: 'How many fingers are there on my left hand?'. Let $\prec$ denote the order of time.[11] I have experienced a plurality of fingers and let $i \prec j \prec k \prec l \prec m$ be the different moments in time corresponding to these experiences. Although the sequence of image-universes is generally not monotone, let me assume that *locally* there is monotonicity such that the partitions of $U_i, \ldots, U_m$ consisting of finger-images are the same from i to m. Alternatively we could assume that the schema belonging to the concept 'finger-on-my-left-hand' is constant. Such a pragmatic assumption seems reasonable.[12] Therefore *any* image from the partition can be used to represent any finger from time i to time m. Let x be such an image. x represents any of the fingers that I experience on my left hand. The minimal requirement making the images different is their location in time.[13]

---

[10] Of course this is an idealization which is perhaps not fully justified. Empirical concepts are vague concepts, and therefore it is perhaps not possible, given a concept P, to form two *disjoint* sets U_i^P and $U_i^{\neg P}$ such that $U_i = U_i^P \cup U_i^{\neg P}$. This problem is, however, not a problem which threatens the interpretation of "the schema of magnitude", and therefore I will make this idealization.

[11] Kant's precise understanding of the order of time is not important for this example. For our example it only matters that the collection of past moments is linearly ordered.

[12] The assumption of monotonicity locally around a concept seems reasonable, given that I in the time from i to m do not discover new essential properties of the concept, and given that I do not forget any of the essential properties either.

[13] We could also imagine another way of making the images different, namely that they are cognized as fingers with coordinates in space. This would also separate the images. But this would only be an additional property making the images different, because we can-

Time "determines the relation of representations in our inner state" (A33/B49-50). Therefore we have *temporized* images: x_i, x_j, x_k, x_l and x_m. This is my interpretation of Kant's "transcendental time-determination", and note that the temporal indexes are *necessary* for my judgment 'there is a plurality of fingers'. Now, by judging unity in plurality I form the set M consisting of the temporized images which I simultaneously count ("summarize"). This is done by enumerating M. Mathematically speaking this is a determination of a set of natural numbers which is equinumerous with M. In other words, we determine a set of natural numbers N and establish an injection f from N to M. The canonical domain for f is, of course, $\{1, 2, 3, 4, 5\}$. Figure 1 represents this mental process.

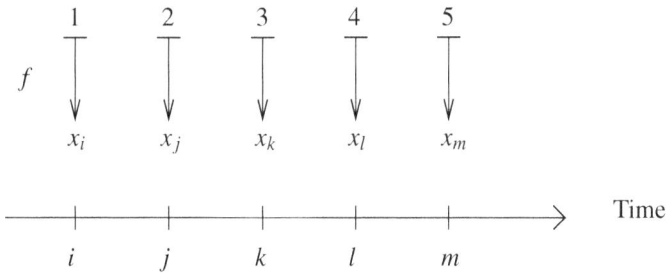

Fig. 1.

On pages A142–3/B182 Kant defines the "schema of magnitude" in one paragraph. I quoted gave most of the first half of that paragraph on page 66. The second part goes:

> Thus number is nothing other than the unity of the synthesis of the manifold of a homogeneous intuition in general, because I generate time itself in the apprehension of the intuition. (A142–3/B182)

In terms of the finger-example my interpretation is the following:

Firstly, I can produce the injection f "because I generate time itself in the apprehension of the intuition"—it is I who locate the images according to inner

---

not help that the images are *constructed* in time. Thus the images with spatial coordinates would be different both with respect to time and coordinates.

sense. This temporization is necessary. Secondly, that "number is nothing other than the unity of the synthesis of the manifold of a homogeneous intuition in general" means that the number 5 describes the unity of my set of fingers in the sense that a bijection (f is of course also surjective) can be established between the set of natural numbers less than or equal to 5 and M. But what precisely does the correspondence between 5 and M consist of?

The order of time $\prec$ induces a natural order on the temporized images; let $\prec$ denote also the induced order. The function f specified in Figure 1 is actually showing that $(\{1,2,3,4,5\}, <)$ and $(M, \prec)$ are isomorphic. Could it be a notion along these lines that Kant has in mind when saying that number 5 *is* "the unity of the synthesis"? The bijection established would then show that the two sets are *equal* up to isomorphism. I do not think this is Kant's idea. We are only interested in the *size (Größe)* of M, not its order. It is therefore 5 as a cardinal number and not as an ordinal number Kant is interested in. The unity Kant is after is the unity which is expressed through *cardinality*. A contemporary mathematical understanding of size is the following. A and B are *equinumerous* or *equal in cardinality*, if a bijection between the two exists:

$$A =_c B, \text{ if and only if, there exists a bijection } f \text{ such that } f : A \to B.$$

Of course, given a set N of temporized images with n elements any bijection g between N and $\{1, \ldots, n\}$ gives rise to an isomorphism between $(\{1, \ldots, n\}, <)$ and $(N, <_g)$, where $<_g$ is the order induced by g—but this additional information is not what Kant has in mind. It is the information which lies *in the process* determining the cardinality of any finite set, but one has to pay attention to this. I understand Kant as saying that this type of information is not what we are after when applying the category of totality, although it can be unwinded from the process. Thus, in the finger example 5 reflects the unity of the synthesis with respect to *cardinality*. The relation of equality is not isomorphism, rather it is $=_c$. The function f depicted in Figure 1 could consequently be any bijection between M and $\{1, 2, 3, 4, 5\}$.[14]

I can now give my general interpretation of the "schema of magnitude". The task of the schema is to give unity to a succession of objects thought under the same concept; objects that we represent as temporized images. In our interaction with the world we temporize images. Before temporization the images belong to the universe of images constructible in principle. For the sake of presentational simplicity let us assume that this universe is constant, i.e., that my empirical schemata generating images are the same over time. Later I will dismiss this restriction. From this universe we form, by "transcendental time-determination" a temporized derivative, namely, the universe U of temporized images.

The general problem concerns the possible unity of a plurality of experiences of objects falling under a concept. Let us in accordance with this purpose form a

---

[14] Of which there are many: 5!.

second-order universe $\mathscr{U}$ consisting of sets of temporized images representing the same concept. Thus the definition of $\mathscr{U}$:

$M \in \mathscr{U}$ iff M is a subset of U and the members of M represent the same concept.

Clearly the notion of *equal in cardinality*, $=_c$ is an equivalence relation on $\mathscr{U}$. The "schema of magnitude", which Kant calls "number", is the general rule which generates this equivalence relation. When we judge unity in a plurality of homogeneous units we determine the equivalence class to which the unit belongs. And if no such class can be found we cannot judge unity: A plurality of experienced units can be judged to be a unit itself *only if* we can determine a cardinality of this plurality. Therefore, the generation of the members of $\mathscr{U}$ proceeds by production of the equivalence classes. An M becomes a member of $\mathscr{U}$, when it becomes a member of an equivalence class of $\mathscr{U}$. Consequently, it is a genuine *constructive* notion of existence.

Let us now see, what the content of the rule "number" is. First of all, it consists of a "transcendental time-determination", as otherwise we cannot distinguish between images representing the same concept. But what is the content of the act which partitions $\mathscr{U}$ in accordance with $=_c$? It is our ability to enumerate finite sets, and the mathematical description of this is the ability to produce bijections between finite sets.

In this way, we see that the "schema of magnitude" is reflective because of reflection on already constructed first-order level images. But it is constructive on a second-order level; the sets of $\mathscr{U}$ are determined (second-order constructed) by the schema, and they live in $\mathscr{U}$ only when this determination has taken place. Moreover, the schema is an *act* of the understanding which itself takes place in time, temporal succession is a transcendental condition:

> No one can define the concept of magnitude in general except by something like this: That it is the determination of a thing through which it can be thought how many units are posited in it. Only this how–many–times is grounded on successive repetition, thus on time and the synthesis (of the homogeneous) in it. (A242/B300)

Let us now dismiss the restriction posed on the universe of images. Thus a certain invariance of meaning of images can happen, as an image can represent a concept at a certain time and not represent that concept at another time. As a consequence of this the universe $\mathscr{U}$ is a dynamical floating universe where the elements of the equivalence classes defined by $=_c$ are not the same over time. Does this pose a problem for the status of the *pure* concept of totality? Certainly not, given any variant of $\mathscr{U}$ the rules determining equivalence classes according to $=_c$ are precisely the same. The underlying universe may vary, as our empirical concepts vary, but the structure imposed on it is the same. Of course there are extensions of empirical concepts which we are unable to determine, of which the sorites paradox provides an excellent example. But this is not a problem for our theory of magnitude, it is (perhaps) a problem for our theory of empirical concepts.

3 Number as concept

It must be admitted that my interpretation is somewhat involved. I am, for instance, using notions like equivalence-class and second order object—notions which were not present in the mathematics of Kant's time. On the other hand, it is also true that the bijections involved are constructively fully meaningful: We are only working with functions from finite sets to finite sets—the essence of that is simply to pair objects from two different finite sets. Moreover, the universes mentioned should certainly *not* be understood in some kind of Platonic sense—rather they are universes constructed by a cognizing human.

Some interpretation *is* needed as Kant's own text is unclear and lacks important details. Kant designates the constructive procedure used when counting as "**number**".[15] But from his text (or texts) alone it is not clear precisely what he means. For a full justification of my interpretation I should be able, however, to explain central themes in the Kantian theory of transcendental schemata. One of the most distinctive ones is, that arithmetic has no axioms. I will take that up in the last section. Another important distinction in Kant's theory is the distinction between 'number as concept' as opposed to 'number as schema'.

Kant notes on A142/B181 that a transcendental schema "is something that can never be brought to an image at all". His point is that there are no generic images of the pure concepts. In contrast to empirical schemata, pure schemata do not provide images representing the pure concept. How would a paradigmatic image of causality look, for instance?[16] Somehow in contrast to this Kant writes, when discussing the difference between image and schema, that

> if I place five points in a row, ..., this is an image of the number five. On the contrary, if I only think a number in general, which could be five or a hundred, this thinking is more the representation of a method for representing a multitude (e.g., a thousand) in accordance with a certain concept than the image itself (A149/B179)

There is, therefore, a concept for the number five, but there is also a schema called "number" which is an act of the understanding or a "representation of a method

---

[15] Note also that in German to count is *zählen*—a derivative of *Zahl*.

[16] One who was looking for a paradigmatic image of causality was Hume. According to Kant this cannot be found.

for representing a multitude". Therefore we have to distinguish between the concept number and the schema number. The number five is *not* the rule synthesizing my finger-images x_i, x_j, x_k, x_l, x_m, but rather a concept of the specific size—the cardinality—of the corresponding set. This cardinality is realized through the enumeration which simultaneously determines set-hood (unity) and cardinality. Thus number is a concept under which a multiplicity is thought. This concept is thought through the schema "number". The schema is the *procedure*, i.e., a rule-governed activity, that we use to determine whether a given collection of sensible things exhibits unity in plurality.[17]

Let $\bar{n}$ denote the class of elements in $\mathcal{U}$ which are equivalent $\{1, \ldots, n\}$ with respect to cardinality, in other words:

$$\bar{n} = \{\, \{x_1, \ldots, x_n\} \in \mathcal{U} \mid \{x_1, \ldots, x_n\} =_c \{1, \ldots, n\} \,\}.$$

Now, I propose to understand the *concept* of a particular number n as a type, in fact more specifically as the corresponding equivalence class $\bar{n}$. We know, however, that the elements of this equivalence class is not constant: An element of $\mathcal{U}$ is constructed as a second-order object at a certain moment in time. Through this construction the element becomes a member of an equivalence class. Therefore, if we understand the concept five as an equivalence class in the set theoretic sense described above, then the equivalence class is *not* determined by its extension, it is rather determined intensionally, namely the "schema of magnitude" which amounts to the capacity of producing bijections. Only this *intensional* aspect can guarantee that number concepts remain the same over time. Therefore, if the $\bar{x}$ and $\bar{y}$ are number concepts, possibly 'found' at different moments in time, then the equality of $\bar{x}$ and $\bar{y}$ is determined not by their extensions but by the bijections—understood as rules—on which they are generated. In other words $\bar{x} = \bar{y}$, if and only if, the two canonical bijections are the same.

We understand a concept of a certain number as a cardinal number being a *type* whose tokens are members of the corresponding equivalence class. The "schema of magnitude" decides the relation between the type $\bar{n}$ and the tokens falling under this type. It is due to the intensional aspect that the properties a particular number concept are independent of time. Moreover, each and all of the (finite) cardinal numbers are founded by the same schema. See Figure 2 for a diagram represent-

---

[17] In the A-deduction Kant writes: " If, in counting, I forget that the units that how hover before my senses were successively added to each other by me, then I would not cognize the generation of the multitude through this successive addition of one to the other, and consequently I would not cognize the number; for this concept consist solely in the consciousness of this unity of the synthesis" (A103).

ing my interpretation.[18] So the schema and the concept of number certainly are different.

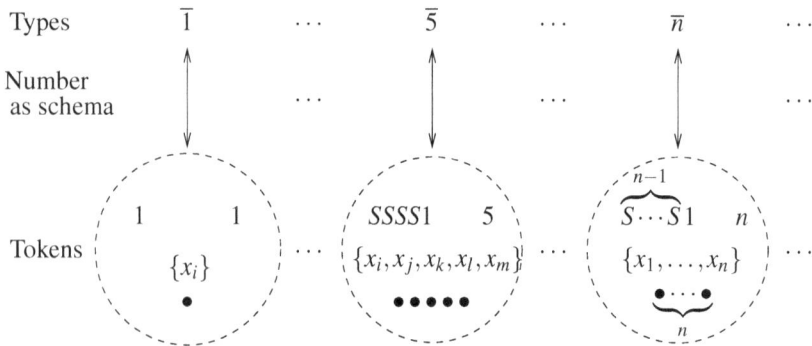

Types $\bar{1}$ $\cdots$ $\bar{5}$ $\cdots$ $\bar{n}$ $\cdots$

Number as schema

Tokens

Fig. 2.

4 The arithmetical schema and universality

In the case of geometry, geometrical schemata make reasoning about say the general triangle possible through reasoning about one particular.[19] My claim is that Kant holds the view that we are in a similar position in the case of arithmetic. The form of universality which emerges in the case of arithmetical schemata can be gained by using "the schema of magnitude". On the face of it, this may seem problematic as the extension of an arithmetical equivalence class is not constant. On the other hand, the members of the geometrical types are pure intuitions and a pure intuition belongs to the type triangle, if and only if, it can be constructed in pure intuition by the schema of triangle. This property is invariant over time—therefore the geometrical types are extensionally determined. Nevertheless, when explaining the synthetic nature of propositions in arithmetic Kant writes:

> The concept of twelve is by no means already thought merely by my thinking of that unification of seven and five, and no matter how long I analyze

[18] Although generally my interpretation is different from the one given by Longuenesse (1998) I think my understanding of the difference between the concept number and the schema is close to her's, according to which "the concepts of number is the concept of a *determinate* quantity, and that number as a schema is the schema of *determinate* quantity" (1998, p. 256).

[19] See for instance (Jørgensen, 2005, pp. 23).

my concept of such a possible sum I will still not find twelve in it. One must go beyond these concepts, seeking assistance in the intuition that corresponds to one of the two, one's five fingers, say, or (as in Segner's arithmetic) five points, and one after another add the units of the five given in the intuition to the concept of seven. For I take first the number 7, and, as I take the fingers of my hand as an intuition for assistance with the concept of 5, to that image of mine I now add the units that I have previously taken together in order to constitute the number 5 one after another to the number 7, and thus see the number 12 arise. (B15–6)

This quote is central for an understanding of Kant's conception of numbers and arithmetic. Kant is claiming two important properties of arithmetic:

1. The simple propositions of arithmetic like $7 + 5 = 12$ are not analytic; "[o]ne must go beyond these concepts, seeking assistance in the intuition" in order to realize that it actually is the case that 5 added to 7 yields 12.
2. For the verification of the correctness of $7 + 5 = 12$ I can use contingent *empirical* representations of 5—fingers on my hand.

I will treat the syntheticity of numbers in the next section. Here I focus on the second point. How can I obtain a necessary proposition about all possible sets with cardinality 5 by using one particular set with cardinality 5? It could happen that I by accident could loose a finger, or that my concept of "finger-on-my-hand" would change over time such that a thumb would no longer be a finger. The answer is of course that due to the "schema of magnitude" we realize the following: When we use the set of (temporized images of the) five fingers we use it *only* with respect to its magnitude: We understand that we could have used any other set with five members, i.e., we could have used any other member of the equivalence class $\bar{5}$ for the verification. The situation is strikingly close to geometrical schemata; only there, however, it is about construction and not reflection:

> The individual drawn figure is empirical, and nevertheless serves to express the concept without damage to its universality, for in the case of this empirical intuition we have taken account only of the action of constructing the concept, to which many determinations, e.g., those of the magnitude of the sides and the angles, are entirely indifferent, and thus we have abstracted from these differences, which do not alter the concept of the triangle. (A713–4/B741–2)

In the case of the set of five finger-images, we realize they represent "the concept without damage to its universality, for in the case of this empirical intuition we have taken account only" of its size, i.e., its cardinality, as determined by the "schema of magnitude". The construction which takes place is on a higher level, namely that *we think* of the five fingers as constituting a set. We realize, however, that we could have used any other set with this magnitude, and the similarity to the geometrical reasoning is striking: When we prove a proposition about a concept, say the number five or a triangle, we take an intuition representing the concept *together* with the rules determining any intuition falling under that concept. The "schema of magnitude" thus ensures that we can operate with specific images of numbers, taking them as representatives of their types, prove properties about the images and be sure that these properties apply, not only to the specific images (i.e., the tokens) but to any image representing the type.

Precisely this aspect of Kant's schematism was simply not understood by G. Frege (1848–1925) in *The Foundations of Arithmetic* (1884/1980). In § 13 Frege ascribes to Kant the position that "each number has its own peculiarities. To what extent a given particular number can represent all the others, and at what point its own special character comes into play, cannot be laid down generally in advance." Well, the schema will take care of this, according to Kant.

It should be noted that in contrast to geometry it is not by first-order construction rather it is by reflection the determination of cardinality takes place. By this I mean the following. The construction which takes place in geometry is construction of pure intuitions which are first-order objects belonging to U. On the other hand, arithmetical reflection is construction of second-order objects, subsets of images, belonging to $\mathscr{U}$.

5 Syntheticity of arithmetical propositions: Space and time

Both space and time ground our notion of number. Kant's point when saying "[t]houghts without content are empty"[20] (A51/B75) is precisely that if, say the equivalence class corresponding to some number is empty, then thinking about that particular number is not possible. The elements of the equivalence classes are mental objects, of which there are two kinds: pure intuitions and images referring to empirical objects. But in either case they refer mediately or immediately to spatial objects. This relation between concepts and intuitions is exemplified in the relation between numbers as types and equivalence classes, by saying "[o]ne must go beyond these concepts, seeking assistance in the intuition" (B15). Pure intuitions, however, gain objectivity through and only through the empirical:

> Now the object cannot be given to a concept otherwise than in intuition, and, even if a pure intuition is possible *a priori* prior to the object, then even this can acquire its object, thus its objective validity only through empirical intuition, of which it is the mere form. Thus all concepts and with them all principles, however *a priori* they may be, are nevertheless related

---

[20] Which means 'concepts without intuitions'.

to empirical intuitions, i.e., to *data* for possible experience. Without this they have no objective validity at all, but are rather a mere play, whether it be with representations of the imagination or of the understanding. One need only take as an example the concepts of mathematics, and first, indeed, in their pure intuitions. Space has three dimensions, between two points there can be only one straight line, etc. Although all these principles, and the representation of the object with which this science occupies itself, are generated in the mind completely *a priori*, they would still not signify anything at all if we could not always exhibit their significance in appearances (empirical objects). [...] In the same science [mathematics] the concept of magnitude seeks its standing and sense in number, but seeks this in turn in the fingers, in the beads of an abacus, or in strokes and points that are placed before the eyes. The concept is always generated *a priori*, together with the synthetic principles or formulas from such concepts; but their use and relation to supposed objects can in the end be sought nowhere but in experience, the possibility of which (as far as its form is concerned) is contained in them *a priori*. (A239–40/B298–9)

If the proposition $7 + 5 = 12$ had been analytic then merely thinking about the concepts of $5, 7$ and $+$ would yield the result 12. But this is not possible. Our concept of 5 is not a collection of marks, rather it is an abstract type, whose semantics is determined by the "schema of magnitude" when the subject is interacting with the world, and therefore numbers have meaning only in connection with intuitions. We have to operate with particular tokens which are found "in the fingers, in the beads of an abacus, or in strokes and points that are placed before the eyes". A consequence of this is that the notion of number is meaningless only if all equivalence classes in $\mathscr{U}$ are empty. In other words, our notion of number is meaningless only if no objects are representable. Thus it becomes an issue to guarantee that these classes are not empty, and this seems to lead to a problem about large numbers. As Frege puts it:

> I must protest against the generality of Kant's dictum: without sensibility no object would be given to us. [...] Even those who hold that the smaller numbers are intuitable, must at least concede that they cannot be given in intuition any of the numbers greater that $1000^{1000^{1000}}$, about which nevertheless we have plenty of information. (Frege, 1884/1980, p. 101)

Tait formulates in the paper *Finitism* a similar criticism, which is extended to a critique of Kant's notion of number through a critique of Kant's dictum (in the words of Tait) that "existence is restricted to what can be represented in intuition" (2005, p. 7). Tait writes:

It is clear—and was so to Kant and Hilbert—that there are numbers, say 10^{10} or 30, which are not in any reasonable sense representable in intuition. Kant seems to have responded to this by saying that at least their parts are representable in intuition [...] The real difficulty, however, is that the essence of the idea of Number is iteration. However and in whatever sense one can represent the operation of successor, to understand Number one must understand the idea of iterating this operation. But to have this idea, itself not found in intuition, is to have the idea of number *independent of any sort of representation*. (Tait, 2005, p. 35)

In should be clear from my exposition of Kant's notion of number that Kant is not in any way affected by the latter part of Tait's criticism. Kant does not hold an empirical understanding of number—rather the concept of a general number is founded on the "schema of magnitude" which flows from the intellect, certainly not from the empirical.

In the former part of the criticism Tait seems to have the same premise as does Frege. Intuition, on their understanding, does not include pure intuition. Granting that we have *pure intuition* Kant would respond by saying that by *the very writing* of $1000^{1000^{1000}}$ or 10^{10} or 30 we in fact *have* intuitions. The very inscriptions provided by Frege and Tait are intuitions. They are understood in terms of the exponentiation function which is basically a primitive recursive function. Therefore, in principle, we can determine the equivalence-class-membership of these inscriptions—they are numbers, as we have an intuition and a rule determining how to operate with this intuition.

Let me give a general solution concerning the meaningfulness of large numbers. As it turns out it is our concept of space which ultimately provides arithmetic with its objects. Our primary geometrical schemata (some equivalents of Euclid's postulates) lead to a production of a sequence of finite spaces,

$$S_1, S_2, \ldots, S_n, \ldots$$

where S_i is strictly smaller than S_j, if $i < j$, and the whole sequence is unbounded. This sequence of pure spaces is a constructive but potentially infinite sequence. Thus given any natural number the equivalence class corresponding to that number is inhabited, at least due to this sequence of increasing finite pure spaces. Therefore, the justification of this argument, which refutes Frege's criticism, rests on Kant's notion of space (see my analysis in Jørgensen, 2005).

Time, however, is also a necessary condition for the concept of magnitude. The concept of iteration is a necessary element in the "schema of magnitude" ("the successive addition of one (homogeneous) unit to another"). Without iteration it would be impossible to determine the magnitude for any given thing. Kant assumes nothing particular about the objects for numbers—they can be anything—but adding unit to unit always takes place in time: "this how–many–times is grounded on successive repetition, thus on time" (A242). Therefore, the natural numbers as a sequence of numbers can only be *represented* as a progression in time. Further-more, also the most simple operations of arithmetic, say addition, takes place in time according to Kant. In a letter to Schultz Kant writes: "If I view $3 + 4$ as the expression of a *problem*" then the results found "through the successive addition that brings forth the number 4, only set into operation as a continuation of the enumeration of the number 3" (Ak. 10, p. 556). So I can take first three fingers together with four fingers and enumerate all of them. This enumeration ends by judging the fingers to constitute a set with cardinality 7.[21]

Under the condition of inner sense (time), number as *schema* generates the synthesis (of representations of) objects subsumed under a concept. As a conse-quence of this cognition, number, pure concept, representations of objects falling under an empirical concept, transcendental time-determination and the transcen-dental imagination are closely related *in inner sense*, as also Figure 1 illustrates: It all takes place under the conditions of inner sense.

But in the same letter to Schultz, Kant claims that:

> Time, as you correctly notice, has no influence on the properties of num-bers (as pure determinations of magnitude) [...] The science of num-bers, notwithstanding the succession that every construct of magnitude requires, is a purely intellectual synthesis, which we represent to our-selves in thought. But insofar as specific magnitudes (*quanta*) are to be determined in accordance with this science; and this grasping must be subjected to the condition of time. (Ak. 10, pp. 556–57)

The natural numbers are pure concepts (types) of the understanding (*Verstand*), in the sense that they are not derived from experience, but from the structure of our representation.[22] Time has no influence on these types, as they are not dependent on time. The rules determining the tokens (members of the equivalence classes) are purely intellectual rules, which remain the same over time. This is the intensional aspect of number. The objects in the equivalence classes, on the other hand, are ultimately empirical objects (and therefore spatial). But reasoning about numbers proceeds necessarily by way of mental images over time. When we do mathemat-ics and examine the properties of numbers, time is a necessary condition for the

---

[21] See also A164/B205.

[22] It is of course an interpretation to say that the numbers are pure concepts of the under-standing, as Kant claims there are only 12 of these categories. It would, however, not make much sense, I think, to regard the numbers as anything but pure. In this respect I completely agree with M. Young (1992, p. 174), and I am generally sympathetic towards his short interpretation of Kant's schematism.

representations of numbers.[23] The interplay between pure concepts, tokens, time and space is summarized by Kant in Dissertation by saying that

> there is a certain concept which itself, indeed belongs to the understanding but of which the actualization in the concrete (*actuatio in concreto*) requires the auxiliary notions of time and space (by successively adding a number of things and setting them simultaneously side by side). This is the concept of *number*, which is the concept treated in ARITHMETIC. (Ak. 2, p. 397)

Therefore both space and time condition our access to and the constitution of the natural numbers: They are constituted by non-temporal schemata but any use will be temporal, and their meaning is ultimately provided by intuitions.

6 What numbers are

Numbers are not characterized extensionally—this would not be meaningful in Kant's framework. Number is rather given an *intensional* characterization in terms a collection of rules. In order to fully appreciate this and to give a coherent account we need Kant's full theory of schemata—all the way through empirical, geometrical and transcendental schemata. On the other hand, we really get a coherent interpretation when the disentangled theory of schemata is taken into consideration. I think that Charles Parsons failed to realize this when writing his article "Arithmetic and the Categories". There he concludes that "Kant did not reach a stable position on the place of the concept of number in relation to the categories and the forms of intuition" (1992, p. 152). In contrast to this, I hope my interpretation of Kant's theory of schemata has shown, precisely how Kant's notion of number relates to the categories, and to the two forms of intuition.

But, according to Kant numbers cannot be objects, as arithmetic has no axioms.

> The self-evident propositions of numerical relation [...] are to be sure, synthetic, but not general, like those of geometry, and for that reason also cannot be called axioms, but could rather be named numerical formulas. (A164/B205)

The numerical formulae Kant is thinking of are propositions like "$7+5=12$". But "[s]uch propositions must [...] not be called axioms (for otherwise there would be infinitely many of them)" (A165/B205). In a sense Kant is very right: *There was no axiomatization of number theory at the time of Kant.* In fact Kant reflects, once

---

[23] Therefore, the situation is not as in mechanics, where time is analyzed together with the alteration of placement in space.

again, the Euclidean paradigm. Euclid has no axioms for numbers. Number theory is treated in Book VII, and already in proposition 2 (the greatest common divisor of two numbers) Euclid uses well-ordering of the natural numbers, but without reference to some first principle.

Now, one could speculate: If Kant had had axioms for number theory, would he have regarded them as genuine axioms? Axioms should be synthetic a priori in analogy to the way that Euclid's postulates reflect schematic spatial procedures.

According to Kant geometrical schemata produce true objects in pure intuition. These are genuine objects in the sense that they are possible objects of experience. Let us use the terminology of a first-order universe U and a second-order universe $\mathscr{U}$, which I introduced on page 73. The objects which the geometrical schemata produce are elements of U. Representations of numbers, however, are elements in the second-order universe $\mathscr{U}$. Thus, they are not really objects to Kant as the only true objects are first-order objects. This also explains why the pure "schema of magnitude" deals more with reflection than construction. The elements of $\mathscr{U}$ are according to Kant not really constructed, they rather reflect a certain relation between objects living in U. I think this is the only good reason Kant has when claiming that the numbers are not objects.

In the course of history we have learned—due to relativity theory—that space is only *approximately* Euclidean. Consequently it seems that the objects we produce in Euclidean geometry are only approximations of possible objects of experience. On the other hand, the number five is still represented by the set of fingers on a normal hand. If we allow second-order objects to be genuine objects, then in fact they are more objective (in the Kantian sense) than the objects of Euclidean geometry.

Today we can give an axiomatization of the natural numbers, which relative to the slight generalization of the Kantian notion of object, is a set of axioms in the Kantian sense.

We can define an arithmetic Q, called Robinson arithmetic, in which there is a constant 0, called zero, a unary operation S, called successor and two binary operations $+$ and $\cdot$ called plus and times, which satisfy the following axioms:

1. $\forall x(0 \neq Sx)$.
2. $\forall x,y\big((Sx = Sy) \rightarrow (x = y)\big)$.
3. $\forall x\big(0 \neq x \rightarrow \exists y(x = Sy)\big)$.
4. $\forall x(x + 0 = x)$.
5. $\forall x,y\big(x + Sy = S(x + y)\big)$.
6. $\forall x(x \cdot 0 = 0)$.
7. $\forall x,y\big(x \cdot Sy = (x \cdot y) + x\big)$.

The first three axioms are realized by the "schema of magnitude". They claim the existence of a concept of iteration S which taken together with a symbol 0 gives rise to a paradigmatic representation of the natural numbers:

$$0,\ S0,\ SS0,\ SSS0,\ldots$$

Due to our ability of producing bijections, these paradigmatic numbers as intuitions put us in a situation where we can use and reason universally about *any* representation of the numbers. Recall, the schema together with any representation, whether empirical or pure, allow for universal reasoning. The sequence is furthermore potentially infinite, and thus for any type $\bar{n}$, the number $\bar{n}$ is meaningful. The realization of the latter axioms is also due to our ability to produce and operate with bijections. Plus corresponds to composition of functions, which we found on page 77 was validated by the "schema of magnitude", and times iterates this concept.

If we therefore allow second-order objects to be objects and accept the above axiom system, then there are numbers, just as much as there are triangles—or perhaps even more. We meet them as tokens and any number $\bar{n}$ is schematisable in the sense that we can produce a representation $S\cdots S0$ such that the representation together with the schema allows for universal reasoning. Finally, the interpretation I have suggested also makes certain elements found the first *Critique* more coherent, as for instance when Kant writes about number images (A149/B179; A240/B299).

References

Frege, G. (1884/1980). *The Foundations of Arithmetic*, Northwestern University Press, Evanston.

Hartnack, J. (1968). *Kant's Theory of Knowledge*, MacMillan, London. Translated by M. Holmes Hartshorne.

Jørgensen, K. F. (2005). *Kant's Schematism and the Foundations of Mathematics*, PhD thesis, Roskilde University, Roskilde.

Kant, I. (1998). *Critique of Pure Reason*, The Cambridge Edition of the Works of Immanuel Kant, Cambridge University Press, Cambridge. Translated and edited with introduction by P. Guyer and A.W. Wood.

Longuenesse, B. (1998). *Kant and the Capacity to Judge*, Princeton Univeristy Press, Princeton, Oxford.

Parsons, C. (1992). Arithmetic and the categoreis, *in* C. J. Posy (ed.), *Kant's Philosophy of Mathematics*, Kluwer, pp. 135–158.

Paton, H. J. (1936). *Kant's Metaphysics of Experience*, Vol. 2, Humanities Press, New York.

Shabel, L. (2003). *Mathematics in Kant's Critical Philosophy*, Routledge.

Tait, W. (2005). *The Provenance of Pure Reason*, Oxford University Press, Oxford, New York.

Tiles, M. (2004). Kant: From general to transcendental logic, *in* D. M. Gabbay and J. Woods (eds), *Handbook of the History of Logic*, Elsevier, Amsterdam, pp. 85–130.

Young, J. M. (1992). Construction, schematism, and imagination, *in* C. J. Posy (ed.), *Kant's Philosophy of Mathematics*, Kluwer, pp. 159–175.

Cardiovascular modelling at IMFUFA

Jesper Kampmann Larsen[1], Viggo Andreasen[1], Heine Larsen[1], Mette Sofie Olufsen[2], Johnny Tom Ottesen[1]

[1] Department of Mathematics and Physics, Roskilde University
[2] Department of Mathematics, North Carolina State University

1 Introduction

Mathematical modelling in life sciences has been a main interest during our professional careers. At the department of mathematics and physics at Roskilde University, IMFUFA, the modelling effort has mainly been on the two related subjects: The human cardiovascular system including its control mechanisms and on fluid dynamics and valveless flow. Stig Andur Pedersen initiated the interest in physiological modelling at IMFUFA in the early nineties through his involvement in the development of training simulators for anaesthesiologists. In the present paper we give a brief description of the anaesthesia simulation and report on some of the early developments in cardiovascular modelling at IMFUFA. As an alternative to the widespread use of electrical analogue models of the cardiovascular models the use of energy-bond graphs (EBG) was considered. We report on the use of this technique for modelling the aorta. We include a brief discussion of the merits of EBG. At one point it was considered to use a commercially available software package for the simulation of cardiovascular flows. We report on the application of an urban sewer system package for the simulation of flow in the aorta. We present the mechanism making the analogy possible. We conclude the paper by presenting some of the current research in cardiovascular modelling at IMFUFA showing how Stig Andur Pedersen's original seed continues to thrive.

2 SIMA

In the early nineties Stig Andur Pedersen was involved in the development of an anaesthesia simulator. For that purpose he needed mathematical models that could predict the physiological response of a human subject to actions taken by the anaesthesiologist under training. Typically for SAP, he insisted that these models be based on real scientific insight from "first principles" rather than relying on fixed scenarios that simply reproduced the response envisioned by experienced anaesthesiologists. This was the start of the SIMA project and the study group at IMFUFA. In this project we developed reliable deterministic mathematical models for blood flow and blood pressure in humans together with models of relevant control mechanisms regulating these flows and pressures. A central demand was that it should be possible to solve the models in "real time" on a computer or that they should be used for validation of simpler implementable models. This research was done in close cooperation between Math-Tech, Herlev University Hospital, S&W Medico Teknik, and the group of researchers from mathematics at Roskilde University constituting the BioMath group. The work resulted in two simulators, SOPHUS, and its successor, SIMA, and in addition a lot of research related to and on modelling various physiological systems. Extensive reviews of the models underlying the simulator SIMA, and in particular on modelling the cardiovascular system and the baroreflex-feedback control mechanism, is given in the monograph (Ottesen, Olufsen and Larsen, 2004).

We have constructed models consisting of systems of nonlinear ordinary differential equations, describing the electro-mechanical transformation, which takes place in the baroreceptor nerves in carotid sinus. These and other models have been coupled to models of how heart rate, contractility, peripheral resistances, venous compliance and the venous unstressed volume of the pulsatile cardiovascular system react to nerve stimuli. Another baroreceptor model was established distinguishing between the sympathetic and parasympathetic nervous activities. The purpose was to include and study the effects of the time delays of the nerve paths with respect to the control actions. All models were validated for instance during simulated acute hemorrhages and bicycle ergometer tests. These models typically benefit from the theory for systems of linear and nonlinear ordinary differential equations with or without time delay. Existence and uniqueness results of equilibrium have been shown and stability has been investigated analytically. A cascade of stability switches (bifurcations) may appear when the time delay is considered as a parameter. The oscillation found corresponds nicely to the well known but previously unexplained Mayer waves frequently observed in humans.

In addition to this we became involved in a critical yet constructive discussion on how mathematical modelling may contribute to forming concepts of theories in theoretical physiology and particular in a debate of how to describe ventricular performance. In this work new mathematical models of the pumping heart have been proposed describing the heart as a pressure source depending on time, volume and flow. The underlying concept is based on a new two-step paradigm that allows separation between isovolumic (non-ejecting) and ejecting heart properties. A characterisation of what constitutes an optimal model has been given and used as a criterion for choosing the optimal model in this family. It was shown that the model exhibits all major features of the ejecting heart, including how ventricular pressure and flow varies in time for various heart rates and how stroke volume and cardiac output varies with heart rate. The modelling strategy presented embraces the same steps and demarcation as those suitable for clinical examination whereby new experiments became possible. It was also shown that the traditional and widely used time-varying elastance concept is disqualified as an independent description of the heart. Based on our earlier work on control mechanisms and on the new heart model, a model describing cerebral blood flow and its control during posture change from sitting to standing has been developed. Since hypertension, decreased cerebral blood flow, and diminished cerebral blood flow regulation are among the first signs indicating the presence of cerebral vascular disease a mathematical model has important applications. The mathematical model uses a compartmental approach to describe pulsatile blood flow and pressure in a number of compartments representing the systemic circulation. To justify the fidelity of the mathematical model and control mechanisms developed, the model was validated against experimental data from a young subject and optimal control theory was used for parameter identification. Parts of the above-mentioned research are documented in the monographs (Ottesen and Danielsen, 2000), and (Ottesen, Olufsen and Larsen, 2004).

3 Cardiovascular modelling

Some of the first development on the simulation of blood flow in the aorta was made in 1993 in a project at the introductory level of the education at Roskilde University. Using energy bond diagram techniques (cf. Christiansen, 2003) a discrete mathematical model of the aorta was developed. The model is one-dimensional and only allows for a description of the flow and pressure of the blood as a function of distance along the length of the aorta and time. By considering the aorta to be built of a number of discrete elements a physically based model of each element was combined into a mathematical model of the whole aorta.

In Figure 1 we show a sketch of the aorta. In formulating the model we disregard the curvature of the aorta and consider elements of length Δx. The total length of the aorta is L and the number of elements are N. Thus $L = N\Delta x$.

The energy bond formalism allows one to create models of a system in a physically consistent way. The formalism operates with storage of either potential or kinetic energy symbolised as a birdcage like icon with an o or an x written in the top centre of the icon, see Figure 2. Connecting the storage to the rest of the system is an energy bond consisting in our context of flow and pressure variables, the product of which is an energy flow. The flow has direction and is symbolised by an

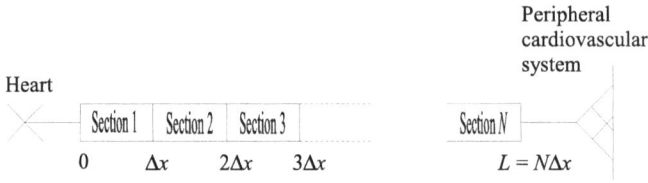

Fig. 1. The model of the aorta. From (Nørgaard et al., 1993)

arrow whereas the pressure is symbolised by a cross mark, see Figure 2. More energy bonds can be connected by junctions that are depicted by tilted squares with either a cross or a circle inscribed designating a flow junction or a pressure junction, respectively. In a flow junction the flows are equal corresponding to a serial connection whereas in a pressure junction the pressures are equal corresponding to a parallel connection.

In addition to the gatherers there are transformers (symbolised by triangles) converting variables and gyrators (that resembles a fish) converting one type variable to another type of variable (see Figure 2). Finally, energy might dissipate, either as a flow out of the system or as pressure drop. The pressure drop due to flow resistance in the system is symbolised by the element at the bottom of Figure 2.

In Figure 2 we have depicted the model of one element away from the boundaries shown in Figure 1. The lay out is staggered such that Q_n is the flow at $x = (n + \frac{1}{2})\Delta x$. In the model the radius, r_n of section n is chosen as the measure of potential energy since it relates linearly with the pressure through $F_n = \Delta x E D(r_n - R_n)/R_n$, where F_n is the force in section n of the wall of the aorta, E is Young's modulus, D is the thickness of the wall and R_n is the radius at zero pressure.

The change of radius of section n can readily be established from Figure 2, i.e.

$$\frac{dr_n}{dt} = \frac{Q_{n-1} - Q_n}{2\pi\Delta x r_n}. \tag{1}$$

The impulse equation is similarly found to be

$$\frac{dp_n}{dt} = \Gamma_n\left(\frac{dp'_n}{dt} - \frac{dp'_{n+1}}{dt} - Q_n\Lambda_n\right) + \Phi_n\frac{dr_n}{dt} + \Phi_{n+1}\frac{dr_{n+1}}{dt}, \tag{2}$$

where

$$\Gamma_n = \sqrt{\frac{r_n r_{n+1}}{R_n R_{n+1}}},$$

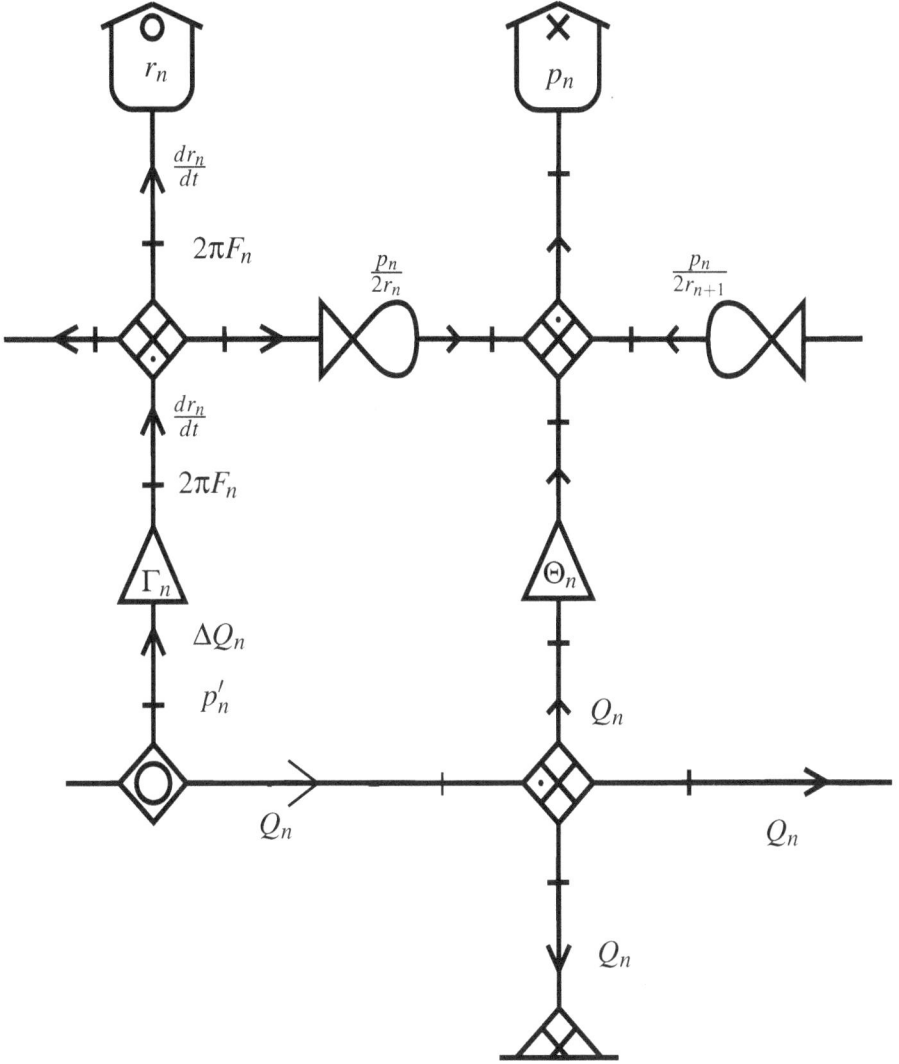

Fig. 2. The EBG of a section of the aorta

$$\Lambda_n = \frac{8\mu\Delta x}{\pi r_n^2 r_{n+1}^2},$$

and

$$\Phi_n = \frac{p_n}{2r_n}.$$

Here μ is the kinematic viscosity of blood. In addition to Equation 1 and Equation 2 we have a relation between flow and pressure,

$$p_n = \Gamma_n p_n' = \Gamma_n \rho \Delta x \frac{Q}{A_n}, \tag{3}$$

where $A_n = \pi r_n r_{n+1}$.

By combining equations for all the n sections of Figure 1 with suitable modifications for section 1 and n, there results a system of $2n$ ordinary differential equations in $2n$ unknowns.

By going to the limit $\Delta x \to 0$ Equation 1 goes into

$$\frac{\partial A}{\partial t} + \frac{\partial Q}{\partial x} = 0, \tag{4}$$

where

$$A = \pi r^2.$$

Similarly, the Equation 2 goes into an Euler equation with an extra Poiseuille term

$$\frac{\partial Q}{\partial t} + \frac{\partial}{\partial x}\frac{Q^2}{A} + \frac{A}{\rho}\frac{\partial p}{\partial x} = -8\pi\nu\frac{Q}{A}. \tag{5}$$

The EBG provides a discrete model of the aorta. A discrete model that can be readily solved by use of standard software for systems of ordinary differential equations. Used correctly, EBG ensures that the physics behind the model is well represented. It also imposes a specific discretisation of the underlying partial differential equation (PDE). In subsequent research in cardiovascular modelling we used PDE's in the formulation of the models.

4 "Life is like a sewer..."

In the fall of 1993 a group of students by use of an analogy between flow in the aorta and flow in a sewer were able to simulate aortic flow by a commercial sewer code. The equations governing flow in a full sewer are the continuity equation:

$$\frac{\partial Q}{\partial x} + \frac{Q}{\rho}\frac{\partial \rho}{\partial x} + \frac{gA_0}{a^2}\frac{\partial Y}{\partial t} = 0, \tag{6}$$

where ρ is the density of the compressible fluid, A_0 is the cross sectional area of the tube at zero pressure, g is the acceleration of gravity and $a^2 = a_0^2/(1 + a_r^2/a_0^2)$ in which a_0 and a_r are the speed of sound in the fluid and the tube wall material, respectably.

The density changes only little with distance,i.e. $\partial \rho/\partial x \approx 0$. Using this and introducing $B = gA_0/a^2$, we rewrite equation 6 as

$$B\frac{\partial Y}{\partial t} + \frac{\partial Q}{\partial x} = 0, \tag{7}$$

where Y is the pressure height.

The equation of motion reads

$$\frac{\partial Q}{\partial t} + \frac{\partial}{\partial x}(\beta\frac{Q^2}{A}) + gA\frac{\partial Y}{\partial x} = gA(I_0 - I_f). \tag{8}$$

Here $0 \leq \beta \leq 1$ is a dimensionless velocity distribution coefficient, $I_0 = \sin\theta$, where θ is the slope of the tube and I_f is a friction term.

If we introduce the tube law $p = P(A)$ in the momentum equation for the aorta, Equation 8 with zero friction, we get

$$\frac{\partial Q}{\partial t} + \frac{\partial}{\partial x}(\frac{Q^2}{A}) + \frac{A}{\rho}P'(A)\frac{\partial A}{\partial x} = 0. \tag{9}$$

If we introduce a narrow slot in the top of the aorta as in Figure 3 we find a relation between the area A and the height of blood $Y(x,t)$ as

$$A = A_r + B \cdot (Y(x,t) - d), \tag{10}$$

where B is the width of the slot, A_r is the area of the tube itself and d is the diameter of the tube. Thus

$$\frac{\partial A}{\partial x} = B\frac{\partial Y}{\partial x}. \tag{11}$$

If we choose the width of the slot as

$$B = \frac{\rho g}{P'(A)}, \tag{12}$$

we can rewrite the momentum equation for the aorta as

$$\frac{\partial Q}{\partial t} + \frac{\partial}{\partial x}(\frac{Q^2}{A}) = -gA\frac{\partial Y}{\partial x}, \tag{13}$$

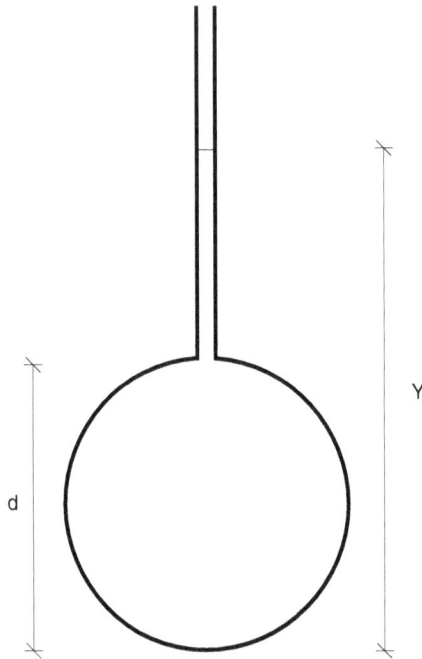

Fig. 3. Cross section of tube with narrow slot

which is the equation of motion for a flow in a channel with a free surface, slope $\theta \approx 0$ and $\beta = 1$. By introducing this slot of width B it is possible to use the commercial well proven code for sewer simulation to simulate the aorta (Marcussen et al., 1993). In order to control all details of the modelling it was decided to build the cardiovascular models from first principles.

5 Recent and current developments in blood flow simulation

Another main focus of our research is reflected in our interest in fluid dynamics and valveless flow. A 1-D model of blood flow in the visco-elastic aorta has been built, described by a system of hyperbolic partial differential equations based on the Navier-Stokes equations and the theory of elasticity. The purpose was to study change in pulse profile during propagation through the highly bifurcated arterial tree and especially how the end-boundary condition could be constructed correctly. Thus the effects of various boundary conditions were studied. This fluid dynamical study was related to the systemic approach discussed above through the question on whether respiration contributes in circulating the blood in the cardiovascular system or not. Inspiration and expiration temporarily change external pressure to the vein, thereby changing its cross-sectional area periodically. When the inertial properties of blood were taken into account (changing the downstream impedance into a complex number), and a complete set of requirements was satisfied, the respiration was found to promote venous return. Hence, it was found that respiration under certain circumstances does promote venous return flow to the heart but under other circumstances doesn't. This is reported in (Ottesen, Steendijk, Karle, Rower, Noordergraaf, Scheffer, Schilders and Noordergraaf, 2004).

This work did inspire us to the more fundamental question in fluid dynamics, namely that of valveless flow in systems of elastic tubes. The phenomenon seems to be explained by effects caused by nonlinear boundary conditions in the study of partial differential equations. Lately, research in this field has resulted in three papers (Ottesen, 2003b,a; Ottesen et al., 2005) and more are in preparation. Valveless flow appears typically if an elastic rubber tube connected with a stiffer rubber tube in series forming two halves of a torus and filled with water. Compressing one of the rubber tubes symmetrically and periodically at a point of asymmetry creates a remarkable unidirectional mean flow in the system. The size and the direction of the mean flow depend on the frequency of compression, the elasticity of the tubes, the compression ratio, the duration and the type of compression with respect to time in a complicated way. The system was modeled by use of the Navier-Stokes equations and the model was analysed partly analytically and partly numerically. A series of experiments on a physical realisation of the system have been described. The theoretical findings and experimental results are compared; they show a remarkable agreement between the experiments and the model predictions. A mechanism underlying the valveless flow is suggested, studied analytically and validated qualitatively. This work also gives rise to some speculations on whether the rather ineffective cardiopulmonary resuscitation method (CPR) used today in case of cardiac arrest could be improved. If the size and even the direction of mean flow are frequency dependent in the CPR treatment it is indeed important to study at what frequency of chest compression the CPR treatment will be most effective. The early stage of the human foetus when heart valves are not yet developed but the heart causes blood to circulate in a unidirectional fashion in the absence of heart valves, constitutes another application. Furthermore, certain invertebrates (e.g. Amphioxus) and some vertebrates (e.g. Tunicate) have a valveless circulatory system. Also variations in human transmural pressure due to respiration may locally be considered as valveless pumping. Finally, non-equilibrium molecular dynamics techniques have been used to study valveless flow as well. These show that microscopic simulations resemble the macroscopic predictions and confirm the findings based on Navier-Stokes' equations. Thus valveless flow appears to be

scale invariant and the phenomenon is expected on nano-scale as well, leaving a
large potential for applications, such as constructing nano-pumps for instance.

6 Concluding remarks

Back in the early nineties Stig Andur Pedersen initiated our research in cardiovascular modelling. The anaesthesia simulator that grew out of this development is
today used for training and education of anaesthesiologists and anaesthetic nurses
in Denmark and abroad. As stated in the introduction the seed that Stig Andur
Pedersen planted continues to thrive.

References

Christiansen, P. V. (2003). Energy-bond-graphs – a semiotic formalization of modern physics, *Technical Report 419*, IMFUFA, Roskilde University.

Marcussen, A., Nilsson, A. C., Michelsen, L. and Hansen, P. M. (1993). "Life is
like a sewer..." et projekt om modellering af aorta via en model for strømning i
kloakrør, *Technical Report 268*, IMFUFA, Roskilde University. In Danish.

Nørgaard, T., Ellegaard, J., Jalving, J., Pedersen, J. B., Gregersen, P., Meyer, S. I.
and Wittus, H. (1993). Aorta-modellering, *Technical report*, Roskilde University. In Danish.

Ottesen, J. T. and Danielsen, M. (eds) (2000). *Mathematical Modelling in
Medicine*, Studies in Health Technology and Informatics, IOS Press.

Ottesen, J. T., Hansen, J. S. and Lemarchand, A. (2005). Molecular dynamics simulations of valveless pumping in a closed microfluidic tube-system, *Molecular
Simulations* **31**(14–15): 963–969.

Ottesen, J. T., Olufsen, M. S. and Larsen, J. K. (eds) (2004). *Applied Mathematical Models in Human Physiology*, Monographs on Mathematical Modelling and
Computation, SIAM, Philadelphia.

Ottesen, J. T., Steendijk, P., Karle, H., Rower, N., Noordergraaf, G. J., Scheffer,
G. J., Schilders, W. H. A. and Noordergraaf, A. (2004). Cardiopulmonary re-
suscitation: Biomedical and biophysical analysis, *The Biomedical Engineering
Handbook*, 3. edn.

Ottesen, J. T. (2003a). Symmetric compressions of a fluid filled torus of asymmetric elasticity generates mean flow of frequency dependent size and orientation,
in V. Capasso (ed.), *Mathematical Modelling and Computing in Biology and
Medicine*, Progetto Leonardo, ESCULAPIO Pub. Co., Bologna.

Ottesen, J. T. (2003b). Valveless pumping in a fluid-filled closed elastic tube-
system – one-dimensional theory with experimental validation, *J. Math. Biol.*
46: 309–332.

Part II

Philosophy and history of science

Einstein on the history and nature of science

Helge S. Kragh

Steno Institute, University of Aarhus

Although Albert Einstein never articulated his views on the history of science, he was keenly interested in the subject and often framed his philosophical reflections on science in the form of historical narratives. His ideas on the history of science cannot be separated from his ideas on the methods and nature of science. Although he came to admire the past masters of physics and see himself as part of a historical tradition, as a young man he had only disdain for theories that were obsolete. Having looked at a book on the ether, published in 1885, he wrote to Mileva Marič, his future wife: "One would think it came from antiquity, its views are so obsolete. It makes one see how fast knowledge develops nowadays." (Letter of 28 December 1901, Einstein, 1987).[1] As he grew older, he also grew wiser: "We must examine old ideas, old theories, although they belong to the past, for this is the only way to understand the importance of the new ones and the extent of their validity." (Einstein and Infeld, 1938, pp. 7–8).

1 Einstein, Mach and the history of physics

The Austrian physicist and philosopher Ernst Mach deeply influenced young Einstein's philosophical views, which tended towards empiricism and sensationalism. He was thoroughly familiar with Mach's celebrated *Die Mechanik in ihrer Entwicklung*, first published in 1883, a critical and historical study of key concepts such as space, time, mass and substance (Mach, 1883).[2] As Einstein wrote in a

---

[1] The book was probably (Hoh, 1885). On Einstein's views on and use of the history of science, see (Byrne, 1980; Kragh, 2005).

[2] On Einstein and Mach, see (Holton, 1988, pp. 237–278). Einstein read *Die Mechanik* in 1899 and subsequently also Mach's *Die Prinzipien der Wärmelehre*, another work in the historico-critical tradition.

letter to Carl Seelig of 1952, "The book exerted a deep and persisting influence upon me ..., owing to its physical orientation toward fundamental concepts and fundamental laws." (Holton, 1988, p. 241).[3] From Einstein's correspondence with Mach we know that this evaluation was no slip of the memory. Thus, in a letter to Mach of 1909, Einstein wrote, "I know, of course, your main publications very well, of which I most admire your book on Mechanics." (Einstein, 1993b, p. 204).

It was primarily Mach as a philosopher of science who appealed to young Einstein, rather than Mach as a historian, but in Mach's thinking history and philosophy of science were so closely intertwined that the two aspects cannot be easily separated. Einstein soon came to reject Mach's epistemology, at first implicitly and eventually explicitly. It was in particular his work on the new gravitation theory which convinced him that scientific insight does not flow from empirical data, but from considerations of inner consistency, unificatory power and logical simplicity. It was a theme to which he would often return, as in a letter to Seelig: "The main significance of the theory [of general relativity] does not lie in the verification of little effects, but rather in the great simplification of the theoretical physics as a whole." (Seelig, 1954, p. 195).

In contrast to Einstein's growing dissatisfaction with Mach's philosophy of science, he continued to value him as a historian. In 1947, Einstein's lifelong friend Michele Besso (who had introduced him to *Die Mechanik* half a century earlier) wrote him: "As far as the history of science is concerned, it appears to me that Mach stands at the center of the development of the last 50 or 70 years." (Letter of 8 December 1947, Speziali, 1974, p. 387). I do not know if Einstein responded to Besso's evaluation, but apparently he agreed. When the physicist Robert Shankland interviewed Einstein in 1950, the conversation included history of science, a field Einstein did not have a high opinion of. "Nearly all historians of science are philologists," he claimed; they "do not comprehend what physicists were aiming at, how they thought and wrestled with their problems." Neither did Einstein have confidence in physicists as historians of their science, because generally they lacked "historical sense."

But there were exceptions, and as a model worth to follow he mentioned Mach's *Die Mechanik*. Mach, Einstein told Shankland, "did not *know* the real facts of how the early workers considered their problems," yet he had sufficient insight to write a book which was very likely correct anyway (Shankland, 1963); (Holton, 1988, pp. 345–346). The critical attitude to historians of science – observers and analysts of historical events, rather than participants in them – also appeared in Einstein's responses to the essays in the 1949 volume celebrating his

[3] Similarly in (Einstein, 1949, p. 21), written in 1946: "This book exercised a profound influence upon me." For Einstein's appreciation of Mach, see also his long and sympathetic obituary (Einstein, 1916) and (Einstein, 1996, pp. 277–282). On young Einstein as a positivist and positivist philosophers' embracement of his works, see (Neidorf, 1963).

seventy-year's birthday. In relation to the essays written by Wolfgang Pauli and Max Born, he opined:

> Only those who have successfully wrestled with the problematic situations of their own age can have a deep insight into those situations; unlike the later historian, who finds it difficult to make abstractions from those concepts and views which appear to his generation as established, or even as self-evident. (Einstein, 1949, p. 665)

Five years after his conversation with Shankland, Einstein had another opportunity to reflect on the history of science. In a conversation with young I. Bernard Cohen near the end of his life, Einstein pointed out the obvious, that history is much less objective than science.[4] In his view, there was an intuitional history and a documentary history, and he found the former to be the most interesting, if admittedly also the less objective. As far as the sources are concerned, he stated to Cohen that "the worst person to document any ideas how discoveries are made is the discoverer" – and he did not exclude himself from the rule. It was Einstein's belief that the trained historian is better suited to understand the creative process of a scientist than the scientist himself. This may call for two comments. First, Einstein's statement concerning the ability of historians to follow processes of scientific creativity contradicts what he said to Shankland just five years earlier (historians "do not comprehend what physicists were aiming at"). Second, his sense of the problem of reflectivity in historical narratives agrees nicely with the warning with which he introduced the Herbert Spencer lecture in Oxford in 1933. In his address "On the Method of Theoretical Physics," Einstein famously said: "If you want to find out anything from the theoretical physicists about the methods they use, I advise you to stick closely to one principle: don't listen to their words, fix your attention on their deeds." (Einstein, 1982, p. 270).[5]

2 Intuitional history

Let us follow Einstein's advice and see how he, explicitly or implicitly, formulated his ideas about science and its historical development. Of course, Einstein was not a historian of science and one should not be surprised that his writings of a more or less historical nature did not agree with the standards of professional, critical history as they were in the 1950s (such as represented by I. B. Cohen and Alexandre Koyré, for instance). They agree even less with current standards.

---

[4] Cohen's interview with Einstein took place two weeks before Einstein's death. See (Cohen, 1955, 1979).

[5] The text was first published as a separate print in 1933 and appeared subsequently in (Einstein, 1934).

For example, his version of the history of science was very much a scientist's history, blatantly ignoring documentation and being unashamedly presentist and non-contextualist.[6]

Nonetheless, Einstein's writings qualify as historical accounts in so far as they were concerned with the emergence in the past of new scientific ideas. Many of Einstein's philosophical reflections were organized along historical lines, as he was not content to show what science *is*, but also wanted to show how it *develops* over time. The temporal or dynamical dimension was an integral part of his view of science and naturally invited historical reflection. It should be noted that although Einstein often wrote and spoke of "science", what he really meant was "physics"; and not just any kind of physics, for his interest was limited to theoretical or fundamental physics. He hardly ever expressed any interest in other sciences, although mathematics was an exception (and then only because of its connection to physics).

After Einstein had become a famous physicist, he wrote numerous articles of a biographical nature, either obituaries, contributions to festschrifts, or memorial articles on great physicists such as Galileo, Kepler, Newton and Maxwell. He believed that biographical accounts of individual scientists must form an important part of history of science, as this is the only approach through which one can understand the creative scientific process. In an article on the German physicist Emil Warburg on the occassion of his retirement in 1922, Einstein wrote:

> The content of a science can undoubtedly be understood and evaluated without paying attention to the development of those individuals who have formed it. But by such a one-sided, objective presentation the steps in the development often appear to be fortuitous. One can only understand why these steps were possible, indeed necessary, if one follows the intellectual [geistigen] development of those individuals. (Einstein, 1922)[7]

As another example, consider an essay Einstein wrote upon the bicentenary of Newton's death in 1927. Characteristically, Einstein was unconcerned with chronology and the context of Newton's life and work. What interested him was Newton's psychological traits, the thought processes of a great natural philosopher. Indeed, much of the essay was not so much about Newton as it was a semihistorical survey of the conceptual development from Newton to contemporary physics.

Einstein often used history of science in this way, to reflect on the general development of scientific thought and how it had led to the present world view. He ended his essay on Newton with reflecting on the situation in contemporary quantum physics and its claim that the law of causality needs to be abandoned.

---

[6] For scientists' history of science, see, e.g., (Kragh, 1987, pp. 150–158) and (Brush, 1995).

[7] On Einstein's many biographies and obituaries, see (Herneck, 1984).

Einstein admitted a certain sympathy for the wave mechanics of Louis de Broglie and Erwin Schrödinger because it "has, in a certain sense, the character of a theory of fields" and resulted "in amazing agreement with the facts of experience." Yet this was not enough to make wave mechanics an acceptable theory: "But it has to dispense with a localisation of the mass-particle and with strictly causal laws. Who would be so venturesome as to decide to-day the question whether causal law and differential law, these ultimate premises of Newton's treatment of nature, must definitely be abandoned?" (Einstein, 1927).[8]

The structure of Einstein's arguments, partly historical and partly philosophical, was essentially the same in an article of 1940 on fundamental physics. In order to illustrate what is the fundamental features of theoretical physics, and how knowledge develops over time, he started with Newton's mechanics and proceeded with the electromagnetic theories of Faraday, Maxwell and Hertz. As usual, his account ended with modern physics, meaning relativity theory and quantum mechanics. As to the general theory of relativity, he stated that it "owes its origin to the attempt to explain a fact known since Galileo's and Newton's time," namely the experimental identity of a body's inertial and gravitational mass. As to quantum mechanics, Einstein made it clear, once again, that he could not accept the theory in its present interpretation: "Some physicists, among them myself, cannot believe that we must abandon, actually and forever, the idea of direct representation of physical reality in space and time; or that we must accept the view that events in nature are analogous to a game of chance." (Einstein, 1982, pp. 323–335).[9]

3 Biographies and surveys

Galileo was another of the scientific pioneers who attracted Einstein's interest, such as is manifest in a careful foreword he wrote to Stillman Drake's translation of the *Dialogo* of 1632. Einstein was particularly interested in Galileo's puzzling rejection of Kepler's planetary ellipses, such as they were introduced in *Nova astronomia* of 1609 but which Galileo chose to ignore. This was a question that Einstein almost took personally. "It has always hurt me to think that Galilei did not acknowledge the work of Kepler," he told I. B. Cohen in 1955 (Cohen, 1955,

---

[8] Reprinted in (Einstein, 1982, pp. 253–261). The essay was first published in *Manchester Guardian*, March 19, and appeared in German in *Die Naturwissenschaften* **15** (1927), 273–276. Einstein had been instrumental in directing Schrödinger's attention to Louis de Broglie's theory and initially had some sympathy for wave mechanics. Shortly before he wrote the essay on Newton, he presented to the Prussian Academy of Sciences a kind of hidden-variable formulation of wave mechanics. However, realizing that his formulation was unsatisfactory he never published it (see Belousek, 1966).

[9] First published in *Science*, 24 May 1940.

p. 69).[10] As usual in his historical writings, Einstein used the history of science exemplarily, in this case to reflect on the false dichotomy between elements of speculation and experiment in creative science. He did not accept the traditional picture of Galileo as the one who had replaced speculative methods with empirical methods and thereby founded modern science:

> I believe, however, that this interpretation would not stand close scrutiny. There is no empirical method without speculative concepts and systems; and there is no speculative thinking whose concepts do not reveal, on closer inspection, the empirical material from which they stem. To put into sharp contrast the empirical and the deductive attitude is misleading, and was entirely foreign to Galileo. (Einstein, 1953, p. xvii)

Apart from his biographical articles, Einstein also wrote several papers in which he surveyed the development of a scientific concept in a semi-historical perspective. As early as 1909 he surveyed in this manner the nature of radiation (Einstein, 1909)[11], and in 1924 he gave a conceptual and historical survey of the ether (Einstein, 1924).

In an article of 1930 on the concepts of space and field, Einstein adopted a historical framework in explaining how prescientific thoughts about space had evolved into the modern notion as found in the general theory of relativity. As steps on the road he mentioned the contributions of Descartes, Newton, Maxwell and Lorentz, but as usual without bothering to include any chronology. According to Einstein's way of thinking, it was possible to identify in history one or more principles that can be used to order historical events in such a way that their intrinsic significance stands out. He likened the task of the conceptually oriented historian of science with that of the archaeologist and believed that a method of "intuitive archaeology" would lead to a clearer, more logical picture of the past. The problem that faced the historian-archaeologist trying to comprehend the prescientific concept of space was this:

> We have, so to speak, forgotten what features in the world of experience caused us to frame those concepts, and we have great difficulty in calling to mind the world of experience without the spectacles of the old-established conceptual interpretation. There is the further difficulty that

---

[10] In (Einstein, 1953), he described Galileo's rejection of Kepler as "a grotesque illustration that creative human beings frequently lack a receptive frame of mind" (p. xv). Einstein's remark inspired the art historian Erwin Panofsky to examine Galileo's motive, which he localized in Galileo's aesthetic sensibilities. On Panofsky and Einstein's "blind spot," see (Bredekamp, 2001, pp. 184–186).

[11] Also in (Einstein, 1989, pp. 564–582).

our language is compelled to work with words which are inseparably connected with those primitive concepts. (Einstein, 1982, p. 277)[12]

Modern intellectual historians will not disagree.

The kind of history that Einstein outlined in his survey of the concept of space
was admittedly a constructed or intuitional history, not a factual history based on
the right chronology and causal sequence of events. As mentioned, in Einstein's
view the recovery of the significance of historical events was more important than
their chronology and contexts. The same can be said about Mach, who used history
of science to illuminate and promote his epistemological views, first of all his basic
claim that all of our knowledge of nature is based in experience.[13] Einstein was
much less a historian than Mach, but also in his case did history of science serve
methodological and epistemological purposes, only were these very different from
those held by Mach.

This approach is quite clear in his 1930 article, which includes a long digres-
sion on Einstein's view of the aim and methods of science, such as exemplified by
the theory of relativity, "a fine example of the fundamental character of the modern development of theoretical science." The ideas included in the article on space
of 1930 were essentially the same as those he would state more fully in his Herbert Spencer lecture three years later. "The grand aim of all science," according
to Einstein, is "to cover the greatest possible number of empirical facts by logical deduction from the smallest possible number of hypotheses or axioms." In the
search for the laws of nature, the scientist cannot rely solely on experiments but
must be guided by purely formal considerations, particularly mathematical arguments. Although "the observed fact is undoubtedly the supreme arbiter" when it
comes to justifying a theory, it is of little help in the context of creation. Contrary
to what Mach and his allies believed, there is no way in which the basic principles
of physics can be obtained by means of induction from phenomena. There is only
"intense, hard thinking." And when such thinking has resulted in a satisfactory law
with a deductive structure, there is always the possibility that it can be falsified by
new experimental data:

> The theorist has to set about this Herculean task fully aware that his efforts
> may only be destined to prepare the death blow to his theory. The theorist
> who undertakes such a labour should not be carped at as "fanciful"; on the

---

[12] See also (Byrne, 1980, pp. 266–269). "The Problem of Space, Ether, and the Field in
Physics" (Einstein, 1982, pp. 276–285) is a translation of the article that appeared in
Mein Weltbild, first published in 1934. The version differs in phraseology from the original article (Einstein, 1930), and on Einstein's request it left out a long account of Einstein's view of unified field theory anno 1930.

[13] On Mach as a historian of science and his use of history for philosophical purposes, see
(Blüh, 1968; Hiebert, 1970).

contrary, he should be granted the right to give free reign to his fancy, for there is no other way to the goal. His is no idle daydreaming, but a search for the logically simplest possibilities and their consequences. (Einstein, 1982, p. 282)[14]

Einstein found this view justified by his own experiences with the general theory of relativity and, more generally, by the historical development of theoretical physics.

4 The Evolution of Physics

Einstein's most extensive work of a historical form was *The Evolution of Physics*, a semipopular book he wrote in 1937 jointly with his collaborator, the young Polish physicist Leopold InfeldS. The background of this hugely popular book (which is still in print) was that Infeld had come to Princeton and there started a collaboration with Einstein on gravitational radiation and other problems of relativistic physics. However, as it proved impossible to support Infeld's stay by means of a grant from the Institute of Advanced Study, Infeld proposed to finance his further stay by writing a book with Einstein, who found it to be "not a silly idea at all." Although the idea to write the book thus was Infeld's, the content of the book was mainly due to Einstein, whose conception of the development of physics it reflected. Initially *The Evolution of Physics* was planned as an elementary and popular work, but as Einstein got involved with it the concept changed and became more ambitious. The two authors now aimed at a more learned work, though written in a straightforward way and with an appeal to readers of a "fairly high intellectual level."[15] Because the book includes a rather full account of Einstein's view on the historical development of physics and its relevance for the philosophy of science, it should not be dismissed as merely one more popular book on physics. It was considered serious enough to be reviewed in *Isis*, the premier academic journal of history of science founded and edited by Georges Sarton.[16]

Einstein and Infeld freely admitted that their account was highly selective, a constructed history which included only events they found to be interesting and

---

[14] On Einstein's mature philosophy of science, see also his letter to his friend Maurice Solovine of 7 May 1952, as analyzed in detail in (Holton, 1998, pp. 28–56).

[15] See (Infeld, 1976, pp. 78–79) and also (Infeld, 1941). On the relationship between Infeld and Einstein, see (Stachel, 2002).

[16] *Isis* **30** (1939, pp. 124–125). The reviewer, the physicist and philosopher Victor Lenzen, was full of praise for the book, except that he objected to Einstein and Infeld's claim that physics started with Galileo. As he pointed out, this was to seriously underrate the contributions made in ancient Greece and in medieval and renaissance Europe.

important: "Through the maze of facts and concepts we had to choose some highway which seemed to us most characteristic and significant. Facts and theories not reached by this road had to be omitted." Nor were they interested in telling how physical theories actually had come into being, for they attempted rather to "reconstruct the line of progress logically, without bothering too much about chronological order." (Einstein and Infeld, 1938, pp. v and 129).[17] Their relaxed attitude to historical facts and contexts permeates much of the book. Although it includes several quotations, some of them quite extensive, dates are almost entirely missing. The lack of context may be illustrated by Einstein and Infeld's account of the early science of electricity, where the reader is told, "Regardless of the truth concerning details, there is no doubt that Galvani's accidental discovery led Volta ... to the construction of what is known as a voltaic battery." From there they proceed to Ørsted's discovery of electromagnetism, without mentioning the background in romantic *Naturphilosophie* which was, after all, a crucial element for Ørsted. And from Ørsted they leap directly to Rowland's experiment, although the two experiments were separated in time by nearly sixty years (Einstein and Infeld, 1938, pp. 88–92).

The Evolution of Physics is an exemplary history, in the sense that many of the chosen cases are used to illustrate Einstein's favourite view of methodology of science. This view included that laws of nature cannot be obtained inductively from experiments, but only by the imaginative establishment of mathematical principles with a maximum amount of logical simplicity. For example, Einstein and Infeld used their history to bring home the point that the law of inertia "cannot be derived directly from experiment, but only by speculative thinking consistent with observation." (Einstein and Infeld, 1938, p. 9). Whereas experimental data are not very important in the creative phase of science, reinterpretation is: "To raise new questions, new possibilities, to regard old problems from a new angle, requires creative imagination and marks real advance in science. The principle of inertia, the law of conversation of energy were gained only by new and original thoughts about already well-known experiments and phenomena." (Einstein and Infeld, 1938, pp. 95–96). These and similar methodological points are illustrated not only by the older history of physics, but also by Einstein's own contributions. Thus, in the account of the anomalous precession of Mercury's orbit and its later explanation by the general theory of relativity, the authors emphasized that this theory was developed "without any attention" to the anomaly. Einstein knew that this was an exaggeration, but chose to ignore the role that the Mercury anomaly had played in the process that led to the field equations of general relativity (see Earman and Janssen, 1993).

---

[17] There is an obvious similarity to Imre Lakatos' concept of a rationally reconstructed "internal historiography," where the historian decides to "omit everything that is irrational in the light of his rationality theory." See (Lakatos, 1976, p. 18).

5 Testing of theories

Einstein and Infeld were not, of course, unconcerned with experiments and observations, but they restricted their role in theory development to testing. Einstein always insisted that if there were *conclusive* evidence against a theory, it had to be abandoned; he found it unacceptable to save a theory by modifying it by means of auxiliary hypotheses. This was the case in particular for fundamental theories such as the theory of relativity, of which he wrote in 1919: "If any deduction from it should prove untenable, it must be given up. A modification seems impossible without the destruction of the whole." (Einstein, 1919).[18] In *The Evolution of Physics* Einstein took up the holistic theme of testing in relation to Newton's celestial mechanics. This theory works wonderfully, but one can easily imagine modifications that work just as well:

> It is really our whole system of guesses which is to be either proved or disproved by experiment. No one of the assumptions can be isolated for separate testing. In the case of the planets moving around the sun it is found that the system of mechanics works splendidly. Nevertheless we can well imagine that another system, based on different assumptions, might work just as well. (Einstein and Infeld, 1938, p. 33)

Einstein was indeed a holist, and his view of a physical theory as a collection of hypotheses, assumptions, background knowledge etc. which can only be tested as a whole has a clear affinity with Pierre Duhem's philosophy of science, as expounded most fully in his *La théorie physique, son objet et sa structure* of 1906.[19] Although Einstein never referred in print to Duhem's book, he most likely knew it, and had possibly read it, by the end of 1909. Don Howard and others have convincingly argued that Duhem exerted a strong and lasting influence on Einstein's epistemology (Howard, 1990).[20] Whereas Einstein found inspiration in Duhem's philosophy of science, apparently he showed no interest in his historical

---

[18] And in (Einstein, 2002, pp. 213–214).

[19] Compare the quotation by Einstein and Infeld with Duhem's statement: "The physicist can never subject an isolated hypothesis to experimental test, but only a whole group of hypotheses; when the experiment is in disagreement with his predictions, what he learns is that one of the hypotheses constituting this group is unacceptable and ought to be modified; but the experiment does not designate which one should be changed." (Duhem, 1974, p. 18)

[20] See also (Hentschel, 1992). *La théorie physique* appeared in German in 1908 as *Ziel und Struktur der physikalischen Theorie*, translated by Einstein's close friend Friedrich Adler and with a sympathetic foreword by Mach. In a letter of 1918, Einstein referred approvingly to "the clear book by Duhem." See (Howard, 1990, p. 368).

and historiographical writings. Einstein's early exposure to Duhemianism is illustrated by his lesson notes in electricity for the term 1910–11, where he discusses the empirical meaning of the electrostatic field. The force can be defined by way of a test charge, "But the force thus defined is no longer immediately accessible to exp[periment]. It is part of a theoretical construction that is true or false, i.e., corresponding or not corresponding to experience, only *as a whole*." (Einstein, 1993a, p. 325, Einstein's emphasis).

It was an important feature in Duhem's conception of physics that a theory cannot, strictly speaking, ever be falsified or verified. Einstein, on the other hand, defended the strong testability of theories. For example, in an address delivered upon Max Planck's sixtieth birthday in 1918, Einstein referred to the problem of empirical underdetermination, namely that there are "any number of possible systems of theoretical physics all equally well justified" to account for some phenomena. But to Einstein, this was not a serious problem that physicists needed to worry about:

> The development of physics has shown that at any given moment, out of all conceivable constructions, a single one has always proved itself decidedly superior to all the rest. Nobody who has really gone deeply into the matter will deny that in practice the world of phenomena uniquely determines the theoretical system, in spite of the fact that there is no logical bridge between phenomena and their theoretical principles. (Einstein, 1982, p. 226).[21]

A similar view can be found in several passages in *The Evolution of Physics*. The role of experiment is to test theory, which according to Duhem presents a problem, at least from a logical point of view. Yet, Duhem realized that there is an ironic contrast between the logical fact of underdetermination and the practise of science, which often leads to unambiguous results. This contrast was more clearly stressed by Einstein and Infeld, who had no problem with the concept of a *crucial experiment*. Thus they considered the Michelson-Morley experiment to be crucial in relation to the pre-relativity ether theories (although they knew, of course, that originally the experiment did not have the effect of ruling out the ether). In general, Einstein had a relaxed attitude to systematic philosophy. As he expressed it in 1949: "The scientist ... cannot afford to carry his striving for epistemological systematic that far. ... He therefore must appear to the systematic epistemologist as a type of unscrupulous opportunist." ("Reply to Criticisms", Einstein, 1949, p. 684).

In their semi-historical account of the creation of the general theory of relativity, Einstein and Infeld wrote as follows: "Every speculation must be tested by

---

[21] The address ("Principles of Research") was first published in *Mein Weltbild* in 1934.

experiment, and any results, no matter how attractive, must be rejected if they do not fit the facts." (Einstein and Infeld, 1938, p. 251). This may sound like an empiricist statement, but a few lines later the important qualification is made: "Even if no additional observation could be quoted in favour of the new theory [of general relativity], if its explanation were only just as good as the old one, given a free choice between the two theories, we should have to decide in favour of the new one." The reason is that the equations of general relativity are based on simpler and more general assumptions than the equations of Newtonian gravitation theory.

Elsewhere in the book, Einstein's light quanta or photons are introduced in the standard textbook way, by pointing out the incompatibility of the photoelectric effect and the wave theory of light. Somewhat surprisingly, given Einstein's role in the development and knowledge of it, it is claimed that the photoelectric effect (and particularly the result that the speed of the photoelectrons are independent of the intensity of light) presents an unsolvable anomaly to the wave theory: "This experimental result could not be predicted by the wave theory. Here again a new theory arises from the conflict between the old theory and experiment." (Einstein and Infeld, 1938, p. 274).[22] What is here described is the traditional role of the experiment, following Popper, namely to disprove a theory and guide the theorist to come up with an alternative hypothesis.

6 Einstein and Popper

As mentioned, there are several Popperian elements in Einstein's methodology and his rationalization of the history of physics, especially when it comes to falsification, the failure of inductivism, and the existence of crucial experiments. Moreover, Einstein, the physicist, was in agreement with Popper, the philosopher, with regard to the distinction between the contexts of discovery and justification. It is well known that young Popper was much impressed by Einstein and his attitude to theory shift, not least his willingness to let the general theory of relativity depend on how its predictions related to observations. In his autobiography, Popper made clear his indebtedness to Einstein in this way:

> What impressed me most was Einstein's own clear statement that he would regard his theory as untenable if it should fail in certain tests. ...
> Here was an attitude utterly different from the dogmatic attitude of Marx,

---

[22] In fact, there need not be any inescapable contradiction between the photoelectric effect and the wave theory of light. Around 1910 several physicists suggested wave-based theories that could accomodate the anomalous result, if only by introducing ad hoc hypotheses. Einstein knew about the theories, but chose to ignore them. See (Stuewer, 1970).

Freud, Adler, and even more so that of their followers. Einstein was looking for crucial experiments whose agreement with his predictions would
by no means establish his theory; while a disagreement, as he was the first
to stress, would show his theory to be untenable. This, I felt, was the true
scientific attitude. (Popper, 1976, p.38)[23]

Einstein did indeed express himself as mentioned by Popper, although at some
other occasions his attitude was less clear. Einstein read *Logik der Forschung* soon
after it appeared in the fall of 1934, and he liked it, such as we know from a
letter he wrote to Popper on 15 June 1935. "Your book has pleased me in many
ways," Einstein wrote. "Rejection of the 'inductive method' from an epistemological standpoint. Also the falsifiability as determining property of a theory of reality.
... You have also defended your position really well and sharpwittedly."[24] He even
offered to bring Popper's work to the attention of his colleagues and asked Popper
of how he could best help him. In another letter to Popper, dealing with the new
Einstein-Podolsky-Rosen thought experiment, Einstein wrote: "I really do not like
the now fashionable 'positivistic' tendency of clinging to what is observable. ... I
think (like you, by the way) that theory cannot be fabricated out of the results of
observation, but that it can only be invented."[25]

The basis of conceptual advances in scientists' creative imagination, rather
than inductions from experiments, was another element common to Einstein and
Popper. Indeed, Popper explicitly pointed out the similarity between Einstein's
view and his own contention that the discovery process is intuitive (or even "irrational").[26] In an essay published in 1919 in the *Berliner Tageblatt*, Einstein argued
from the history of science ("the real development") that the creative researcher
always starts with a hypothesis, some preconceived intuitive view. "Galileo could
never have discovered the law of freely falling bodies, had he not maintained the
preconceived opinion that the circumstances which we really encounter are complicated by the effects of air resistance so that one has to focus on cases in which
air resistance plays as marginal a role as possible."[27] Einstein further remarked that
a theory can be proven wrong in two ways, either logically (if there is an error in

---

[23] Einstein's importance for Popper is also illustrated by *Logik der Forschung*, Popper's
main work in philosophy of science. In the first English edition of 1959, Einstein is
the second-most cited author, whether scientist or philosopher. Only the logician and
philosopher Rudolf Carnap received more references. See (Popper, 1959).

[24] Quoted in (Van Dongen, 2002, p. 39). See also (Adam, 2000).

[25] Letter of 11 September 1935, reproduced in (Popper, 1959, pp. 457–460).

[26] See (Popper, 1959, p. 32), where he cites Einstein's 1918 address: "[The laws of physics]
can only be reached by intuition, based upon something like an intellectual love of the
objects of experience." Cf. (Einstein, 1982, pp. 224–227).

[27] The article is translated into English in (Adam, 2000, pp. 34–35). Having realized the
close affinity between Einstein's and Popper's views, John Stachel asked in 1983 Popper

its deductions) or empirically (if a prediction disagrees with facts). However, "the *truth* of a theory can never be proven. For one never knows that even in the future no experience will be encountered which contradicts its consequences."

The Einstein-Popper view of the construction and development of scientific theories is not as unusual as it may appear. Similar views have been expounded by other theoretical physicists, both before and after Einstein. Consider as one example young Fred Hoyle and his astronomer colleague Raymond Lyttleton, who in 1948 wrote a survey article on stellar physics. Neither Hoyle nor Lyttleton were philosophically inclined, and there is no reason to assume that they were inspired by or knew of the ideas of Einstein and Popper. Yet their spontaneous philosophy of science reflected the same emphasis on intuition as an essential element in theory building. Referring to the inductivist view of science, where hypotheses are suggested only after a careful consideration of empirical evidence, they wrote:

> This traditional view, however, is largely incorrect, for not only is it absurdly impossible of application, but it is contradicted by the history of the development of any scientific theory. What happens in practice is that by intuitive insight, or any inexplicable inspiration, the theorist decides that certain features seem to him more important than others and capable of explanation by certain hypotheses. Then basing his study on these hypotheses the attempt is made to deduce their consequences. The successful pioneer of theoretical science is he whose intuitions yield hypotheses on which satisfactory theories can be built, and conversely for the unsuccessful (as judged from a purely scientific standpoint). (Hoyle and Lyttleton, 1948, p. 90)

At the end of *The Evolution of Physics*, Einstein stated with force what he believed the history of physics had demonstrated: "Science is not just a collection of laws, a catalogue of unrelated facts. It is a creation of the human mind, with its freely invented ideas and concepts." (Einstein and Infeld, 1938, p. 310). He had entertained such an anti-positivist view for many years and it can be found in several of his writings since about 1920. For example, when he was interviewed by the journalist and author Alexander Mozkowski in 1921, he stressed the difference between the concepts of "discovery" and "construction". Whereas the former concept refers to finding something preexisting in nature, to construct a new theory of physics involves an inventive or creative act. "Discovery is not really a creative act," Einstein argued (Mozkowski, 1922, p. 100). In agreement with this view, he never referred to the theory of relativity as a discovery, but preferred to call it an invention.

---

if he knew about the 1919 article in the *Berliner Tageblatt*. Popper replied that he did not and that the essay was new to him.

It was clear to Einstein that, as he expressed it in *The Evolution of Physics*, "Physical concepts are free creations of the human mind, and are not, however it may seem, uniquely determined by the external world." (Einstein and Infeld, 1938, p. 33). Imagination and intuition play an important and legitimate role, and Einstein often justified his beliefs in terms of such concepts. Nowhere is this better illustrated than in his dismissal of the cosmological constant which he added to his field equations in 1917 in order to secure a static model of the universe. Einstein soon came to see the constant as unwarranted, not because it lacked observational support but because it reduced "the formal beauty of the theory," as he wrote in 1919. In a letter to the Belgian cosmologist Georges Lemaître of 1947 he repeated that the constant was "very ugly indeed" and consequently had no place in physical theory. He explained: "About the justification of such feelings concerning logical simplicity it is difficult to argue. I cannot help to feel it strongly and I am unable to believe that such an ugly thing should be realized in nature."[28]

Science was for Einstein a highly personal and emotional activity, a kind of "cosmic religion" that could not possibly be accomodated by empiricist philosophy, whether in Mach's version or in the later version of logical empiricism. Or, for that matter, by any formal system of philosophy.

7 Scientific revolutions

Einstein believed that the history of science gave support to the view that there are no final or eternal theories in physics.[29] Science is a never-ending process (as also Popper stressed), and even the crowning achievement of his own creativity, the general theory of relativity, was no exception. One day it would be replaced by a better theory. In connection with older theories of heat and electricity, he wrote:

> There are no eternal theories in science. It always happens that some of the facts predicted by a theory are disproved by experiment. Every theory has its period of gradual development and triumph, after which it may experience a rapid decline. ... Nearly every great advance in science arises from a crisis in the old theory, through an endeavour to find a way out of the difficulties created. (Einstein and Infeld, 1938, p. 77)

---

[28] Quoted in (Kragh, 1996, p. 54). Einstein abandoned the cosmological constant in 1931, after the expansion of the universe had shown that it was unnecessary.

[29] He did, however, exclude classical thermodynamics from the rule because of its high degree of simplicity and generality. In his autobiographical notes, he wrote: "It is the only physical theory of a universal content which I am convinced that within the framework of the applicability of its basic concepts, it will never be overthrown." (Einstein, 1949, p. 33).

While this passage does not disagree with Popper's ideas, it also has a Kuhnian ring, as has some of his observations later in the book and in other writings. Einstein's picture of the long-term evolution of physics is in broad agreement with Thomas Kuhn's views as developed in *The Structure of Scientific Revolutions* from 1962. Thus, Einstein pictured the history of physics since Galileo as a sequence of conceptual revolutions in which new physical ideas were born in the struggle with old ideas. On the other hand, the subsequent development, following the revolutions, was evolutionary and cumulative, or what Kuhn called paradigm-governed or normal science. "The initial and fundamental steps are always of a revolutionary nature," Einstein said. "The continued development along any line already initiated is more in the nature of evolution, until the next turning point is reached when a still newer field must be conquered." (Einstein and Infeld, 1938, p. 28).

Yet, Einstein's use of terms such as "revolution" and "crisis" does not make him a Kuhnian. In Kuhn's original view, revolutions mark separations between incommensurable world views, whereas Einstein's revolutions retain a close connection with past science. In describing the transition from the mechanical world view to field theory – a crucial phase in the realist Einstein's version of the history of physics – he expressed his view as follows. "The new theory shows the merits as well as the limitations of the old theory and allows us to regain our old concepts from a higher level. This is true not only for the theory of electric fluids and field, but for all changes in physical theories, however revolutionary they may seem." To bring home his point, he illustrated it in the following way:

> To use a comparison, we could say that creating a new theory is not like destroying an old barn and erecting a skyscraper in its place. It is rather like climbing a mountain, gaining new and wider views, discovering unexpected connections between our starting-point and its rich environment. But the point from which we started out still exists and can be seen, although it appears smaller and forms a tiny part of our broad view gained by the mastery of the obstacles on our adventurous way up. (Einstein and Infeld, 1938, p. 159)[30]

Needless to say, this is a most un-Kuhnian view. Einstein never claimed that his theory of relativity was a revolution, and certainly not in the strong sense associated with Kuhn's early philosophy of science. On the contrary, he saw it as a natural extension of the classical physics developed by Newton, Maxwell and Lorentz.[31] As he said in a lecture he delivered at King's College, London, in 1921: "We have here no revolutionary act, but the natural continuation of a line that can

---

[30] In his biography of Duhem, Stanley Jaki refers to the quotation, which he finds is "uncannily Duhemian in its ring" (Jaki, 1984, p. 436).

[31] In spite of Einstein's disclaimer, his physics, and the theory of relativity in particular, was generally hailed as a revolutionary break with the past. The revolution metaphor,

be traced through centuries." (Einstein, 1982, p. 246). In fact, he felt that the notion of revolutions in science had become inflated and might produce a false impression of how science develops. It would be quite wrong to believe, he said in a press statement of 1947, "that every five minutes there is a revolution in science, somewhat like the coups d'état in some of the smaller unstable republics." No, by and large science progresses cumulatively, it is a process to which scientists "of successive generations add by untiring labor" and which "slowly leads to a deeper conception of the laws of nature." (Klein, 1975, p. 113)

So much for the (in)compatability of Einstein's view and the picture of science as a series of discontinuing revolutions, as presented in Kuhn's *The Structure of Scientific Revolutions*. On the other hand, the later Kuhn did not maintain this radical conception. In the 1980s he came to see revolutions in science as gradual to a much greater extent and also modified his view of incommensurability accordingly. He described, rather abstractly, the central characteristic of scientific revolutions to be, "that they alter the knowledge of nature that is intrinsic to the language itself and that is thus prior to anything quite describable as description og generalization, scientific or everyday." (Kuhn, 2000, p. 32)[32]. It is possible that this version of Kuhn's view is in better harmony with Einstein's dynamical philosophy of science, although I doubt it. At any rate, it is a question that lies outside the scope of the present essay, which is primarily concerned with Einstein's view of history of science and the progress of scientific knowledge.

References

Adam, A. (2000). Farewell to certitude: Einstein's novelty on induction and deduction, fallibilism, *Journal for General Philosophy of Science* **31**: 19–36.

Belousek, D. W. (1966). Einstein's 1927 unpublished hidden-variable theory: Its background, context and significance, *Studies in the History and Philosophy of Modern Physics* **27**: 437–461.

Blüh, O. (1968). Ernst Mach as an historian of physics, *Centaurus* **13**: 62–84.

Bredekamp, H. (2001). Gazing hands and blind spots: Galilo as draftsman, *in* J. Renn (ed.), *Galileo in Context*, Cambridge University Press, Cambridge, pp. 153–192.

Brush, S. G. (1995). Scientists as historians, *Osiris* **10**: 215–232.

Byrne, P. H. (1980). The significance of Einstein's use of the history of science, *Dialectica* **34**: 263–276.

---

as associated with Einstein, was equally popular among scientists and non-scientists. Thus, Mao Zedong referred explicitly and repeatedly to the Einsteinian revolution as a philosophical justification for his politics in the 1950s and 1960s (Friedman, 1983).

[32] The essay was first published in 1982.

Cohen, I. B. (1955). An interview with Einstein, *Scientific American* **193**: July, 68–73.

Cohen, I. B. (1979). Einstein and Newton, *in* A. P. French (ed.), *Einstein. A Centenary Volume*, Harvard University Press, Cambridge, Mass., pp. 40–42.

Duhem, P. (1974). *The Aim and Structure of Physical Theory*, Atheneum, New York.

Earman, J. and Janssen, M. (1993). Einstein's explanation of the motion of Mercury's perihelion, *in* J. Earman and J. D. Norton (eds), *The Attraction of Gravitation: New Studies in the History of General Relativity*, Birkhäuser, Boston, pp. 129–172.

Einstein, A. and Infeld, L. (1938). *The Evolution of Physics*, Cambridge University Press, Cambridge.

Einstein, A. (1909). Entwicklung unserer Anschauungen über das Wesen und die Konstitution der Strahlung, *Physikalische Zeitschrift* **10**: 817–825.

Einstein, A. (1916). Ernst Mach, *Physikalische Zeitschrift* **17**: 101–103. Reprinted in Einstein (1993b), pp. 277-282.

Einstein, A. (1919). Time, space and gravitation, *The Times* **28**: November, 13–14.

Einstein, A. (1922). Emil Warburg als Forscher, *Die Naturwissenschaften* **10**: 823–828.

Einstein, A. (1924). Über den Äther, *Verhandlungen der Schweizerische naturforschende Gesellschaft* **105**: part 2, 85–93.

Einstein, A. (1927). Isaac Newton, *The Observatory* **50**: 146–153.

Einstein, A. (1930). Raum, Äther und Feld in der Physik, *Forum Philosophicum* **1**: 173–180.

Einstein, A. (1934). On the method of theoretical physics, *Philosophy of Science* **1**: 162–169.

Einstein, A. (1949). *Albert Einstein. Philosopher-Scientist*, edited by Paul A. Schilpp. Library of Living Philosophers, New York.

Einstein, A. (1953). *Foreword*, pp. vi-xix in Galileo Galilei, *Dialogue Concerning the Two Chief World Systems*, translated by Stillman Drake. University of California Press, Berkeley.

Einstein, A. (1982). *Ideas and Opinions*, Three Rivers Press, New York.

Einstein, A. (1987). *The Collected Papers of Albert Einstein*, Vol. 1 The Early Years: 1879-1902, Princeton University Press, Princeton. Edited by J. Stachel, D. C. Cassidy and R. Schulmann.

Einstein, A. (1989). *The Collected Papers of Albert Einstein*, Vol. 2 The Swiss Years: Writings, 1900-1909, Princeton Univerity Press, Princeton. Edited by J. Stachel, D. C. Cassidy, J. Renn, R. Schulmann and D. Howard.

Einstein, A. (1993a). *The Collected Papers of Albert Einstein*, Vol. 3 The Swiss Years: Writings, 1909-1911, Princeton Univerity Press, Princeton. Edited by M. J. Klein, A. J. Kox, J. Renn and R. Schulmann.

Einstein, A. (1993b). *The Collected Papers of Albert Einstein*, Vol. 5 The Swiss Years: Correspondence, 1902-1914, Princeton Univerity Press, Princeton. Edited by M. J. Klein, A. J. Kox and R. Schulmann.

Einstein, A. (1996). *The Collected Papers of Albert Einstein*, Vol. 6 The Berlin Years: Writings, 1914-1917, Princeton Univerity Press, Princeton. Edited by A. J. Kox and M. J. Klein and R. Schulmann.

Einstein, A. (2002). *The Collected Papers of Albert Einstein*, Vol. 7 The Berlin Years: Writings, 1918-1921, Princeton Univerity Press, Princeton. Edited by M. Janssen, R. Schulmann, J. Illy, C. Lehner and D. K. Buchwald.

Friedman, E. (1983). Einstein and Mao: Metaphors of revolution, *China Quarterly* **93**: 51–75.

Hentschel, K. (1992). Einstein's attitude towards experiments: Testing relativity theory 1907-1927, *Studies in the History and Philosophy of Science* **23**: 593–624.

Herneck, F. (1984). Albert Einstein als wissenschaftlicher Biograph, *in* Herneck (ed.), *Wissenschaftsgeschichte. Vorträge und Abhandlungen*, Akademie-Verlag, Berlin, pp. 92–101.

Hiebert, E. (1970). Mach's philosophical use of the history of science, *in* R. H. Stuewer (ed.), *Historical and Philosophical Perspectives of Science*, University of Minnesota Press, Minneapolis, pp. 184–203.

Hoh, T. (1885). *Die Stellung der Atomlehre zu Physik des Aethers*, W. Gärtner, Bamberg.

Holton, G. (1988). *Thematic Origins of Scientific Thought. Kepler to Einstein*, Harvard University Press, Cambridge, Mass.

Holton, G. (1998). *The Advancement of Science, and its Burdens*, Harvard University Press, Cambridge, Mass.

Howard, D. (1990). Einstein and Duhem, *Synthese* **84**: 363–384.

Hoyle, F. and Lyttleton, R. (1948). The internal constitution of the stars, *Occasional Notes of the Royal Astronomical Society* **no. 12**: 89–108.

Infeld, L. (1941). *Quest. The Evolution of a Scientist*, Chelsea Publishing Company, New York.

Infeld, L. (1976). *Leben mit Einstein: Kontur einer Erinnerung*, Europa Verlag, Wien.

Jaki, S. (1984). *Uneasy Genius: The Life and Work of Pierre Duhem*, Martinus Nijhoff Publishers, The Hague.

Klein, M. J. (1975). Einstein on scientific revolutions, *Vistas in Astronomy* **17**: 113–120.

Kragh, H. (1987). *An Introduction to the Historiography of Science*, Cambridge University Press, Cambridge.

Kragh, H. (1996). *Cosmology and Controversy. The Historical Development of Two Theories of the Universe*, Princeton University Press, Princeton.

Kragh, H. (2005). Einstein as a historian of science, *in* J. Renn (ed.), *Albert Einstein, Chief Engineer of the Universe: One Hundred Authors for Einstein*, Wiley-VCH, Berlin, pp. 358–361.

Kuhn, T. (2000). What are scientific revolutions?, *in* J. Conant and J. Haugeland. (eds), *The Road Since Structure*, University of Chicago Press, Chicago, pp. 13–32.

Lakatos, I. (1976). History of science and its rational reconstruction, *in* C. Howson (ed.), *Method and Appraisal in the Physical Sciences*, Cambridge University Press, Cambridge, pp. 1–40.

Mach, E. (1883). *Die Mechanik in Ihrer Entwicklung, historisch-kritisch dargestellt*, Barth, Leipzig.

Mozkowski, A. (1922). *Einstein. Einblicke in seine Gedankenwelt*, Fontane, Berlin.

Neidorf, R. (1963). Discussion: Is Einstein a positivist?, *Philosophy of Science* **30**: 173–188.

Popper, K. R. (1959). *The Logic of Scientific Discovery*, Basic Books, New York.

Popper, K. R. (1976). *Unended Quest. An Intellectual Autobiography*, Fontana, London.

Seelig, C. (1954). *Albert Einstein*, Europa Verlag, Zurich.

Shankland, R. S. (1963). Conversations with Albert Einstein, *American Journal of Physics* **31**: 47–57.

Speziali, P. (ed.) (1974). *Albert Einstein, Michele Besso: Correspondence 1903-1955*, Hermann, Paris.

Stachel, J. (2002). Einstein and Infeld: Seen through their correspondence,, *in* Stachel (ed.), *Einstein from 'B' to 'Z'*, Birkhäuser, Boston, pp. 477–497.

Stuewer, R. H. (1970). Non-Einsteinian interpretations of the photoelectric effect, *in* Stuewer (ed.), *Historical and Philosophical Perspectives on Modern Physics*, University of Minnesota Press, Minneapolis, pp. 246–264.

Van Dongen, J. (2002). *Einstein's Unification: General Relativity and the Quest for Mathematical Naturalness*, Faculty of Science, University of Amsterdam.

How to recognize intruders in your niche

Hanne Andersen

Steno Institute, University of Aarhus

One important problem concerning incommensurability is how to explain that two theories which are incommensurable and therefore mutually untranslatable and incomparable in a strictly logical, point-by-point way are still competing. The two standard approaches have been to argue either that the terms of incommensurable theories may share reference, or that incommensurable theories target roughly the same object domain as far as the world-in-itself is concerned. However, neither of these approaches to the problem pay due respect to the incommensurability thesis' insights. In this paper I shall first show the inconsistency between the basic premises underlying Kuhn's incommensurability thesis and the two standard responses to the thesis. I shall then argue that if one adopts Kuhn's position, the response must build on a notion of overlap between phenomenal worlds. Finally, I shall argue that overlap between complex structures of features can provide the basis for such a notion, and that this makes it possible to explain how incommensurable theories may compete.

1 The problem: Intruder or next-door neighbour?

At the beginning of the 1990s Thomas Kuhn introduced a new metaphor in his account of the development of science: speciation.[1] According to this view the sciences develop like an evolutionary tree in which new subspecialties emerge and gradually get isolated from each other and from the specialties from which they proliferated. The mutual isolation of the subspecialties is brought about by a growing conceptual disparity between the developed tools, hence, the specialized

---

[1] Although Kuhn had compared the development of science to biological evolution in his earlier works, his main metaphor in these works had been selection, not speciation.

scientist with his highly adapted tools, refined to serve the purposes of the sub-specialty, inhabits a niche isolated from the niches of other subspecialties (Kuhn, 1992, p. 20).[2] Occasionally, the exploitation of the niche by its inhabitants reaches its limit, and usually reorganization and new proliferation are the result. The inhab-itants of the old, exhausted niche may die out, and with them the niche vanishes. In this respect speciation resembles the revolutions which Kuhn previously used as a metaphor for certain phases of the development of science. However, in other respects the new metaphor of speciation is fundamentally different from the old metaphor of revolutions.[3] One of the most striking differences is the role played by incommensurability. In the process of speciation incommensurability is claimed to play the role of making the subspecialties distinct and keeping them apart:

> ... what makes these specialties distinct, what keeps them apart and leaves the ground between them as apparently empty space ... is incommensu-rability, a growing conceptual disparity between the tools deployed in the two specialties. Once the two specialties have grown apart, that disparity makes it impossible for the practitioners of one to communicate fully with the practitioners of the other. (Kuhn, 1992, p. 19f.)

Ironically, this new role ascribed to incommensurability emphasizes one of the most serious problems implied by the original incommensurability thesis: how to make sense of the idea that incommensurable theories are actually competing.

Intuitively, one would say that the fact that there is no communication between the different niches – like astrophysics and immunology – reflects only that they address "something different", that they are not "about the same thing". On the contrary, for such theories as, for example, oxygen theory and phlogiston theory one would say that they are indeed "about the same thing" and therefore within this shared niche compete on offering the better account of their common domain. When formulated in terms of the speciation metaphor this amounts to the differ-ence that scientists inhabiting a given scientific niche live in peaceful co-existence with scientists from *other niches*, but competition and combat is the immediate re-sult if intruders, or betrayers from inside, start exploiting their *own niche*, changing it slightly as they go.

The distinction between intruders in a given niche and inhabitants of another is blurred by Kuhn's formulation of the speciation metaphor when he claims incom-mensurability to hold between different subspecialties and not just between a new subspecialty and an old one whose problems the new is expected to solve. In other

---

[2] Wray (2005) describes how most accounts prior to Kuhn's have focused on social changes as the cause of the formation of new specialties, whereas Kuhn brings the epis-temic dimension into focus. However, as Wray notes, Kuhn's work on scientific special-ization has been largely neglected by the science studies research community.

[3] For a discussion of some differences between the two metaphors, see (Chen, 1997).

words, there seems to be a difference for scientists working within a given subspecialty between intruders or betrayers who threaten to change their own niche, and
those who are simply inhabitants of another, neighbouring niche.

The problem of how to maintain this distinction has accompanied the notion of
incommensurability from the outset. Provoked by Kuhn's repeated use of wordings
like "when paradigms change, the world itself changes with them" (Kuhn, 1970a,
p. 111), "after a revolution scientists are responding to a different world" (Kuhn,
1970a, p. 111), or "the proponents of competing paradigms practice their trades
in different worlds" (Kuhn, 1970a, p. 121), critics have argued that there seems
to be simply no point at issue between incommensurable theories – that if they
are formulated in untranslatable languages, any kind of conflict between them is
precluded. Hence, incommensurable theories cannot be rivals. As Shapere put it in
his review of *The Structure of Scientific Revolutions*: "if they disagree as to what
the facts are, and even as to the real problems to be faced and the standards which a
successful theory must meet – then what are the two paradigms disagreeing about?
And why does one win?" (Shapere, 1964, p. 391).

In the following I shall first summarize the main points of the two standard
responses to the problem. Next, I shall show that these responses build on realist premises of a kind which are incompatible with Kuhn's non-realist position.
Finally, I shall suggest an alternative response that is placed between traditional
realism and non-realism[4] and that does not suffer from the same problem as the
two standard responses.

2 The referential stability approach

One standard response to the incommensurability thesis is the referential stability
approach. This approach was first proposed by Scheffler (1967) who argued that
even if two theories are mutually untranslatable, as long as their terms share reference it is possible for statements from the two theories to conflict, hence, for the
theories to be rivals and comparable.

Such a solution to the incommensurability problem would have to draw upon
a theory of reference which could account for reference determination in a way
that secured referential stability during theory change. The classical descriptive
theory of reference according to which the reference of a term is determined by its
associated descriptive content clearly did not fulfil the requirement of referential
stability: change of theory entails new descriptions that may not be true of the

---

[4] An abbreviated version of this argument can be found in (Andersen, 2001). For further
details, see also (Andersen, 2000, 2004).

same things as previous descriptions, and as description determines reference such changes of description will entail change of reference.[5]

The causal theory of reference may at first sight seem more suited to solve the problem of securing referential stability.[6] According to the causal theory, the extension of a term consists of objects that bear a special same-kind-as relation to each other. Thus, if a term has been introduced in an ostensive act it will in subsequent use refer to other objects of the same kind. This same-kind-as relation as a theoretical relation that is determined by the internal structural traits of the objects to which the term refers, and is therefore a relation that may discovered by scientific research (e.g. Putnam, 1975; Boyd, 1979; Sankey, 1994). Claiming that later theories are "in general, *better* descriptions of the *same* entities that earlier theories referred to" (Putnam, 1975, p. 137, italics in the original), causal theorists base their referential stability thesis on the realist premise that the objects are theory-independent entities that remain unaltered during theory-change (cf. Putnam, 1975). However, as I shall argue below, this is exactly the kind of realism questioned by the incommensurability thesis.

3 Kuhn's theory of world constitution I

Kuhn's ontological viewpoint may be difficult to extract from his writings which suffer from a tension between two different meanings of key terms such as "world" in passages like "though the world does not change with a change of paradigm, the scientist afterward works in a different world" (Kuhn, 1970a, p. 121). However, a reconstruction which dissolves this tension has been provided by Hoyningen-Huene (1989, 1993). According to this reconstruction, two concepts of "the world" have to be distinguished in Kuhn's philosophy: the *phenomenal world* which is a "perceived world" (Kuhn, 1970a, p. 128), and the *world-in-itself* which is a "hypothetical fixed nature" (Kuhn, 1970a, p. 118), cf. (Hoyningen-Huene, 1993, ch. 2.1).

The phenomenal world is "a world already perceptually and conceptually subdivided in a certain way" (Kuhn, 1970a, p. 129). This subdivision is not read off

---

[5] To use a classical descriptive theory in the referential approach was suggested by Scheffler (1967). The inability of the classical descriptive theory to secure referential stability has been pointed out by various scholars, most notably Putnam (1973), but also e.g. Hacking (1983); Newton-Smith (1981); Nola (1980a); Sankey (1994).

[6] This theory was originally introduced by Kripke (1972) to cover proper names and by Putnam (1973) to cover physical magnitude terms and extended to cover natural kind terms in general. Various modifications to the causal theory have been suggested by, among others, Boyd (1979); Enç (1976); Kitcher (1978, 1983); Nola (1980b); Sankey (1991, 1994).

from the world itself, but is a structure which is imposed on the world by means of the concepts applied to it. The structure is established by relations of similarity and dissimilarity between objects.[7] These relations are thus decisive for the determination of reference and the constitution of the conceptual structure, but it is important to note that that they are not necessarily based on underlying structural traits, and that there are no restrictions on *which* characteristics can be used to judge the objects similar or dissimilar: "in matching terms with their referents, one may legitimately make use of anything one knows or believes about those referents" (Kuhn, 1983, p. 681). This means that there is no distinction between defining and contingent features.[8] By the same token, different members of a given language-community may use different features to identify referents and non-referents of the concept.

4 The inadequacy of the referential stability approach

According to the causal theory, entities and the same-kind-as relation that hold between them are expected to exist objectively in the world prior to our concepts. By the same token, the internal structural traits of the entities that determine the same-kind-as relation can be discovered by scientific investigations. Kuhn, on the contrary, denies the realist assumption of the existence of theory-independent entities, and that gives the similarity and dissimilarity relations used in Kuhn's theory a different status than the same-kind-as relations used in the causal theory. Absent the world's real joints, the similarity and dissimilarity relations cannot be determined by them. Instead, Kuhn rhetorically asked if it makes

> better sense to speak of accommodating language to the world than of accommodating the world to language. Or is the way of talking which creates that distinction itself illusory? Is what we refer to as "the world"

---

[7] "Object" shall here be understood as *perceived* objects and not as entities given by an observer-independent world. This may be disturbing to a traditional realist. However, as pointed out by Hoyningen-Huene et al. (1996) and Oberheim and Hoyningen-Huene (1997) many of the central terms used in the realism debate have different meanings in realist and non-realist contexts. The same use of the term object can be found in, for example, Putnam's internal realism is based on a similar view that "'objects' do not exist independently of conceptual schemes. We cut up the world into objects when we introduce the one or another scheme of description" (Putnam, 1981, p. 52).

[8] The latter point is similar to Shapere's rejection of the essentialism inherent in the causal theory: "it is not just one property of set of properties – the 'essential' ones – that determines or affects how scientists will apply terms in new situations; all the (true) properties may ... play a role, and furthermore, the properties and behaviour of other entities (substances, etc.) may also play a role" (Shapere, 1982, p. 7).

perhaps a product of a mutual accommodation between experience and language? (Kuhn, 1979, p. 418)

Hence, instead of being determined by the world's real joints, the relations of similarity and dissimilarity are constitutive of the structure of the phenomenal world, that is, of which objects exist in this world. As a consequence, different sets of similarity and dissimilarity relations may constitute different ontologies – "carve different joints".

This difference between Kuhn's view and the causal theory is the subject of Kuhn and Putnam's discussion of Putnam's Twin Earth argument. In this argument, Putnam had imagined a Twin Earth which is exactly like our own Earth except for the single difference that what is called "water" on Twin Earth is not the chemical compound H_2O. Instead, it a different compound with a very long and complicated chemical formula, abbreviated as XYZ. Nevertheless, this Twinearthian water is indistinguishable from the Earth's water at normal temperature and pressure. On Putnam's view, when the first astronauts from Earth came to visit Twinearth and did not know the chemical formula, they would simply take Twinearthians' water to be water. However, with the development of chemical theory they would at some point discover that they only *mistook* Twinearthian water for water. Thus, Putnam argued that although the astronauts first supposition might probably be that the term "water" has the same meaning on Earth and on Twin Earth, after doing some chemical analyses they would report back to Earth that on Twin Earth the word "water" means XYZ.

Kuhn disagreed to this part of the scenario, arguing that the astronauts' report would rather be something like "Back to the drawing board! Something is badly wrong with chemical theory" (Kuhn, 1989, p. 27); (Kuhn, 1990, p. 310). Kuhn argued that modern chemical theory is incompatible with the existence of a substance with properties very nearly the same as water but described by an elaborate chemical formula. Therefore, the discovery that Twinearthian water is not H_2O but XYZ is not merely a discovery of the underlying traits of a particular substance, but the discovery of an anomaly to chemical theory as such. Resolving that anomaly might lead to a restructured chemical lexicon, and only with this "differently structured lexicon, one shaped to describe a very different sort of world, could one, without contradiction, describe the behaviour of XYZ at all, and in that lexicon 'H_2O' might no longer refer to what we call 'water'" (Kuhn, 1989, p. 27).[9]

---

[9] Further, Kuhn attacked Putnam's claim that the term "water" simply refers to substances with the chemical formula H_2O, pointing out that "H_2O" not only picks out water, but also ice and steam. If one wants to specify the underlying trait of water more than just the chemical constitution is necessary, such as information about packing and relative motion of the molecules. The conjunction of this set of properties picks out a smaller class of objects than the properties taken individually, and that raises the question which

In sum, by drawing on the existence of some real joints in the world to provide the referential stability, the referential approach solves the problem how to establish sameness of object-domain for incommensurable theories, but without taking seriously that what the incommensurability thesis denies is exactly the existence of such world's real joints. Thus, on this point the causal theory and the incommensurability thesis are based on incompatible premises. Consequently, if the problem inherent in Kuhn's thesis – how can theories that are mutually untranslatable and therefore incomparable in a strictly logical, point-by-point way still compete – shall be solved within its original non-realist framework and not just be dismissed by changing to a realist framework, another approach free from traditional realist assumptions is requested.

5 The double-world approach

The reconstruction of Kuhn's position in terms of a perceived world constituted by relations of similarity and dissimilarity relations was introduced to capture Kuhn's rejection of traditional realism. At first sight, it may appear as if the similarity and dissimilarity relations can be freely invented to constitute any arbitrary structure of the phenomenal world. However, this is not the case: "nature cannot be forced into an arbitrary set of conceptual boxes. On the contrary ... the history of the developed sciences shows that nature will not indefinitely be confined in any set which scientists have constructed so far" (Kuhn, 1970b, p. 263).

This "resistance" against giving arbitrary structures to the phenomenal world is ascribed to the world-in-itself. Kuhn first tried to avoid the use of Kantian things-in-themselves (Kuhn, 1979, p. 418f.), and the term world-in-itself was introduced by Hoyningen-Huene in his reconstruction of Kuhn's philosophy (Hoyningen-Huene, 1993, sec. 2.1.a). However, Kuhn later endorsed a similar view:

> Underlying all these processes of differentiation and change, there must, of course, be something permanent, fixed, and stable. But, like Kant's *Ding-an-sich*, it is ineffable, undescribable, undiscussible. Located outside of space and time, this Kantian source of stability is the *whole* from which have been fabricated both creatures and their niches. (Kuhn, 1991, p. 12, italics added)

One here notes the difference between Kuhn's position and both extreme constructivist anti-realism and traditional realism. According to Kuhn's position, the "resistance" appears in the form of *anomalies*, i.e. situations in which it becomes clear that something is wrong with the structure given to the phenomenal world by

---

property or conjunction of properties should be interpreted as cutting the world at it's joints (cf. Kuhn, 1989).

our concepts – that objects do not behave or situations do not develop as prescribed by the current conceptual structure. However, the anomalies only show that there is something wrong with the structure of the phenomenal world, but not how the phenomenal world has to be structured instead. If the resistance fully determined the structure of the phenomenal world there would be no need to introduce the phenomenal world as a changeable perceived world and the position would reduce to traditional realism, if there was no resistance it would be extreme constructivist anti-realism – and Kuhn's position is somewhere in-between.[10]

6 The inadequacy of the double-world approach

Based on his double-world interpretation of Kuhn, Hoyningen-Huene has attempted to solve the key problem of this paper – how can incommensurable theories be rivals – by recourse to the world-in-itself: "Incommensurable theories ...target *roughly the same* object domain, *as far as the world-in-itself is concerned*, though this 'object domain' isn't graspable in any theory-neutral way, since different lexica must always produce different object domains" (Hoyningen-Huene, 1993, p. 219, italics in the original). Although this solution may have some intuitive appeal, this attempt to solve the problem within a Kuhnian non-realist framework seems inconsistent. Thus, the alleged solution consists of two parts: a suggestion to establish sameness of "object-domain" as far as the world-in-itself is concerned, and a qualification to the concept of "object-domain" that it *cannot* be seized in terms of the world-in-itself. Evidently, this is problematic. The two parts together amount to a quest for establishing (a rough) identity between that to which we do not have any epistemic access. That venture is inconsistent. Either we must take seriously that the world-in-itself is a hypothetical fixed nature to which we do not have any epistemic access – but then we cannot establish the required (rough) identity. Or we may establish the identity – but then we are not dealing with a world-in-itself. The world-in-itself here seems to serve as an intuitive realist foundation in the anti-realist building.[11]

---

[10] In Hoyningen-Huene's terminology, the structure of the phenomenal world, constituted by the relations of similarity and dissimilarity, has both object-sided and subject-sided moments. These are inseparable; because the resistance offered by the world-in-itself does not fully determine the structure of the phenomenal world, we cannot get rid of the subject-sided moments, and hence cannot gain full epistemic access to the purely object-sided (Kuhn, 1979, p. 418). See also (Hoyningen-Huene, 1993, sec. 2.3 and Epilogue).

[11] Similar arguments have been raised against Hoyningen-Huene's double-world reconstruction of Kuhn's position by Sankey (1997).

7 Kuhn's theory of world constitution II

Since incommensurability cannot be based on identity between object-domains as far as the world-in-itself is concerned, what must be required instead is some sameness of object-domains as far as the phenomenal world is concerned. However, since incommensurable theories apply to different phenomenal worlds, sameness of object-domain as far as the phenomenal world is concerned can only be established if there is some *overlap* between the phenomenal worlds in question. To clarify how phenomenal worlds may overlap we must again consider the constitution of a phenomenal world and the intertwining of object-sided and subject-sided moments.

As described in section 3 above, the phenomenal world is constituted by a web of similarity and dissimilarity relations that are immediate in the sense that the relation is not based on a similarity-conferring third. Kuhn claims that this is possible because of an "empty perceptual space between the families to be discriminated" (Kuhn, 1970a, p. 197, fn. 14). This can be exemplified by Kuhn's favourite example of the child who learns to recognize ducks, geese and swans. The successful recognition of three different categories presupposes that there are no intermediate forms, no duck-swans that fall between the categories of ducks and swans. If the features involved in recognizing ducks, geese and swans are thought to span the dimensions of perceptual space, there must be discontinuities along some of these dimensions. For example, while the necks of ducks are quite short, the neck of swans are relatively long, and no duck-swans with intermediate lengths are found in nature. It is such empty perceptual space between the families to be discriminated that makes discrimination possible.

Perceptual space and the relations of similarity and dissimilarity seem here to be mutually dependent. At first sight, the mutual dependence of the phenomenal joints and the relations of similarity and dissimilarity may seem circular: the phenomenal joints – categories such as ducks and geese – secure the immediacy of the relations of similarity and dissimilarity,[12] but at the same time the relations of similarity and dissimilarity are constitutive of the phenomenal joints. However, adopting a developmental view this problem of circularity dissolves.[13] The important point is that the phenomenal world is never structured from scratch by its inhabitants. Instead, the inhabitants of any phenomenal world inherit it from their predecessors. A phenomenal world is therefore always provided by the historical situation and may from there again be reshaped by introducing new relations of

---

[12] In this way, Kuhn can be said to draw on the joints of the *phenomenal world* to substantiate the claim of the immediacy of the similarity and dissimilarity relations, but contrary to the purely objective "world's real joints" of the causal theory, the joints of the phenomenal world have both subject-sided and object-sided moments.

[13] A more detailed version of this argument can be found in (Andersen, 2000).

similarity and dissimilarity and abandoning old ones, thus providing a different phenomenal world with different phenomenal joints to the generations to come.

8 Joints in the phenomenal world

In constituting the structure of the phenomenal world, Kuhn ascribes a special importance to the features which differentiate between instances of contrasting concepts. This is due to the taxonomic character of conceptual structures constituted by similarity and dissimilarity relations. Such conceptual structures are necessarily taxonomic, that is, hierarchical structures in which a general concept decomposes exhaustively into a group of more specific, non-overlapping concepts called a contrast set (Chen et al., 1998; Andersen, 2000).[14] Each of the contrasting, subordinate concepts may again decompose into yet more specific concepts, and so forth, thereby forming a taxonomic tree.

The decomposition of a superordinate concept into a group of contrasting concepts is determined by the features: "To each node in a taxonomic tree is attached a name ... and a set of features useful for *distinguishing* among creatures at the next level down. ... Attached features are not shared by named creatures. They function as differentiae for the next level down" (Kuhn, 1990, p. 5, emphasis in the original). Hence, differentiae are sets of features attached to a superordinate concept.[15]

It is important to note that the plurality of differentiating features serves two purposes: First, it will often be necessary to draw on several features organized in some pattern in order to distinguish the instances of the concepts in a contrast set. For example, considering Kuhn's favourite contrast set – ducks, geese and swans – neither the feature rounded beak/pointed beak nor the feature short neck/long neck alone suffice to distinguish instances of the three concepts, but taken together they do suffice to distinguish instances of ducks, geese and swans. Secondly, the possibility of basing the similarity and dissimilarity relations on different features presupposes an empirical correlation between them. As only some of those co-occurring features are necessary to use a given concept correctly, adding further

---

[14] As argued by Nersessian and Andersen (1997), whereas normic concepts are taxonomic, for nomic concepts it is the complex problem situations in which the nomic concepts are used that show taxonomic structures (similarly Hoyningen-Huene, 1993, sec. 3.6.e). Hence, for nomic concepts the following argument on overlapping joints applies to the problem situations rather than to the individual concepts.

[15] Formulated in terms of perceptual space, Kuhn's emphasis on the plurality of differentiating features implies that the categories to be discriminated are separated by empty space in multiple dimensions. As described above, neck length may be one dimension in perceptual space along which discontinuities may be found, but others could be body size, or shape or colour of the beak.

features will say something not just about how to pick out instances of the concept, but also something about how an object already picked out will behave. For example, when distinguishing between ducks, geese and swans some speakers may use the combination of the two features neck and beak, while others use the combination of the two features colour and beak. Each of these sets of features are jointly sufficient to identify instances of the contrasting concepts, but none of the features are individually necessary. This equivalence of different sets of jointly sufficient features presupposes an empirical correlation between *all* features and makes it possible, for example, for the former group of speakers not only to identify ducks on the basis of neck and beak but further for them to expect a bird such identified to be brown.

The conjunction of all features can therefore be seen as a hypothesis about the behaviour of the instances of the corresponding concept – in the example above, the hypothesis that fowls classified as ducks due to neck and beak will be brown. In this way conceptual structure is linked to *projectibility*. Drawing on Goodman's definition of a projectible hypothesis as a hypothesis that has some undetermined, some positive and no negative cases (Goodman, 1965, p. 90), one can say that all previously examined instances of the concept which have established that the co-occurrence of the bundled features is an empirical regularity provide the positive instances of the hypothesis, while all hitherto unexamined instances of the concept provide the unexamined cases.[16]

Hence, concepts are projectible because they imply expectations of how instances of the concept already identified behave. For a concept to be projectible, the expectation must exist that the bundle of features involves more features than needed just to pick out instances of the concept, and that the classifications which the different sets of jointly sufficient features give rise to must be coextensive.

It is a key premise of Kuhn's position that nature cannot be forced into any arbitrary set of conceptual boxes. Whether the bundle of features are actually co-extensive is for Kuhn an objective matter. Anomalies will inevitably appear if a set of features cannot be bundled.[17] However, as I have argued elsewhere, several qualifications are necessary here (Andersen, 2004, forthcoming): Not all anomalies are equally severe, they need not be discovered at once, and different scientists

---

[16] A qualification to the requirement of no negative cases will be necessary, as anomalies, that is, instances that do not conform to the knowledge of regularities, do not always lead to conceptual change. I shall come back to this point later.

[17] This possibility that previous classifications may later turn out wrong is also one of the key points in e.g. Shapere's rejection of essentialism: "science is constantly open to the possibility that doubt may (though it need not) arise, that our present views, including the ways we 'conceptualize' objects and kinds, and name and describe them, may have to be revised or rejected and replaced" (Shapere, 1982, p. 14).

may judge the same anomaly very differently.[18] Projectibility does therefore not offer any simple algorithm for theory comparison. Due to some historically determined emphasis on specific features, the possibility of bundling a few emphasized features may be considered far more important than the possibility of bundling several other, less emphasized features. Such differences must also be taken into account when comparing theories.[19]

Further, the claim that it is an objective matter whether features can be bundles in a particular way does not rule out that features can be bundled in different ways. As argued previously, different sets of similarity and dissimilarity relations may bundle different sets of features which carve different joints in the phenomenal world. Thus, the claim is a purely negative claim that not any arbitrary bundling of features is possible; it is not a positive claim about the existence of a privileged set of features bundles constituting the world's real joints.[20]

9 The alternative response: Overlapping phenomenal worlds

Given Kuhn's theory of the constitution of phenomenal worlds and their joints, *identical* phenomenal worlds are those which are carved into the same joints, that is, phenomenal worlds which share ontology. For phenomenal worlds to be carved into the same joints presupposes shared conceptual structure, and that again presupposes shared relations of similarity and dissimilarity. However, as argued above, the similarity and dissimilarity relations constitutive of the conceptual structure need not be attached by means of the same features by all speakers. Hence, it is shared structure, not shared features, that yields shared ontology. The only requirement is that in so far as features are not shared, it is expected that they are all compatible:

---

[18] For detailed treatments of the possibility of different reactions to anomalies, see also (Kuhn, 1970a, ch. VI-VIII); (Hoyningen-Huene, 1993, sec. 7.1); (Chen et al., 1998), and (Nersessian and Andersen, 1997).

[19] Although there is here a certain similarity to Goodman's notion of entrenchment (Goodman, 1965, ch. IV), the emphasis on features introduced here goes beyond mere recurrence and more in the direction of recent cognitive theories on the role of causal status in determining the importance of features. For a historical case study, see (Andersen, forthcoming).

[20] It may be questioned whether this negative claim suffices for a realist position. However, comparing to Sankey's (2001) description of the principles by which scientific realism can be characterized it can be noted that this position can well be characterized by the principles that the world investigated by science is an objective reality that exists independently of human thought, that it is the external world that renders our claims about the world true or false, and that truth consists in a correspondence between a claim about the world and the world.

homologous structures, structures mirroring the same world, may be fash-
ioned using different sets of criterial linkages. ... What members of a language community share is homology of lexical structure. Their criteria
need not be the same, for those they can learn from each other as needed.
(Kuhn, 1983, p. 683)[21]

Hence, in so far as the features for a given set of contrasting concepts are not
shared, it is expected that they can in principle all be bundled, that is, that all the
different sets of jointly sufficient features used by different members of the language community carve the same joints in the phenomenal world. Only in this
case will members of the language community who use different features to distinguish instances of a set of contrasting concepts categorize these instances in the
same way.

   The question now is how to develop a notion of *overlap* between the phenomenal worlds which can serve to identify incommensurable theories. Still lacking of
a notion of object-domain, I shall start from a developmental perspective and later
generalize to cover the ahistorical perspective as well. On a developmental view,
incommensurable theories are (usually) the result of a historical process which has
required changes of conceptual structure that go beyond mere additions and re-
finements (Kuhn, 1991, p. 7); (Kuhn, 1992, p.16).[22] Such changes are made as the
response to severe anomalies, that is, they are made when an object has been discovered which does not fit into the similarity classes of the conceptual structure.
More specifically, since conceptual structures constituted by relations of similarity and dissimilarity are characterized by a hierarchy of contrast sets, any object
which challenges the non-overlapping division of the contrast set to which it belongs calls the conceptual structure into question. This is what Kuhn introduced as
the no-overlap principle:

   no two kind terms, no two terms with the kind label, may overlap in their
   referents unless they are related as species to genus. There are no dogs that
   are also cats, no gold rings that are also silver rings, and so on: that's what
   makes dogs, cats, silver, and gold each a kind. Therefore, if the members
   of a language community encounter a dog that's also a cat (or, more real-
   istically, a creature like the duck-billed platypus), they cannot just enrich
   the set of category terms but must instead redesign a part of the taxonomy.
   (Kuhn, 1991, p. 4)

   Hence, conceptual structure is challenged when an object is encountered which
on the basis of different differentiating features can be ascribed to different con-

---

[21] Similarly, (Kuhn, 1990, p. 7).

[22] For historical case-studies of such developments, see e.g. (Andersen, 1996, forthcoming;
Barker et al., 2003).

trasting categories. In this case, the anomaly reveals that the previously assumed bundling of features is not projectible after all.

When this happens, conceptual structure must be changed. However, as noted previously, an anomaly reveals only that the previous bundling of features was not projectible after all, but it does not determine any alternative way to bundle the features. The only requirement is that the conceptual structure must be brought to comply with the no-overlap principle again, that is, the similarity classes must be changed in such a way that also the anomalous object will be ascribed to only one of the contrasting concepts. Such a change implies changes in the relations of similarity and dissimilarity and hence changes of the bundlings of features, where the only guideline is that projectibility must be reestablished by bundling the features such that the new combinations of features can be seen as hypotheses with some positive, but no negative cases. This requirement that projectibility must be reestablished therefore provides *reasons* for conceptual change. For a conceptual structure that develops gradually through several incremental changes, each step will be based on such considerations of how concepts remain projectible. From step to step, these considerations provide reasons for conceptual change, and as the conceptual structure gradually develops, these new considerations about projectibiliy are linked to the old as links in a chain. Adopting an expression from Shapere, the bundles at different stages in a theory's development can be said to be related by chains-of-reasoning (Shapere, 1982) (similarly Nersessian, 1984).

However, it is possible to imagine several different changes of the bundling of features which all have some positive, but no negative cases. If these different bundlings overlap, that is, share some but not all features in the bundle, and if these different bundlings lead to differing similarity classes, then the different bundlings can be seen as incompatible hypotheses, which are nevertheless all projectible. As the different bundlings carve different joints in the phenomenal worlds to which they give rise, although they are all projectible, they are projectible in different phenomenal worlds.

But these phenomenal worlds do not differ in any arbitrary sense. They are not unrelated. The important point is that the different bundles are incompatible due to the shared features. That means that they give rise to different phenomenal worlds that are also mutually exclusive. If you adopt one, you simultaneously reject the others. Hence, while it is shared structure and not shared features that yields shared ontology, when structure is no longer shared, it is shared features that provides the overlap between different phenomenal worlds necessary for them to compete in offering the better account of the world in the form of more successful or more promising bundlings. Thus, it is shared features that yield incommensurability in the original sense of combat and competition. Or, to put it in terms of niches, it is the shared features which provide the overlap between portions of the two niches necessary for their inhabitants to see each other not as inhabitants of another niche, but as intruders wanting to change their own niche.

This argument was built on the developmental view, that usually incommensurable theories are *successive* theories. However, in principle nothing prevents the result being generalized in such a way that it is applicable independently of how the theories in question have developed. Chinese astronomy may overlap – and hence compete – with Western astronomy by sharing features, despite its origin in a totally separate tradition. Shared features bundled differently yields incommensurability whether or not the shared features are inherited from the same predecessor; the developmental perspective simply make more plausible that, among the staggering amount of possible bundlings of features, two different theories will actually overlap in their bundlings because they are connected through chains-of-reasoning.

10 The incommensurability thesis revisited

As Kuhn has repeatedly pointed out since the publication of *The Structure of Scientific Revolutions*, the incommensurability thesis is much more modest than many of its critics have supposed (e.g. Kuhn, 1983, p. 671). The problems related to incommensurability only relate to a small part of the scientific lexicon, while most of the lexicon functions the same way in both theories (e.g. Kuhn 1983, p. 670f.). Hence, most bundles of features are shared by the incommensurable theories. It is only a few, overlapping bundlings of features establishing differing similarity classes which are responsible for the incommensurability. Further, it is important to remember that incommensurability arise because these different bundles provide incompatible hypotheses. Many pairs of theories may share large parts of their conceptual structure, but they only compete on explaining the same object-domain when some bundlings of features overlap.

Astrophysics and health physics, for example, may share large parts of their conceptual structure – parts dealing, for example, with nuclei and radiation. They may also develop unshared parts – the interaction between radiation and tissue in health physics, or the production of heavy elements in astrophysics – but none of the bundlings of features in these unshared parts overlap. They do not offer incompatible hypotheses about some object which can be classified differently in the two theories, leading to different expectations of its behaviour. Such theories have nothing to compete about, but live in peaceful coexistence, each in their own niche. The situation is different for Ptolemaic astronomy and astronomy after Kepler. These theories may also share large parts of their conceptual structure, but the unshared parts provide incompatible hypotheses due to overlapping bundling of features. In this case scientists adopting different theories agree on some features, but bundle them in different, incompatible ways.

Hence, a second no-overlap principle can be formulated that has to hold between different niches. For two niches to be different and their inhabitants to live in peaceful co-existence, each in their own niche, all bundlings of features must be either identical, that is, totally overlapping, or not overlap at all. Of course, if *all* bundlings of features are identical, the two niches are identical, and its inhabitants are scientists who work within the same subdiscipline, but with the same concepts and hence without any disagreement about which tool is the better one for investigating the niche. But if only *some* bundlings of features are identical, the non-identical bundlings are not allowed to overlap. As long as they do not overlap, the different sets of bundlings are directed towards different niches, and hence, again, there is no disagreement about which tool is the better one because they use it in investigating different niches.

If, however, this new no-overlap principle for features is violated, that is, if some bundles of features overlap, the niches are neither identical, nor distinct. Instead, the niches overlap in the sense that their inhabitants compete on which tool to use in exploring the niche, and how the niche shall be developed as a result. Hence, it is through the overlapping bundles of features that intruders can be recognized. The different bundling may arise as the result of a development *within* the niche among the scientists of which some have become betrayers who attempt to simultaneously change the niche and develop a tool for investigating it in its changed form, or it may be a result of intruders from *outside* who have discovered a niche sufficiently overlapping with their own to attempt an invasion, but in both cases the result is competition and combat among the scientists and incommensurability between the theories they employ.

References

Andersen, H. (1996). Categorization, anomalies, and the discovery of nuclear fission, *Stud. Hist. Phil. Mod. Phys.* **28**: 463–492.

Andersen, H. (2000). Kuhn's account of family resemblance: A solution to the problem of wide-open texture, *Erkenntnis* **52**: 313–337.

Andersen, H. (2001). Reference and resemblance, *Philosophy of Science* **68**: S50–S61.

Andersen, H. (2004). A dynamic account of conceptual structures, *in* L. Ping, X. Chen, Z. Zhilin and Z. Huaxia (eds), *Science, Cognition, and Consciousness – The Proceeding of the International Conference of Philosophy and Cognitive Sciences*, JiangXi People's Press, Guanzhou, pp. 118–140.

Andersen, H. (forthcoming). Unexpected discoveries, graded structures, and the difference between acceptance and neglect, *in* J. Meheus and T. Nickles (eds), *Models of Discovery and Creativity*, Springer, Dordrecht.

Barker, P., Chen, X. and Andersen, H. (2003). Thomas Kuhn and cognitive science, *in* T. Nickles (ed.), *Thomas Kuhn*, Cambridge University Press, pp. 212–245.

Boyd, R. (1979). Metaphor and theory change: What is 'metaphor' a metaphor for?, *in* A. Ortony (ed.), *Metaphor and Thought*, 2nd edn, Cambridge University Press, Cambridge, pp. 481–532. 2nd edition published 1993.

Chen, X., Andersen, H. and Barker, P. (1998). Kuhn's theory of scientific revolutions and cognitive psychology, *Philosophical Psychology* **11**: 5–28.

Chen, X. (1997). Thomas Kuhn's latest notion of incommensurability, *Journal for the General Philosophy of Science* **28**: 257–273.

Enç, B. (1976). Reference of theoretical terms, *Noûs* **10**: 261–282.

Goodman, N. (1965). *Fact, Fiction and Forecast*, 2nd edn, Bobbs-Merril, Indianapolis.

Hacking, I. (1983). *Representing and Intervening – Introductory Topics in the Philosophy of Modern Science*, Cambridge University Press, Cambridge.

Hoyningen-Huene, P., Oberheim, E. and Andersen, H. (1996). On incommensurability, *Stud. Hist. Phil. Sci.* **27**: 131–141.

Hoyningen-Huene, P. (1989). Idealist elements in Thomas Kuhn's philosophy of science, *History of Philosophy Quarterly* **6**: 393–401.

Hoyningen-Huene, P. (1993). *Reconstructing Scientific Revolutions – Thomas S. Kuhn's Philosophy of Science*, University of Chicago Press, Chicago.

Kitcher, P. (1978). Theories, theorists and theoretical change, *Philosophical Review* **LXXXVII**: 519–547.

Kitcher, P. (1983). Implications of incommensurability, *PSA 1982* **2**: 689–703.

Kripke, S. (1972). Identity and necessity, *in* M. K. Münitz (ed.), *Identity and Individuation*, New York University Press, New York.

Kuhn, T. S. (1970a). *The Structure of Scientific Revolutions*, 2nd edn, University of Chicago Press, Chicago.

Kuhn, T. S. (1970b). Reflections on my critics, *in* I. Lakatos and A. Musgrave (eds), *Criticism and the Growth of Knowledge*, Cambridge University Press, Cambridge, pp. 231–278.

Kuhn, T. S. (1983). Commensurability, comparability, communicability, *PSA 1982* **2**: 669–688.

Kuhn, T. S. (1989). Possible worlds in history of science, *in* S. Allen (ed.), *Possible Worlds in Humanities, Arts and Sciences*, de Guyter, Berlin, pp. 9–32.

Kuhn, T. S. (1991). The road since structure, *PSA 1990* **2**: 3–13.

Kuhn, T. S. (1992). The trouble with the historical philosophy of science, Robert and Maurine Rotschild Distinguished Lecture, Nov. 19th, 1991. Department of the History of Science, Harvard University.

Kuhn, T. (1979). Metaphor in science, *in* A. Ortony (ed.), *Metaphor and Thought*, Cambridge University Press, Cambridge, pp. 409–419.

Kuhn, T. (1990). A historians theory of meaning, Talk to the Cognitive Science Colloquium, UCLA, 4/26/90.

Nersessian, N. and Andersen, H. (1997). Conceptual change and incommensurability: A cognitive-historical view, *Danish Yearbook of Philosophy* **32**: 111–151.

Nersessian, N. (1984). *Faraday to Einstein: Constructing Meaning in Scientific Theories*, Martinus Nijhoff, Dordrecht.

Newton-Smith, W. H. (1981). *The Rationality of Science*, Routledge & Kegan Paul, Boston.

Nola, R. (1980a). 'Paradigms lost, or the world regained' – an excursion into realism and idealism in science, *Synthese* **45**: 317–350.

Nola, R. (1980b). Fixing the reference of theoretical terms, *Philosophy of Science* **47**: 505–531.

Oberheim, E. and Hoyningen-Huene, P. (1997). Incommensurability, realism and meta-incommensurability, *Theoria* **12**: 447–465.

Putnam, H. (1973). Explanation and reference, *in* G.Pearce and P. Maynard (eds), *Conceptual Change*, Reidel, Dordrecht, pp. 199–221. Reprinted in: Putnam, H.: Mind, Language and Reality. Philosophical Papers Volume 2, Cambridge: Cambridge University Press, 1975, pp. 216–271.

Putnam, H. (1975). The meaning of 'meaning', *in* K. Gunderson (ed.), *Language, Mind and Knowledge*, Vol. VII of *Minnesota Studies in the Philosophy of Science*, University of Minnesota Press, Minnesota. Reprinted in: Putnam, H.: Mind, Language and Reality. Philosophical Papers Volume 2, Cambridge: Cambridge University Press 1975, pp. 215–271.

Putnam, H. (1981). *Reason, Truth and History*, Cambridge University Press, Cambridge.

Sankey, H. (1991). Translation failure between theories, *Studies in the History and Philosophy of Science* **22**: 223–236.

Sankey, H. (1994). *The Incommensurability Thesis*, Avesbury, Aldershot.

Sankey, H. (1997). Incommensurability: The current state of play, *Theoria* **12**: 425–445.

Scheffler (1967). *Science and Subjectivity*, Bobbs-Merrill, Indianapolis.

Shapere, D. (1964). The structure of scientific revolutions, *Philosophical Review* **LXXIII**: 383–394.

Shapere, D. (1982). Reason, reference, and the quest for knowledge, *Philosophy of Science* **49**: 1–23.

Wray, K. B. (2005). Rethinking scientific specialization, *Social Studies of Science* **35**: 151–164.

Science and reality

Jan Faye

Department of Media, Cognition and Communication, University of Copenhagen

Since the heyday of logical positivism, the dominant view in philosophy of science has been Realism. But over the last two or three decades its prominence seems to decline. No one wants to return to the excesses of logical positivism, but as the dust after the battle settled, it became more and more clear that not everything the defeated part stood for was without merit. And, as we shall see, Realism has its excesses and problems too. Hardcore instrumentalists believed that the scientific theories are mere tools for predictions and calculations and that they contain no content telling us how the world really is, being conceptual tools that are neither true nor false. Theories help us to organize empirical data in virtue of the claim of theoretical entities, but theoretical entities are, and always will be, fictitious mental constructions because their alleged existence would transcend anything that could be established by sense experience.

Realism grows out of the practical and observational success of science itself. Instrumentalism, in contrast, is generated by a philosophical desire to strip metaphysics of any veil of legitimacy and to dress science in armour of epistemic warrant. As long as astronomy, physics, chemistry and biology dealt mainly with macroscopic objects which could be observed, as was the case to the end of 19th century, the acceptance of the instrumentalist view had no far-reaching implications, neither with respect to the number of theoretical entities explained away, nor with respect to possible technological consequences of a belief in these entities. But with the development of new theories about invisible entities, forces and processes such as electric and magnetic fields, molecules, and atoms, and together with the rapid increase in technology based on our beliefs in such entities and processes, it seems pointless to push the claim that we do not possess knowledge of that part of reality which is not directly accessible to the naked eye. It is, the realist would say, only because scientific theories provide us with knowledge of the hidden structure behind phenomena that we have been able to change nature, design new organisms, and improve the material and technological level of modern society. Science does not merely yield theories that predict how well-existing phenomena may change. It also fosters theories that give us insight into the laws of nature – thus allowing the creation of quite new phenomena never seen before. As Hilary Putnam once declared: realism is the only philosophy which does not consider the empirical success of science a miracle.

In this paper I shall take issue with some of the most common arguments in favour of scientific realism. My aim is to show that "theory realists" who advocate *semantic realism* have not presented convincing arguments for their thesis that currently accepted theories must be true or approximately true if we shall be able to explain their empirical success. Similarly, I hope to demonstrate that an alternative form of scientific realism, *structural or syntactic realism*, which is very much in vogue, is no way out for the realist. Rather than being a realist concerning theories I share company with those philosophers who are realists concerning entities.

1 Ontological commitments

Realism is a possible position in many different fields. In case one believes that the external world exists independently of our consciousness regardless of whether one believes in its existence or not, one is a realist with respect to the surrounding reality. Or in case one is in favour of the idea that there are moral facts which are not, in some way or another, determined by people's sentiments and emotions, one is a realist with respect to what is right and wrong. Or if one takes the view that abstract entities such as numbers exist, even though they are not provable or constructible, one will be a realist concerning mathematical quantities. We can also be realists when it comes to kinds, universals, modalities, and possible worlds. Common to every realist concerning these different areas is that what he is a realist about is taken to be real, regardless of whether he himself or other human beings had existed. But it is not a requirement that if somebody is a realist in one area, he must be so in every other area. Thus, there is no implication between a belief in the objective existence of the external world and a belief in, say, the independence of moral values.

Nonetheless, since one can be a realist with regard to truth too, the obvious question is whether or not one can be a realist in some areas without being a realist with respect to truth. Before answering this question, we shall throw more light on the realist view that entities exist objectively, independent of our knowledge of them. For the matter of focus we shall restrict the discussion to the problem of the reality of the external world.

As a start, let us turn to the realist claim of mind-independence. Here the realist may have two ideas in mind. The first is that the external world exists objectively, which must be taken to mean that the world is what it is independently of human consciousness. The objective world is not constituted by our knowledge of it; space, time, things, events, properties, and laws of nature may exist whether we believe that they do or not. These entities may be real, even though they are not objects of our perception. The second idea is that the objective world is a physical world. It does not consist of experiential objects like sense-data or other mental objects. The realist hereby also makes the external world physical, or mostly physical. Indeed, a realist is not prevented from submitting that the mental is different from the physical, nor therefore from claiming that the mental is objectively real, independent of whether someone believes it or not. Realism does not rule out objects like minds, but it claims that the existence of minds and their specific nature are what they are regardless of the way one actually may conceive and apprehend them, and regardless of whether they are objects of anybody's apprehension. Thomas Nagel, for instance, believes that there are subjective facts which are unattainable to human knowledge (Nagel, 1974). The requirement of logical independence of human knowledge also means that things, events and laws can exist even if they cannot be known, that is, even if they are, in principle, empirically inaccessible. The realist must agree upon the possible reality of such entities. The world may be inconceivable to our mind. Nothing in his metaphysical point of departure excludes the existence of unknowable entities as a genuine possibility.

Another aspect of the realist's thesis is the question of existence. What is it that is real? Assuming that the realist is bound to assume that the external world exists the way it does, irrespective of whether it is empirically accessible or not, it means at least that the world is what it is in itself. Whether or not we are capable of understanding the external world such as it is in itself, is not a question which excludes that it is what it is in virtue of itself. Reality is not just what it is as a result of our way of apprehension. The external world is both structured and ontologically determinate or unstructured and ontologically indeterminate, but whatever it is, it is what it is prior to our knowledge of it. But the realist is not required to believe more than that. He is not forced to believe anything specific about the world's organization. He may, for instance, contend that the world in itself consists in those things which surround us in our ordinary life. The physical world as we perceive it is the world as it is in reality. The world in itself consists of stuff and objects like gold, water, human beings, animals, cars and refrigerators. This view could be called the *everyday version* of Realism. The realist may also hold that the common sense view of reality has to be supplemented with the scientific story about laws and unobservable things and properties, a position which shall be called the *tolerant version* of Realism. Finally, the realist can take a step further. He may deny completely that reality is what it is considered to be on the basis of our ordinary experience. Instead he can argue that the real world is as science tells us. The later formulation may be called the *intolerant version* of Realism. This is the view Kant scornfully called transcendental realism. Whether the realist adopts the tolerant or the intolerant version, he holds a view to which scientific theories narrate about a reality hidden from our immediate senses: the world is furnished with different kinds of particles and forces impossible to see with our naked eye and which do not possess the same properties as those being ascribed to perceptual things.

Setting the various versions of Realism aside for the moment, what arguments can be levelled in support of Realism in general? Many will probably agree with Thomas Nagel when he points out that if we look at our history, we see that at some time our ancestors did not know, or were not able to conceive, aspects of reality which we know or can conceive today (Nagel, 1986, ch. 6). Similarly, there are things we cannot now grasp, but will be able to later. From these observations, most people will accept an inference to the conclusion that there may be things we cannot conceive of at a particular time in the future, and therefore never ever come to understand. The decisive factor is, of course, whether this means that there are things of which we have no conception because of the way we and these things are, and not because we are at too early a stage of our history. Here the waters divide between realists and antirealists. For the realist would argue that even now some people lack a capacity to conceive of colours or sounds if they are born blind or deaf. And some people don't have the mental power to understand quantum mechanics or the general theory of relativity. Analogously, we can imagine that there are aspects of the world which nobody, in principle, is able to think or know about. The antirealist, on the other hand, would dispute this argument by saying that our thought cannot reach beyond the conditions for the possibility of thoughts. We can make sense of the examples of the disability of the blind, the deaf, and the person with a low mental ability to see, hear or understand aspects of the world only because we realize that other people have the ability to know or conceive them. In other words, the antirealist believes that the examples make sense since we already have a language in which these features are fully specifiable. We cannot, according to him, claim to have a general concept of reality based on what we know or comprehend already, and then meaningfully apply it to something which is incomprehensible.

This dispute cannot be addressed further until we know more about what sets the boundaries of our thought and how truth relates to sentences expressing our thoughts. But Nagel mentions that in our notion of a universal or an existential quantification, the value of a variable need not have to be the referent of a specific name or description in our language (Nagel, 1986, p. 98). The reason is that we already have a general concept of everything which comprises both what we can name or describe, and what we can't. Consequently, we can speak of 'All the things we can't describe imagine or conceive of owing to our very nature.' For this claim to become a way out for the realist, it seems as if he must admit that such a sentence can be true only if there is a negative fact making it true. So long as the realist talks negatively about something which is known, say, 'The Eiffel Tower is not made of wood', a statement like this does not require the existence of a negative fact that the Eiffel Tower is not being made of wood to be true. What makes it true is the positive fact that it is made of steel. If it is completely made of steel, it cannot also be made of wood. In the case of the sentence concerning everything we can't describe, the realist does not have the same opportunity to state which positive fact makes the negative sentence true. Thus, if this consideration is true, it raises serious doubts about the realist's claim that the general concept of reality he applies to what humans cannot understand is the same as the one he uses for what is conceivable by us.

The metaphysical account of realism as regards the external world has so far provided us with three more precise claims: 1) physical things which we experience immediately through our senses exist objectively in some way or another irrespective of our beliefs in them; 2) theoretical entities which are not objects of direct sense experience, but which are related to our best scientific theories, are real and not merely mental constructions; and 3) the best scientific theories tell us how the world is. Nevertheless, it is not uncommon to hear an objection against this metaphysical account of realism. The complaint at this is point is that realism in terms of a mind-independent world is obscured by metaphorical language. Is it possible to specify the realist's position further? Perhaps not. A possible supplement would be to say that realism with respect to the external world also implies a semantic formulation: if the world does not necessarily square with our cognitive resources, then sentences about physical laws and objects are not reducible to sentences about mental states. The former type of expressions has a meaning which cannot be translated into expressions of the latter type. For example, according to common sense realism, sentences about the external world are not translatable into sentences about sense-data, the truth of physical-object statements cannot be expressed in terms of the truth of statements concerning mental states or subjective experiences. I am not claiming that this semantic formulation is logically equivalent with the ontological formulation of the mind-independent thesis. What I am saying is that for the realist the mind-independent thesis has to be associated with the untranslatable thesis to be intelligible, and this holds for scientific realism as well as for common sense realism. Even though the realist would admit the possibility of some unknowable entities, he cannot claim without serious difficulties that the reality-in-itself is completely unknowable, and therefore that our language does not concern such a mind-independent world. Though logically possible it is difficult for the realist to argue positively for the existence of a reality *an sich* and at the same time hold that this reality could be cognitively inaccessible in principle. Because how could he ever know its existence? The common-sense realist would most likely assume that we are not prevented from having knowledge of the reality of the things in themselves and that this knowledge can be expressed in physical-object sentences. Consequently, the scientific realist can semantically be characterized as one who argues that: (i) statements about theoretical entities cannot be reduced with respect to truth conditions to statements about what we can perceive, and (ii) sentences concerning laws of nature cannot be reduced with respect to truth conditions to sentences about their physical manifestation.

Based on the above discussion, we may define Realism as a general metaphysical doctrine consisting of four components. First, there is the *ontological* component of the view: whatever there is is what it is regardless of how we think of it. A real entity, or a law of nature, has full, concrete specificity and determinateness, or lacks both, independently of our mental powers. The realist is not forced to argue that determinateness holds good for the world as a whole. For instance, instead of maintaining that the future (and the past) is ontologically determinate, he could claim that the future (and the past) is ontologically indeterminate or simply unreal. Likewise he could argue that quantum objects are vague or fuzzy entities which have indeterminate attributes. This leaves, apparently, the realist with three different options concerning the nature of the mind-independent world. First, he can hold that everything real is ontologically determinate in the sense that it has concrete specific attributes; second, he can hold that at least a part of what is real is ontologically indeterminate in the sense that it lacks actuality and attribute specificity; and third, he can argue that parts of the world are unreal in the sense that nothing exists corresponding to certain thoughts or imaginations.

Although reality in itself according to the realist exists entirely detached from our cognitive capacities, it is generally assumed that those physical-object statements and/or scientific statements we use in our communicative discourse refer to such a mind-independent world. Thus the claim of the existence of a mind-independent world is associated with a thesis that the true common sense account and/or the true scientific account concerns the objective reality as it is regardless of our senses, opinions, and emotions. An important consequence of the thesis is that statements about the world are not reducible to statements about anything else, especially not to statements about our subjective experience or mental states of the mind. However, according to the intolerant version of scientific realism, as defined above, it is possible to reduce the ordinary physical-object language to the language of science without any loss of meaning.

Second, there is the semantic aspect of the view: the meaning of statements about the external world must be analyzed by reference to the notion of truth conditions whose specification in principle may reach beyond any possible empirical justification. A sentence is true or false independently of whether or not we have any means to verify or ascertain its truth value. What determines these truth conditions is an alleged natural and mind-independent relation between a statement and the objective world. A set of common sense descriptions or a scientific theory is true only if it is related to the world in a way describing the world as it really is.

The third element is the epistemic component: we have objective knowledge of the world as it is. Knowledge in the objective sense is independent of anybody's beliefs or anybody's claims of knowledge. Thus, the epistemic realist maintains that objective knowledge exists in the form of propositions and scientific theories. In other words, propositions and theories concerning the reality-in-itself are held to be true independently of whether we have proven, or might prove, them or not. As Karl Popper states this position: Objective knowledge is knowledge without a knowing subject (Popper, 1972, p. 109).

Since reality *an sich* for God would be one with his understanding of it, he does not, according to such a viewpoint, need reliable methods to prove his possession of objective knowledge. The world-in-itself would be inseparable from God's knowledge of it, or reality *an sich* would at least be congruent with his conception of it. For God as an infinite mind would not be bound by a distinction between the subject and the object. But mortal human beings, in contrast, need reliable procedures to determine whether or not their mental representations are in accordance with reality *an sich*. Thus, the fourth element of realism is the methodological component: in the right circumstances ordinary people or scientists are able to provide warranted judgement about the truth of all kinds of beliefs regardless of whether they are about observable or unobservable entities or are formulated in terms of singular or universal sentences. This is due to the fact that some objective methods or procedures exist such that their application yields a true belief that something is the case if and only if it is the case. Beliefs about the external world, according to the methodological realist, are ascertainable by reliable means: nevertheless, there are procedures which, when followed, yield only good, and not certain, grounds to believe that something is the case. Such a procedure provides us with a rational method by showing that the appropriate statement is likely to be true or false.

In order to defend his position, the realist is bound to explain what kinds of fact make statements about the external world true. He must give us a metaphysical account of how the truth value of statements about ordinary things, about unobservable objects and about natural laws is procured. Furthermore, the realist must explain how we can have epistemic access to ordinary things, the realm of an unobservable reality, and universal truths. He must point out which truth-conducive procedures of inquiry are at our disposal for gaining such knowledge. He must also identify under which circumstances we can know that truth conditions are in fact fulfilled, and in general, what conditions have to be fulfilled for a meaningful use of the sentence in question. Indeed, the realist's position becomes precarious if his metaphysical analysis of the truth conditions means that scientific facts lie beyond the empirical domain.

Having laid out these various forms of realism, it must be emphasized that some philosophers see themselves as both realists and empiricists. This is true of Karl Popper and Hans Reichenbach to mention only a couple. At the same time others, such as Bas van Fraassen, call themselves empiricists and antirealists. Whether one prefers to call oneself a realist or an antirealist is more or less inconsequential, so long as one holds most of the realist's presuppositions as one's own. More important than such labels is it that a given view is characterized unequivocally and exhaustively. However, there seems to be a tendency among those empiricists who consider themselves as epistemic optimists that they believe in the existence of some methods that can provide us with a rational belief in the claims of science; methods, that is, which makes scientific statements more or less probable. On the other hand, epistemic pessimists focus on an assumption that there are no reliable procedures of inquiry yielding the truth of scientific theories.

The opposition to realism with regard to theoretical entities of the invisible world has traditionally been marked by the instrumentalist doctrine. It entertains the view that theoretical concepts are merely heuristic tools for organizing the scientist's observations. Instrumentalists take a nominalist stand on theoretical entities. Common names and natural kind terms of unobservable entities don't refer to anything in reality; hence statements about these entities should not be considered literally true. All concepts of unobservable things, events and properties are nothing but logical constructions from observables. Accordingly, the backbone of this view is that invisible things like forces, fields, atoms, molecules, genes, and viruses are not real, and that the names of these things proclaimed are merely a unifying designation of concrete experimental results. This contention leads to the claim that scientific theories containing sentences about such imperceptible things do not express proper knowledge; instead they are inference schemes which can be utilized for predictions of future experiences on the basis of past experiences.

Instrumentalism is an ontological position about theoretical entities closely associated with the application of empiricist or phenomenalist constraints on what can possibly exist. Only things with which we are directly acquainted can be said to exist by any justification. Embracing such strong epistemic requirements on ontology, instrumentalism can be regarded as a form of non-cognitivism about what we cannot directly perceive. Similar non-cognitivist views have been asserted within other areas of human cognition: discussions about the reality of tenses, moral values, causality, probability and possible worlds can in many cases be seen as a continual battle between realists and nominalists. The question is therefore whether the instrumentalist has better arguments against the existence of theoretical entities than those of the phenomenalists against the existence of ordinary physical objects.

The language of science is full of terms that refer to invisible entities and properties. One therefore seems to be ontologically committed to entities and properties that we cannot see; unless the instrumentalist can prove, for instance, that all sentences concerning them can be translated without loss of meaning into sentences of a language in which each and very term concerns visible objects. Few instrumentalists, other than operationalists, would argue that a given theoretical sentence has the same intension as any observation sentence, that the truth conditions for a sentence 'X is F' containing terms for an unobservable object X and a similarly unobservable property F are identical with the truth conditions of an appropriate observation sentence, or a set of sentences, 'Y is O_1, O_2, O_3, O_4, ..., O_n', which only contains the terms for an observational object Y and the observational properties Os. An instrumentalist does not have to argue that these two sentences necessarily have the same meaning.

Another option for the instrumentalist would be to say that he does not claim the synonymity of such sentences but merely considered them coextensive. One way to vindicate such a consideration is to do like Ramsey and substitute existentially bounded variables for predicates and names. He proved that all theoretical predicates of a theory, i.e., terms of unobservables, can be treated as existentially quantified variables to the effect that the axioms of the theory links the predicate variables to each other and a dictionary links them to observables (Ramsey, 1930/1990). The result is that all problematic predicates are eliminated but the structure and observational consequences remain. If the so-called Ramsey sentence is true, it tells us to what we are ontologically committed. Therefore, Ramsey-sentences have been used in the attempt to get rid of theoretical terms and replace them with observational terms. In fact this was not Ramsey's own purpose. Rather, he used his method to define the observational terms of observational language in terms of the theoretical terms of theory.

The instrumentalist disapproval of the fact that the language of science presupposes the existence of unobservables in order to be true is only one of two questions about the ontology of unobservables the realist must deal with. It is simply not enough for the realist to prove that the language of observables cannot express all our scientific beliefs. The other question rises from the fact that the language of logic and mathematics, for instance, requires the existence of abstract entities to be true. In Peano arithmetic we are committed to holding that natural numbers exist; and in Zermelo-Fraenkel set theory we have the same obligation towards sets. So, as Rudolf Carnap once pointed out, whenever we adopt such linguistic frameworks we are ontologically committed to the reality of numbers, sets, propositions, and so on (Carnap, 1950). He argues that whenever we wish to talk about some kind of being, we must do so within a linguistic framework. Such a framework is constituted by 1) a set of concept definitions, 2) some principles for governing the syntax between these concepts, and 3) some principles for testing the truth values of statements within the framework. In case of a rational (as opposed to an empirical) framework, 2) and 3) are coextensive.

The commitment is internal with respect to the framework. Carnap, however, argues that no metaphysical question can be answered inside the framework; thus it cannot have a truth value and is as such meaningless. When we ask if something *really* is, we are asking a question that goes beyond the conventional criteria for establishing whether something is. In his terms it is an *external* question to which there can be given no real meaning because it concerns reality considered outside a linguistic framework.

The plausibility of Quine's famous dictum hinges on a similar dichotomy between internal and external commitments: To be is to be the value of a bound variable (Quine, 1969, p. 91, ff). Existence is what existential quantification expresses. Thus the ontological commitment of a given theory can be found by identifying the entities over which the quantification of the theory is made. And Putnam's internal realism rides on the same ticket: "'Objects' do not exist independently of conceptual schemes. We cut up the world into objects when we introduce one or another scheme of description. Since the objects and the signs are alike internal to the scheme of description, it is possible to say what matches what" (Putnam, 1981, p. 52).

The realist's commitment is much stronger: the reality of numbers, sets, and propositions is a question about what *really* exists independently of any linguistic framework. A similar external commitment holds for the scientific realist with respect to unobservables. Thus, he must be prepared to argue for the correctness of the assumption that atoms, quarks, fields, and so on, exist objectively regardless of our way of conceptualizing the world. The realist is forced to show that his beliefs in unobservables can be warranted in some other ways than just by appealing to a given linguistic framework.

For instance, classical mechanics relies on everyday concepts like solidity, motion, and position in the observational description of macroscopic objects. But the usually crude determination of these attributes was not entirely satisfactory with the recognition of the renaissance that they could be measured and therefore become objects of mathematics. They could be turned into quantities. From then on a precise determination of their magnitude would involve instruments. Rulers, clocks, and levers were the basic instruments, and thereby mechanics got a new set of observables which were instrument readings. Such pointer readings must be connected with mass, position, and velocity through operational rules: meter sticks gauge the scale of distances, clocks record the elapse of times, levers measure the weight of masses, and velocity is uniform distances covered by equal times.

Newton's mechanics ascribes unobservable properties to observable entities. The ascription can be done through those of their properties we can experience. Quantum mechanics, however, deals with theoretical objects which cannot be object of direct perception; hence none of their properties can be attributed to them on our visual acquaintance with any of their other properties. Nevertheless, William Craig (Craig, 1956) and Carl Hempel (Hempel, 1965, pp. 173-226) have shown with respect to any such theory which can be axiomatized that it is always logically possible to construct an equivalent theory which entirely leaves out theoretical terms and expressions and replaces them with observational terms and expressions. Thus, theoretical terms are construed as meaningless auxiliary marks that serve as inferential devices between observational statements. Indeed, it has severe costs to choose a theory without theoretical terms such as lack of explanatory power, simplicity, and heuristic fertility.

The realist seeks the ontological commitments of our best scientific theories. The view that the physical world consists of a natural, pre-given and pre-descriptive set of laws, entities, properties, and relations is usually called scientific realism. And, according to the realist, the aim of science is to give a literal and objective description of such a world, and its present success can be seen as a token of the performance of these efforts. He holds that science eventually secures more and more knowledge about the world as it is in itself, and hence knowledge about a world of invisible things and properties. Likewise, the realist position is very often identified with the thesis that the theories that at the present time are considered the best are closer to the truth than earlier ones, and that the central terms of our best current theories are genuinely referential. This means, of course, that the truth of theoretical sentences about invisible objects and attributes are not reducible to the truth of a finite set of sentences about empirically accessible things and properties. As a reason for his position, the realist will point out that only if modern scientific theories are regarded as approximately true can we explain their predicative success.

2 Scientific realism

The scientific realist feels committed to a world of unobservables. But what counts as imperceptible entities and properties? How many or how few of the scientific terms stand for observables? Apparently, it varies from one science to another which physical entities or quantities we consider as observables. In general, macroscopic objects and events can be seen by the naked eye, and their visual properties like size, shape, form, solidity, colours, position, and motion are what distinguish them from each other. Some of these visual properties are ignored in a certain intended description of the object, since they are treated as secondary and mind-dependent properties, or because experience tells us that they don't play any role in the description of the object and its kinematical or dynamical behaviour. In classical mechanics, for instance, an object's position, velocity, rotation, and acceleration are the intended properties which are immediately accessible to the senses. Its mass is also a property we sometimes experience directly as the solidity of matter and feel by the weight. All other mechanical entities and properties like force, momentum, and kinetic energy are not observables; however, they can all be specified in terms of observables: $F = ma$, $p = mv$, and $E = \frac{1}{2}mv^2$. Classical mechanics ascribes certain non-observable properties to a physical object on the basis of observable ones. But the realist would say that these unobservable properties are something over and above the various relationships between the observable properties.

In his defence of realism, Michael Devitt presents us with the following train of thought: A person p is ontologically committed to an object a (or a property F) in uttering assertively a sentence token S if a (or F) must exist to make S true. Though Devitt will not deny the validity of this semantic criterion, he believes that there is another, more basic criterion, according to which a person is so committed if, in asserting S, that person says that a, or an F, exists (Devitt, 1984/1991, sec. 4.6). The first criterion requires that we possess a semantic theory for S to tell us what must exist to make S true, before we can say anything about a person's commitments; whereas the second criterion merely presupposes that we understand S as speakers of a certain language to know what commitments a person has. If someone asserts 'The electron is an atomic particle', this sentence is not true unless there exists something to which 'electron' refers and to which 'atomic particle' applies. But, says Devitt, the commitment of this to electrons and to atomic particles is the same as the one following from the assertions 'the electron exists' and 'the atomic particle exists'.

Devitt's argument is, I think, correct as long as it is taken to establish that no semantic theory is needed to know what existence really means. The word 'exists' in a sentence like 'the electron exists' does not have a meaning different from the one it has when we claim that the electron must exist for the sentence to be true. Had there been any difference between its meaning in the object language and the meta-language, we could decide to replace the meaning in the object language with its meaning in the meta-language, or if not, we might be involved in an infinite regress. But the fact that there is no difference does leave us without an argument to the effect that our commitments are external to the linguistic framework. Moreover, if the Craig-Hempel thesis holds and any theoretical sentence can be proven to be coextensive with a set of observation sentences, the realist is deprived of a strong reason to claim that our ontological commitments are external to the theory. For if a theoretical sentence cannot express a fact which cannot be expressed by a certain appropriate set of observation statements, why should we be justified in our beliefs that unobservable entities and properties are real?

The realist likewise sees the success of science as a strong backing of his thesis that scientific theories are typically approximately true. This success is also taken as evidence for the contention that theoretical terms within our best theories refer to whatever they are supposed to refer to. Sometimes it is even said that realism is the only conceivable view which can explain why science has been so successful, because the prediction of observable phenomena would be a cosmic coincidence or a miracle if theoretical terms only have instrumental value (Smart (1963, p.

39) and Putnam (1978, pp. 18-19)). Without the realist's explanation it would be especially incomprehensible how new and unforeseen phenomena can be predicted by a theory. The discovery of the element hafnium succeeded its prediction on the basis of Bohr's reorganization of the periodic system according to physical features of the atoms. As a consequence of his relativistic theory of the electron, Dirac announced the existence of a positive electron before Anderson discovered it. As an explanation of the continuous spectrum from beta decays, Pauli suggested the existence of an escort particle, the neutrino, which was not directly confirmed until many years later. The exchange of virtual mesons in a nuclear field was an essential part of Hideki Yukawa's theory of the strong nuclear force before these particles were discovered about ten years later. The W bosons and the neutral Z meson were first tracked down after they had for a while figured in Steven Weinberg and Abdus Salam's theory about the amalgamation of the weak and the electromagnetic force. All such examples make it highly unlikely, the realist contends, that theoretical terms making these predictions possible are not standing for entities other than those phenomena which can be observed.

The realist, however, also adduces other arguments for his thesis that theoretical terms refer to something real and we therefore are ontologically committed to imperceptible entities in a strong external sense. In searching for a systematization of his experience with the purpose of explanation and prediction, the scientist needs to operate with hypothetical entities that are not directly observable. As long as the scientist confines his effort to observable entities, the realist argues, he is merely able to formulate empirical generalizations. But, generally, the scientist is not content with the amount of integration which empirical generalizations alone furnish him. What he wants is a further integration of laws that bases itself on a few scientific principles, something that requires a further unification and development of concepts covering a broader domain of experience. The way to pass beyond the empirical generalizations must therefore be accomplished by introducing more general concepts not corresponding to anything observable. And, says the realist, the scientist eventually gets a better and better grasp of the world through his acquaintance with these principles, as he becomes able to expose the laws or mechanisms underlying the phenomena.

But how can this be an argument for the reality of unobservables or invisible entities? What the realist argues is that when the scientist aims at making an integration of concepts, he thereby justifies the ontological commitments entailed by our scientific theories. For the scientist seeks such unification only partly because of pragmatic reasons; that is, he wants to work with as few conceptual tools as possible. Rather the scientist believes that our concepts reflect something in the world. So if he can manage to narrow down the general concepts in his description of a certain domain to a very small number, he has reason to believe that that part of nature has been described in its most basic form. The realist's line of thought is that whenever science is capable of describing the world with all its difference and complexity, given very few concepts, it is most likely to be true because these concepts have dissolved the complexity into its most simple constituents.

This argument, however, suffers from two serious shortcomings. The first one is due to the fact that the conclusion is not consistent with the history of science. Many discharged theories, once used to explain an entire domain of experience in virtue of few general concepts, are not taken seriously anymore. Think, for instance, of the Aristotle's theory of motion. At its time it seemed to give a coherent account of our everyday experience of motions based on a few simple concepts. Vertical movement was considered dependent on the gravity of the body; dense things like rocks and water went downwards, more ephemeral things like air, vapour, and fire, upwards. Horizontal movement of a wagon, a stone, or an arrow required the presence of a moving force in the form of oxen, horses, or man power. All other motions could be described as a combination of these two principles. Likewise, the ancient idea of the world as being built up of the four basic elements, earth, water, air, and fire, contains much fewer elements than any contemporary theory. It is therefore doubtful, at least, that we today should have reached the right categories once and for all, just because we have been able to isolate a few concepts for explanatory purposes. The argument only shows that we always feel internally committed to those entities and properties which our currently best theories presume – it cannot prove that we are externally committed to such things.

The second objection is even more fatal to the realist's argument. For how can we be so certain that a scientific theory with fewer concepts is more likely to be true than one with more concepts? There are really no metaphysical grounds for believing that the world should consist of only few basic entities instead of multiple such. Similarly, nothing proves that these entities have fewer properties rather than more. Even if we grant the realist the existence of such proofs, it is impossible to see how that could help him to establish his belief that there are just those entities or properties which a certain scientific theory prescribes. For such a theory may turn out to be too simplistic in its assumptions about the basic number of entities or properties constituting its domain. Theories can start out by postulating very few entities and properties, and eventually have to go through a lot of conceptual extensions in order to cope with more and more experimental evidence for further entities or properties. Clearly, we do not particularly want a theory that posits superfluous entities or properties. But rejecting superfluousness is not the same as embracing simplicity. In my view, the ideal of simplicity is overrated, both when it comes to the number of entities and properties and to the structure of natural laws. Realists have nothing to gain from pursuing such an ideal.

In addition to the arguments discussed above, further reasons have been advanced in the support of the realist claim of real counterparts. Closely related to the latter argument is the question of abduction or inference to the best explanation. Against the instrumentalist it is said about scientific theories operating with unobservable structures and mechanics: because only some of them can explain all relevant facts in a coherent and convincing way, we have grounds to assume that those theories which are able to do so tell us how the world really is, or at least how it approximately is. However, we have to distinguish between at least two kinds of claims which may motivate the embracement of the inference to the best explanation. On the one hand, the realist may hold that the inference to the best explanation leads us to the objective laws of nature, and in such a case he could be called a realist concerning scientific theories; on the other hand, he may just urge the idea that the inference shows what is the most likely entity causing the effect, and in that case he could be said to be a realist concerning entities.

Abduction as well as induction plays an important role in formulating appropriate theoretical laws of science. But the realist will have a hard time if he wants to defend the view that inference to the best explanation is guidance to truth. Historically, this inference has fallen far behind the production of infallible knowledge, and we have little basis for believing that the situation will change in the future. What is considered to be the best explanation at any given time is whatever theory or assumption that seems to cover all chosen phenomena in the most satisfactory way. For more than a thousand years the Aristotelian theory of motion was the best explanation on the market. Then followed the impetus theory, which again was succeeded by the Galilean theory, the Cartesian theory, and by the Newtonian theory of motion – all of which were considered as the most convincing and adequate explanation of motion for a certain period of time. In the beginning of our century, Einstein provided the latest suggestion.

The realist may attempt to be modest, saying that the abductive inference only provides us with good reasons for an explanation more likely to be true. One may wonder, however, how to establish such a likelihood other than by saying that the theory is in agreement with all phenomena considered to be relevant at a given time. A correlation test, for instance, provides us with a measure of how good the correspondence is between the observed values and the expected values a given hypothesis predicts. Thus, if the measure of the likelihood is nothing but this external virtue, the realist must face the serious question of empirical underdetermination of theories. Usually, though, the realist will trade on internal virtues of a theory, like simplicity and coherence, as what characterizes the best explanation. But how can such internal virtues establish that the unobservables are real regardless of the conceptual framework?

To repeat: simplicity will not do the job. But perhaps coherence might? It could be argued that the idea of a world-in-itself is associated with the conception of everything being connected with everything else, and therefore somehow related to the idea that a hypothesis capable of explaining the facts is better if it agrees with other hypotheses than if it doesn't agree with any. At face value there is, however, a problem with such an argument. For the realist, a true hypothesis may or may not adhere with most other assumptions considered to be true. When the view of the truth-values of these other hypotheses eventually has changed, the hypothesis might be in agreement with the majority of commonly accepted assumptions. A good example of something like this would be the history of the heliocentric theory of Aristarchus of Samos. But the realist can avoid this problem by arguing that a claim is not scientifically interesting, even if it is true, before we have independent warrant for believing it. And he could continue by saying that so long as the hypothesis is not coherently connected with other commonly accepted assumptions about the world, it is not independently justified as true.

Also, the realist could emphasize that a hypothesis does not only have to agree with other reliable hypotheses to be better than its alternatives. It also has to agree with certain ontological principles forming the arrangement of the world, one of which I once named the principle of the unities of time, space, and cause after the classical drama (for instance Faye, 2002, p. 93). For instance, the realist may argue that an explanation has an a priori probability of being true if it accounts for a certain phenomenon in terms of other phenomena which are spatially and temporally connected with the phenomenon under discussion, all of which fit into the same ontic scheme of categories that can possibly enter into a causal relation. Nobody, to put it vividly, would dream of explaining today's hole in the ozone layer over Antarctica by the assassination of crown prince Franz Ferdinand in Sarajevo eighty years ago, because we regard such an explanation as entirely irrelevant. And the reason for this claim of irrelevancy is that the explanation suggested does not respect the unities of time, space, and action. Still, the realist must supply arguments that establish the validity of such a principle and which therefore show that coherence with this principle is necessary for an objective description.

The other way of looking at the inference to the best explanation is to say that it leads us to those entities which are causally responsible for the observed phenomena to be explained. By assuming that the existence of unobservable entities is causally responsible for what we can observe in the laboratory, realism yields the best explanation of why these physical phenomena are stable and occur in a regular way. They don't pop up by mere chance but are caused by underlying entities. A theory that explains different phenomena according to a common cause is also better than one which explains the same phenomena according to various independent causes. For example, as Wesley Salmon has pointed out, the determination of Avogadro's number, i.e., the number of molecules in a mole of any substance, was the decisive achievement in convincing the scientific community of the reality of atoms and molecules. What is crucial is not so much the fact that Jean Perrin succeeded in achieving a precise experimental value of Avogadro's number as the fact that within a few years, he and others reached the same number based on several independent methods and carried out on a variety of phenomena. Among those phenomena were Brownian movement, alpha decay, X-ray diffraction, black body radiation, and electrochemistry. Thus, ruling out the question of a striking coincidence, this remarkable agreement among the results of experiments, which seem to be quite independent of one another, can be taken as strong evidence of the hypothesis that behind the different phenomena there is something common causing their appearances (Salmon, 1984, pp. 214-227). Nevertheless, the history of science also seems, once again, to teach us another and different lesson. As long as the discussion is kept on the empirical level, there are historical cases where theories were regarded as the most prolific explanations available, but where the explanatory success wasn't enough to establish the reality of the entities proposed. The theories of phlogiston and caloric are just two overriding examples. Apart from this fact, the antirealist is always in a position to argue, as Bas van Fraassen does, that a case of the type Salmon mentions merely shows that our best theories are empirically adequate (van Fraassen, 1980, ch. 1). Such a case does not by itself establish philosophically that our theories of molecules have to be true, or that molecules are real.

What is wrong with the realist's argument for the inference to the best explanation is not that no such inferences are used in science. But it fails to prove that we are ontologically committed to those entities or laws of nature which are made subject of our best explanation. The argument works only in favour of the realist's point of view, after he has proven that we do have ontological commitments to the entities and properties postulated by those theories that are empirically adequate.

3 The success argument

I propose that we distinguish between two sorts of scientific success: One kind being related to science's ability to conceptualize the so-called unobservable world in terms of categories and principles in a rigorous fashion, which in turn allows us to make substantially correct prediction of numerous observable phenomena. Let us call this *theoretical or predictive success*. The other being related to our technological conquests of the unobservable world and our ability to manipulate it to create new effects. This kind can be called *practical or manipulative success*.

Theoretical success amounts to the fact that science until now has worked, that scientific theories have passed many empirical tests without being refuted, and that they yield coherent explanations of many otherwise unconnected phenomena. It therefore seems justified, the argument continues, to consider those unobservable entities postulated by a theory as real if they can be used to account for a large number of observable phenomena. So because a concept like 'field' enters into a theoretical explanation of gravitational and electromagnetic phenomena, the realist believes that we have sufficient reasons to assume that this concept stands for an objective feature of reality. If, on the other hand, the unobservable entity in question has been introduced only for the benefit of a certain and rather specific calculation, it is not reasonable to assume that the term by which it is introduced refers to anything in the world, unless, of course, it helps the scientist to predict a new phenomenon.

The practical success makes science successful in virtue of our ability to construct an advanced technology on the basis of the insight in nature we gain from applying scientific theories on practical problems. However, even though science by and large can be said to be successful in both of the above senses, the fact that science can be ascribed theoretical success hardly counts as a strong argument for scientific realism.[1] Theoretical success should merely be taken as evidence that current scientific theories are what they are supposed to be, namely, empirical adequate. For explanatory success depends here entirely on predictive success. It seems as if a causal theory cannot have explanatory success without having predictive success. But does it hold the other way around?

Sometimes it is claimed that predictive success does not imply explanatory success as, for instance, in the case of quantum mechanics. It is held to be an example of a theory with very little explanatory power but with a lot of predictive force. Obviously, in this case the denial of the converse implication happens to rest on premises that are very sensitive to what kind of notion of explanation one subscribes to. However, with respect to the present discussion of what can be inferred from the success of scientific theories, it is not useful to make a distinction between predictive and explanatory success.

In the history of science, and even in science today, there are many examples that theories may be used to predict future phenomena, theories which are either not true, or whose central terms do not refer to something real – e.g., the Ptolemaic system for the motion of the planets and Newton's theory of gravitation. In principle the Ptolemaic theory could still be used for predicting the course of the planets on the vault of heaven, in spite of the fact that nobody any longer believes that the planets are satellites moving around the earth. Such predictions have become even more achievable today because of the calculative power of current computers. Nevertheless, nothing in reality corresponds to 'epicycles' and 'geocentric orbits', the most central terms within the theory. Analogously, the world cannot be as we are told by Newton's theory of gravitation, if Einstein's general theory of relativity gives us the correct description on a grand scale. The central term of the theory, 'gravitational force', does not refer to something in reality; instead it has been replaced with geodic curves in spacetime. But the Newtonian theory is indispensable

---

[1] Several philosophers share the view that theoretical success implies scientific realism. See, for instance, Boyd (1973, 1985, 1990); Newton-Smith (1978, 1981); Niiniluoto (1977).

for calculations of many astronomical and technological problems in connection with space research, tidal movements, etc.

The conclusion is therefore that predicative success implies neither truth nor referential success. Scientific realism cannot make capital out of the fact that science has strong predicative success. What predicative success proves is that the world works as if there were the entities. Rather, the fact that some theories have useful predicative power without being true or having referential success can be seen as a confirmation of certain version of antirealism.

But what about the converse implication: Do truth and referential success imply predicative success? As Larry Laudan brings to light, scientific theories may be genuinely referential without being successful (Laudan, 1982, p. 223). The examples he mentions are Dalton's theory of atoms, the Proutian theory that the atoms of heavy elements are made up of hydrogen atoms, and Bohr's early theory of the electron. All of these were apparently genuinely referring theories, in spite of fact that they made a lot of flawed claims about atoms and their constituents, and hence in the end turned out to be unsuccessful. Laudan also rejects a possible realist retreat, according to which it is said that a theory whose central terms refer will usually be successful. He does so because, as he says, it is always possible by the use of negation to generate 'indefinitely many unsuccessful theories, all of whose substantive terms are genuinely referring'. And he compares this logical point with the many unsuccessful theories of atoms which have been proposed during the two millennia of speculations about the nature of matter. If Laudan were correct, it would imply that the realist's argument at this point is badly damaged.

Nevertheless, I don't think that Laudan gives the realist sufficient benefit of the doubt. I believe that a realist with perfect justice may claim that various historical theories were not successful because some of their central terms did not designate anything. Some of them did, of course, since scientists had correctly identified those entities in question. But Laudan seems to imagine that the realist position involves only that substantive terms are referring. Against this, the realist could argue that the most important predicative terms should also have to be genuinely satisfied for a theory to be successful. For example, a sentence like 'Electrons move around the nucleus in stationary, but classical orbits' expresses one of the fundamental assumptions Bohr made. Here the realist could argue that the terms 'electron' and 'nucleus' refer, whereas predicates like 'move around in stationary but classical orbits' and 'have a determinate position and a determinate momentum' are not satisfied. And for this reason Bohr's theory was wrong: It ascribed the wrong attributes to the right entities. So what made some of the theories mentioned unsuccessful was in fact that some property terms of the theories failed to be satisfactory defined or turned out to be empty.

The above example also reveals how truth and reference are related for the realist. Usually, the truth of a theory is taken to imply the genuine reference of its theoretical terms, while genuine reference does not imply truth. A theory can only be true or approximately true if its terms have real counterparts. In other words, whereas truth is, even according to realists, assumed to be sufficient for successful reference, reference is merely supposed to be necessary for truth. This is not the place to take a more careful look at the realist notion of truth. But we still have to finish our discussion of whether scientific success is a parasite on genuine reference.

In addition to the putatively theoretical success of explanation and prediction, science is connected with practical and technical success. Maybe successful predictions are not a consequence of the fulfilment of the referential aspect of the theoretical terms employed, assuming that all what observation can provide us with are the genuine reference of the observational terms and hence empirically successful theories. Nevertheless, in science we are able to experiment with things which we cannot see with the naked eye; things which afterwards may, on the basis of the knowledge of their causal properties we gain from these experiments, be used in technical apparatuses and instruments. Thus, the realist could say that because we can manipulate with what we cannot see and bring about the observable effects we want to produce, this shows that the theoretical terms of both the causal description of the experiment and of the function of the involved apparatuses genuinely refer. It is an undeniable fact that we incessantly, with greater and greater success, create and construct new technologies by using such unobservable entities and processes as direct tools in the construction and operation of these technologies. But this fact would not be understandable unless our best current theories were genuinely referential. If we, for instance, were able to move around with individual genes in a cell, taking some out and putting some others in, thereby creating new organisms, it would be beyond any rational ground to suggest that genes are not real merely because we cannot see them.

As pointed out by Ian Hacking, the fact that electrons can be used as tools is the strongest evidence for scientific realism (Hacking, 1983, ch. 16). In his opinion it is not because one can make experiment with them that one is committed to believing in their existence. Nor is it because of electrons can be used to experiment on something else. What matters is that by understanding the causal properties of electrons we can use our knowledge to build devices in which the electrons will behave in a certain characteristic manner, whenever we want them to do so. Electrons can be prepared in such a way that they can be employed in the creation of phenomena we wish to investigate in some other domain of nature.

For the realist this amounts to holding that practical success implies referential success, although the converse entailment is not true; theoretical terms may indeed have reference without the referent being an entity that can be used technologically. The basic premise is that you may see something which doesn't exist, and wrongly believe things are real which you cannot see; but you can never manipulate anything which isn't there. And even less can you manipulate an entity to cause an effect unless it exists. The realist's conclusion, therefore, is that a theory of knowledge which confines knowledge to what can be seen *ad oculus* is not very convincing. Our power to manipulate unobservable things justifies the assumption that we finally have knowledge of the physical world as it exists in virtue of itself.

A fine example illustrating some of these points is the discovery of Hafnium.[2] The periodic system of the elements was not established until around 1870. When this happened, it was done only on the basis of the *chemical features* of the elements, and most chemists regarded it as a purely empirical classification of the elements. In 1897 J.J. Thompson suggested a connection between atomic structure and the periodic system; however, it was not until Niels Bohr's second theory of the atom that anybody was able to give a physically satisfactory account of all the elements from hydrogen to uranium, including the transition groups and the rare earths. The theory was a result of a mixture of ill-defined general principles and empirically based concepts coupled with an exceptional physical intuition. Among the principles and theoretical concepts were the construction principle (Aufbauprinzip), the correspondence principle, penetrating orbits, and symmetry concepts. On the empirical side was chemical evidence in the form of ionic colours, magnetic properties, ionization potentials, atomic volumes, polarizability, and physical evidence in the form of optical spectra. Relying on these data and forming principles, Bohr gave a physical description of the atomic structure of the various elements and of how the electrons build up in shells from one element to the next. This description was able to reproduce many of the characteristics of the old periodic system.

After the formulation of Bohr's theory it was soon strongly supported by its ability to incorporate evidence from X-ray spectroscopy made by Dirk Coster. This evidence was in agreement with the predictions that included the right number of curves for the absorption edges, indicating the possible configuration based on levels of three quantum numbers; the curves of absorption edges showed that the building up of electrons started out roughly where it was expected: and finally the curves almost visualized those parts of the periodic system in which the building up occurs at the intermediate, but still incomplete level. Likewise the theory predicted new results for the optical spectra of the elements which were successfully confirmed by Paschen and Fowler.

Nevertheless, Bohr's theory was overthrown a few years later, partly because J.D. Main Smith and E.C. Stoker changed it in order to cope with the structure and the existence of simple chemical compounds, and partly because Wolfgang Pauli could support their changes by his introduction of the exclusion principle as an

---

[2] My knowledge about the discovery of Hafnium rests entirely on an excellent study by Kragh (1979) and Kragh (1980).

explanation of the electron distribution in a single atom. In spite of that, Bohr's theory still had one big victory to claim. While he was working on his model, the element with atomic number 72 had not been satisfactorily identified. It was generally believed to be an element that belonged to the rare earths, and chemists were looking for it in ytterbium minerals. In 1911, Urbain claimed to have isolated this new element by the method of fractionations. He called it celtium. Eleven years later Urbain, together with the X-ray spectroscopist Dauvillier, announced that, based on a few X-ray lines, they finally had identified element 72 in agreement with Urbain's earlier chemical discovery. If, however, this claim had been correct, it would have been fatal for Bohr's theory, according to which element 72 should be considered to be a homologue of zirconium, and therefore have no chemical similarities with the rare earths as celtium was supposed to. Knowing this and unhappy with the quality of Urbain's and Dauvillier's X-ray lines, Coster and G. Hevesy succeeded within half a year to find the new element, called hafnium, among zirconium minerals. They, too, used X-ray spectroscopy to track down the new element, and on the basis of two excellent lines Coster identified them as part of its L-spectrum.

So far as one focuses only on the predictive success of Bohr's theory, one could, as van Fraassen would do, argue that the theory merely provided us with an empirically adequate account of the correlations of the various optical spectra of the elements and of the various X-ray spectra, and a similar account of the mutual correlations between these two kinds of spectra.

4 Constructive empiricism

A theory of elements is empirically adequate if the world is observationally as if there are elements. Van Fraassen distinguishes between the acceptance of a scientific theory and the belief in its (partial) truth, claiming that the acceptance involves only the idea that the theory saves the phenomena, not that it is true (van Fraassen, 1980, p. 8 and 12). Nevertheless, the acceptance of a theory about S means to take all its claims literally, both claims about observable and unobservable entities. His idea is that by acceptance we commit ourselves to using the entire potential of the theory as if S exists in giving explanation and doing research. Still, we should be agnostic about the claims a theory makes about unobservable entities because they cannot be observed. Consequently, according to van Fraassen, the confirmation of Bohr's theory would not force us to embrace a belief of atoms as real. The theory was accepted for a while, simply because it was considered to be empirically adequate in virtue of yielding successful predictions.

But is it possible to account for the discovery of hafnium without believing that Bohr's theory of periodic system is true regarding the assumption of atoms?

In more general terms: is it possible to accept a theory without being externally committed to the theoretical entities it is a theory about? The fact that Coster and Hevesy were able to isolate and produce hafnium in quantities so large that everybody directly could see the stuff seems to justify a belief in atoms. As scientists eventually accepted the reorganization of the periodic system on physical ideas, they had ways to identify the different elements on the atomic level, which, I hold, at the same time established the referent of hafnium, even before this element emerged for their eyes. Elsewhere I have argued for a criterial theory of meaning according to which the evidential criteria for identifying each element is part of the meaning of the name of that natural kind (Faye, 2002, pp. 72–78). There is a causal connection between the use of the name and its bearer. The causal connection is determined by the criteria we have elected to use to identify the bearer of the name; in the present case the evidence was in the form of chemical data and particular lines in the optical spectra and in the X-ray spectra. These evidential criteria are satisfied by the bearer's sortal properties, and they enter into the definition of a particular name 'hafnium' and determine the reference of that name.

The mere fact, however, that Coster and Hevesey could manufacture a new visible element by extracting unperceivable atoms hidden inside zirconium minerals seems unintelligible if we only think of the periodic system as an empirically adequate classification. The last point can be stated even more dramatically. A couple of elements between hydrogen and uranium do not occur in nature as, for instance, technecium. It is a metallic element that can be obtained by bombarding molybdenum with deuterons or neutrons. Now, if the only thing you do is to change one visible element into another visible element by adding invisible things to it, are you not vindicated in a belief that these invisible things exist?

When micro-physical processes can be deliberately manipulated in a purposeful and constructive manner, do we not then have strong and justified reasons to assume that our belief in the existence of atoms, deuterons, and neutrons is true? It seems to be impossible to explain the success of our technological innovations, unless we were able to refer to microphysical entities and to tell a causal story about them. This we are able to do only because we understand their causal properties, and we therefore can use that knowledge in designing experiments and doing measurements. In general, technological success requires that beliefs about what we are doing have to be true, and these beliefs can only be true if we are capable of identifying the entities involved and have knowledge about their causal behaviour.

Explaining that the use of unobservables implies beliefs, and not merely acceptance, as van Fraassen suggests, Sam Mitchell has concocted a functional argument for why it has to be so (Mitchell, 1988). First he lays down a condition which should be acceptable for an empiricist like van Fraassen: only if somebody would act differently towards two kinds of entities does it make sense to argue that he or she harbours different kinds of epistemic attitudes towards these entities; that is, having a belief in one kind and being agnostic about the other. Then he points out that observables and unobservables play no discernible different role in the design of experiments or construction of apparatus. Van Fraassen must therefore either claim that we should be agnostic about observables too, or that we should believe in unobservables too. But since van Fraassen seeks to found our attitudes toward unobservables on our justification for accepting them (namely that claims about them are part of an empirically adequate theory), then the justification for believing in the observables of the theory should be sufficient for believing in the unobservables of the theory. In my opinion, however, there are no obvious epistemic grounds on which to draw a demarcation between observable or unobservable entities.[3]

The criterial theory of meaning on which the causal relationship between the name and the bearer of the name is a result of identifying criteria allows the change of these criteria. The use of a natural kind term is always open to revision because the criteria are fallible. Whenever science discovers that what is regarded as identifying criteria does not refer to sortal properties, we may skip some of these criteria and replace them with new ones, or we may enlarge the number of remaining criteria, or in the worst case scenario, we may give up the idea that a certain set of criteria establish a reference to a genuine entity as it happened with caloric, phlogiston, etc.

5 Structural realism

No doubt, the scientific realist has a strong case if he refers to the technological spin-off from science as something that is sufficient to explain the referential success of scientific theories. The practical success of science supports the external commitment of the language of science. Notice, furthermore, the difference at this point between theoretical success and practical success: it is only the latter which is sufficient for referential success, whereas only the former is necessary for referential success. Technological progress is a result of our power to act and intervene into physical processes. It shows that there is an objective reality which we cannot immediately see with our unaided eyes but which we have cognitive access to through instrumental observations. But, taking this for granted, it still remains to be proved that this kind of progress could not be explained on the assumption that the manipulated reality always exists as a conceptually grasped set of entities, properties and relations, and that these might perhaps be described in another way if the cognitive abilities of human beings had been different.

The kind of realism we have opposed takes the present scientific theories to be true or approximately true about the nature of things. Due to the optimistic no-

---

[3] See, for instance, (Faye, 2000).

miracle argument it holds that only true theories can explain the success of science. Laudan has, in contrast, introduced the pessimistic meta-induction argument: the existence of theory-change in the past seems to supply good inductive grounds for holding that presently accepted theories sooner or later will be replaced by new theories. Therefore predictive success does trade on neither truth nor reference. The physical content of a theory permits it to be true or false, but then if a theory eventually is overturned by a new one, truth cannot be what explains the empirical success of a theory. In the attempt to stay clear of this dilemma, some realists argue instead that theories have empirical success because of the structure of mathematical formulation of a theory. This view, which John Worrall attributes to Poincaré, but which he was first to explicate, is called structural or syntactic realism (Worrall, 1989, p. 112). This form of realism, he argues, can account for the existence of no miracles and meets Laudan's objection that scientific realism is unable to explain the transition from an older theory to a newer theory in which the latter is inconsistent with the former. Structural realism gives us the best of both worlds and still explains why succeeding theories have empirical success.

Structural realism is not a full-blown realism. The idea is that science may completely misidentify the nature of things as they are described by the metaphysical and physical content of our best theories but still attribute the right mathematical structure. Worrall says, "The rule on the history of physics seems to be that, whenever a theory replaces a predecessor, which has however itself enjoyed genuine predictive success, the 'correspondence principle' applies" (Worrall, 1989, p. 120). This requires retention of structure across the change of theory in the sense that the mathematical equations of the old theory reappear as limiting cases of the mathematical equation of the new theory. Worrall's historical case is the transition from Fresnel's to Maxwell's theory of light. Fresnel's theory made correct predictions because it accurately identified certain relation between optical phenomena which depend upon something or other undergoing periodic change at right angles to the light. But what more specifically is a structural realist a realist about?

It cannot be that a realist interpretation of the meaning of scientific theories yields the understanding of the physical content of the laws of nature. In his discussion of this problem James Ladyman points out that structural realism may take the form of two alternative positions: an epistemological refinement and a metaphysical approach (Ladyman, 1998, p. 410). The epistemic structural realism holds that there are epistemic constraints on what we can know about the world. We are justified in believing that we possess objective knowledge if there happens to be a mathematical continuity across theory change and revolutions. This idea requires a clear-cut distinction between the structure and the content of our theories; that is, a distinction between the mathematical equations and the theoretical interpretation of the formalism.

It is possible to find some support for this view in Bohr's methodology of quantum mechanics. Bohr introduced the principle of correspondence, and no other physicist has made such an explicit use of the correspondence principle as a guiding principle in the formation of a new theory. Bohr realized that according to his theory of the hydrogen atom, the frequencies of radiation due to the electron's transition between stationary states with large quantum numbers, i.e. states far from the ground state, coincide approximately with the results of classical electrodynamics for a free electron. But his own model of the atom eventually failed to predict some of the spectroscopic phenomena which were observed in the years to come, and in the beginning of the 1920's it was quite obvious to Bohr and other leading physicists that they still had to look for the final theory. Hence, in the search for a consistent mathematical formalism that could predict all observations, it became a methodological requirement to Bohr that any further theory of the atom should predict values in domains of large quantum numbers that should be a close approximation to the values of classical physics. The correspondence rule was a heuristic principle meant to make sure that in areas where the influence of Planck's constant could be neglected, the numerical values predicted by such a theory should be the same as if they were predicted by classical radiation theory.

The correspondence rule was an important methodological principle. In the beginning it had a clear technical meaning to Bohr. It should guarantee that calculations based on the mathematical formalism of classical electrodynamics gave the same result as a new mathematical formalism in the limit. The way for the correspondence principle to secure such a result was to connect the frequencies of radiation on an atomic spectrum with the Fourier components of the motion of an electron in orbit and then "compare the radiation emitted during the transition between two stationary states with the radiation which would be emitted by a harmonically oscillating electron on the basis of electrodynamics" (Bohr, 1920/1976, p. 51). So Bohr considered quantum mechanics as a mathematical generalization of classical mechanics in which structural elements are preserved. Matrix mechanics fulfilled the promise of the correspondence principle in its retention of the forms of classical equations (Bohr, 1925/1984, p. 852). Accordingly, we can explain the predicative success of classical physics if we take into account that it agrees with quantum mechanics in the domain where the quantum of action did not play any significant role.

In contrast to modern structural realists, however, Bohr realized at the time he became involved in the interpretation of quantum mechanics that it did not suffice to preserve some structural features in order to get to the meaning of quantum mechanics. The formalism cannot be understood unless we continue to use classical concepts in describing the experimental result and we therefore have to apply these while interpreting the mathematical formalism[4]. I think Bohr was right. It is obvious, I believe, that it makes no sense to compare the numerical values of the theory of atoms with those of classical physics unless the meaning of the physical terms in both theories is somehow commensurable. So in Bohr's opinion the use of the correspondence principle in developing the new quantum mechanics substantiated the metaphysical idea that classical concepts, like position, momentum, and energy, are indispensable for our understanding of physical reality, and only when classical phenomena and quantum phenomena are described in terms of the same classical concepts does it make sense to compare the predictive results of different mathematical formalisms. I therefore take the example to show that the structural realists' attempt to draw an interesting philosophical distinction between structure and content, i.e., between formalism and interpretation is futile. For as long as Worrall's structural realism focuses on mathematical structure as separated from interpretation, it is unable to explain the predicative success of theories. To explain predicative success requires attribution of some substantive properties to the phenomena in question.

Ladyman also rejects the epistemological form of structural realism. It does not represent any advantage over traditional scientific realism. His objection concentrates on two possible understandings. One way is to look at a theory as a Ramsey structure in the sense that a Ramsey-sentence for the theory replaces the conjunction of all theoretical constants with distinct variables bound by existential quantifiers. The result is that theoretical terms are eliminated but that the observational consequences are being preserved. It is a mistake, however, to think that the theoretical terms are entirely eliminated. They are still being referred to, not directly with theoretical terms, but indirectly via their Ramsey descriptions whose direct referents are known by acquaintance. The idea is here that the world consists of unobservable entities between which observable properties and relations obtain. Thus the relations form the structure of the world, the structure itself is the abstract form of a set of relations that hold between these entities, and the relations are those which can be known. The problem with this understanding is, as Demopolous and Friedman have pointed out, that any structure of a set of relations can obtain from any (sufficiently large) collection of objects. But if that is the case, a given structure does not pick out a unique set of relations of the world. Therefore we should reject a Ramseyian understanding of the structure of a theory.[5]

Another understanding is proposed by Stathis Psillos, a reading which makes structural realism indistinguishable from traditional realism (Psillos, 1995, 1996). He argues that Worrall's mathematical continuity is not sufficient to answer the

---

[4] A preliminary attempt along these lines can be found in (Giere, 1988, 1999). For a criticism of his semantic view on theories, see (Faye, 2006).

[5] See (Newman, 2004) for a criticism of Ramsey sentence realism posed by Cruse and Papineau (2002).

pessimistic meta-induction; we need a positive argument which connects mathematical formalism as being responsible for the predictive success, an argument which shows that mathematical formalism represents the structure of the world. He also doubts that it is possible to discriminate between our ability to know the structure and our ability to know the nature of the world. Instead he thinks that structure and nature are inseparable; properties are defined by laws in which they feature, and the nature of something consists in its basic properties and their relation as they are structurally described in mathematical equations.

Ladyman advocates an ontic or metaphysical version of structural realism because only this can explain ontological discontinuity. The ontological commitment of structural realism is more than to the empirical content of a theory but less than to the full ontology of scientific realism. He also thinks that the ontic approach to mathematical structures fares well with the semantic or model theoretic view on theories because "theories are to be thought of as presenting structures or models that may be used to represent systems, rather than as partially-interpreted axiomatic systems" (Ladyman, 1998, p. 416). The predictive success of science, such as star light being bent near the Sun as predicted by general relativity, is possible to understand if we assume that the most abstract mathematical structures go beyond a correct description of actual phenomena and represent modal relations between them. He opts for an elaboration of structural realism that takes "structure to be primitive and ontologically subsistent" (Ladyman, 1998, p. 420). He then draws attention to Weyl's view on objectivity according to which the status of objectivity can be bestowed only on relations that are invariant under particular transformations. So ontic structural realism takes structures and relations to be real rather than objects and properties.

Some philosophers have raised objections to the ontic version of structural realism, but I do not have room for presenting these in any detail.[6] My own disagreement rests on the following considerations: First, the semantic view on theories is not necessarily a benefit for the structural interpretation. Not all proponents of the semantic theory of theories consider themselves realists. Bas van Fraassen is one example. Moreover, the semantic view on theories is beset with some of the same problems as structural realism. Both rely on assumptions which are difficult to bring to term. On the one hand, the immediate interpretation of a theory is taken to be a model of abstract objects; and on the other hand a theory consists of a set of descriptive sentences, each of which has a certain truth value.[7] According to an ontic structural realist who focuses on structure rather than content, theo-

---

[6] See, for instance, Pooley (2005): "The main thesis of this paper is that, whatever the interpretative difficulties of generally covariant spacetime physics are, they do not support or suggest structural realism." (Pooley, 2005, p. 2).

[7] Cf. (Faye, 2006) for further criticism of the semantic view on theories. See also (Faye, 2002, ch. 8).

ries represent concrete structures, which means that a scientific theory is true or false with respect to some concrete relations and structures in nature. But how can we assign a truth value to a mathematical equation in virtue of actually existing structures if we understand its meaning in virtue of knowledge of abstract objects and relations? The structure of a theory does not correspond directly to some real structure but to the structure of some models which constitute the interpretation of the theory; i.e., a mathematical expression is structurally coherent with its models, and one of them may then be isomorphic with a real structure. It remains a puzzle to me how we can understand a theory's structure by having access to the abstract structure of the models.

Second, realism in terms of metaphysical structuralism seems to represent a naïve view on the relationship between mathematics and reality familiar from Wittgenstein's old picture theory of language in *Tractatus*. The metaphysical structuralist sees mathematics first and foremost as a means to representing the world in thought. The function of mathematical formulas is to represent how the world is structured. This is possible only in so far as the meaning of a mathematical equation is established in virtue of a corresponding structure which, if it is realized, makes the mathematical formula true. As Wittgenstein argued with respect to language, any combination of sentences consists of a relation of logical structures of atomic sentences, and these atomic sentences stand in a direct relation to the corresponding possible facts so that the sentences are isomorphic with the atomic states of affairs they picture. Similarly, a mathematical formula forms a structure itself, and this structure gets its meaning by saying that the world is structured in the same way as the formula in order for it to become true. In this sense the mathematical structures are logical pictures of possible real structures. The mathematical structure of theory mirrors or pictures the structure of factual relations. Thus our currently best scientific theories and reality exhibit a mutual isomorphism by having the same structural form.

Setting side the later Wittgenstein's criticism of the picture theory, there is, I think, an important difference between his attempts to grasp the function of language in terms of the atomic sentences that picture possible facts and the ontic structural realists' attempts to understand the function of scientific theories in terms of mathematical structures that are isomorphic with some possible factual structures. Wittgenstein's idea was combined with an idea that we have direct empirical access to the facts which were pictured by a language; say, the cup is on the table. But structural realists cannot have a similar empirical knowledge of the *modal* relations of the world, since these structures are ontologically independent of the entities that participate in them. The object of theories is mathematical structures, real counterparts to our mathematical equations, but we have no plausible way to get to know their existence by traditional empirical inquiry. All we can observe and manipulate are objects and their properties.

Third, it does not suffice for the structural realist to point to the ontological commitments of structures given to us by theories. The commitment to a certain structure is always internal to the mathematical framework. The structural realist needs to point to some external commitments. Again, I think that Bohr pointed to some fundamental problems concerning the mathematical structure of our current physical theories to the effect that no such external commitments subsist. In both quantum mechanics and relativity theory we meet complex numbers in the formulation of some of the basic questions such as the commutation rule and the four-interval invariant relation. He therefore rejects the idea that theories give a 'pictorial' representation of the world (Bohr, 1999, p. 86, 105). His reasons seem to be that mathematical structures, which appear as a result of the use of imaginary numbers, can never be real and thus be object of our experience because the existence of imaginary numbers is due to a mathematical abstraction from real numbers. This deprives us from having any external commitments with respect to the structure of such theories.

The final objection I briefly want to present is this. Scientific theories are in general empirically underdetermined. Theories may therefore be empirically equivalent without having the same content or structure. The mere fact that it is possible in principle to construct such theories that have different content and structure should make us suspicious of the ontological claims of structural realism. For if the same observable facts can be described satisfactorily by structurally different theories, we have no reason to argue that mathematical equations represent objective relations and therefore no objective grounds to prefer one particular formulation rather than another.

In my opinion, ontic structural realism relies on an indefensible position on the relationship between mathematically formulated theories and the world: There exists an isomorphic coherence between the mathematical structures, which exist independently of the world, and the real structure of the world as it exists independently of mathematics. This assumption makes sense only if both mathematics and the world are designed according to the same principle of reason that allows a "picture" or "translation" of the logical relations between the elements of the world into logical relations between mathematical elements. In this way, a universal logic functioning as a superior principle for both mathematics and the world guarantees epistemological objectivity. This is all fairly mystic. In contrast, I believe that a less speculative and more practicable approach to an understanding of mathematically formulated theories and their relations to the world does not go via syntax and formal semantics, but through a more cognitive approach to science which may involve ideas from cognitive semantics[8].

---

[8] A preliminary attempt along these lines can be found in (Giere, 1988, 1999). For a criticism of his semantic view on theories, see (Faye, 2006)

6 Conclusion

Invisible entities exist. We do not need scientific theories to be true or approximately true in order to discover the existence of invisible entities. Entities can be, and often are, discovered without scientists having any developed theory at their disposal. We are committed to their existence whenever we are able to interact with them in a constructive way. The truth of scientific theories is not needed because the relation between theory and entities are mediated by models. The entities such as planets, stones, pendulums, light, atoms, electrons, photons, and quarks are not, and will not be, the direct objects of any theory. I have elsewhere argued that fundamental laws, like Newton's laws, Maxwell's laws, and Schrödinger's equation, function as definitions by stating relations between set of quantities (Faye, 2005); see also (Faye, 2002, ch. 8). A theory consists of a vocabulary of certain idealized properties which are then defined as quantities in some mathematical equations. The equations interrelate quantitative terms by defining some of them in terms of the others. Not until a mathematical model is established, which is an abstract representation of some concrete objects, will these quantities become identified with the properties of specific entities. We can then use this abstract model to explain the behaviour of the corresponding physical entities. The upshot is that since past and present theories do not deal with concrete entities but only define idealized attributes, scientific theories may change without affecting our ontological commitment of the entities involved.

References

Bohr, N. (1920/1976). On the spectra of the elements, *in* L. Rosenfeld (ed.), *Collected Works*, Vol. 3, North Holland, Amsterdam. (1976).

Bohr, N. (1925/1984). Atomic theory and mechanics, *in* L. Rosenfeld (ed.), *Collected Works*, Vol. 6, North Holland, Amsterdam. (1984).

Bohr, N. (1999). Causality and complementary, *in* J. Faye and H. Folse (eds), *Niels Bohr's Philosophical Writings*, Vol. 4, Ox Bow Press, Woodbridge, Conn.

Boyd, R. N. (1973). Realism, underdetermination, and a causal theory of evidence, *Nous* **7**: 1–12.

Boyd, R. N. (1985). Lex orandi est lex credendi, *in* P. M. Churchland and C. A. Hooker (eds), *Images of Science*, The University of Chicago Press, Chicago.

Boyd, R. N. (1990). Realism, approximate truth, and philosophical method, *in* S. W. Savage (ed.), *Scientific Theories*, Vol. 14 of *Minnesota Studies in the Philosophy of Science*, University of Minnesota Press, Minneapolis.

Carnap, R. (1950). Empiricism, semantics, and ontology, *Revue internationale philosophie* . Reprinted in the second edition of Meaning and Necessity (1958). The University of Chicago Press, Chicago.

Craig, W. (1956). Replacement of auxiliary expressions, *Philosophical Review* **65**: 38–55.

Cruse, P. and Papineau, D. (2002). Scientific realism without reference, *in* M. Marsonet (ed.), *The Problem of Realism*, Ashgate Publishing Company, Aldershot.

Devitt, M. (1984/1991). *Realism & Truth*, Blackwell, Cambridge, MA.

Faye, J. (2000). Observing the unobservable, *in* E. Agazzi and M. Pauri (eds), *The Reality of the Unobservable. Observability, Unobservability and their Impact on the Issue of Scientific Realism.*, Vol. 215 of *Boston Studies in the Philosophy of Science*, Kluwer Academic Publishers, Dordrecht, pp. 165–177.

Faye, J. (2002). *Rethinking Science*, Ashgate Publishing Company, Aldershot.

Faye, J. (2005). How nature makes sense, *in* J. Faye, P. Needham, U. Scheffler and M. Urchs (eds), *Nature's Principles*, Vol. 4 of *Logic, Epistemology, and the Unity of Science*, Springer Press, Dordrecht, pp. 77–102.

Faye, J. (2006). Models, theories, and language, *in* D. Conci, M. Marsonet and F. Minazzi (eds), *Essays in honor of Evandro Agazzi*, Presidenza del Consiglio dei Ministri, Rome: Poligrafico e Zecca dello Stato (in press).

Giere, R. N. (1988). *Explaining Science. A Cognitive Approach*, The University of Chicago Press, Chicago.

Giere, R. N. (1999). *Science Without Laws*, The University of Chicago Press, Chicago.

Hacking, I. (1983). *Representing and Intervening*, Cambridge University Press, Cambridge.

Hempel, C. G. (1965). *Aspects of Scientific Explanation*, The Free Press, New York, pp. 231–43: The theoretician's dilemma: A study in the logic of theory construction. Article first published 1958 in H. Feigl *et al.*: *Minnesota Studies in the Philosophy of Science*, University of Minnesota Press.

Kragh, H. (1979). Niels Bohr's second atomic theory, *in* R. McCormmach *et al* (ed.), *Historical Studies in the Physical Sciences*, The Johns Hopkins University Press, Baltimore and London, pp. 123–185.

Kragh, H. (1980). Anatomy of a priority conflict: The case of element 72, *Centaurus* **23**: 275–301.

Ladyman, J. (1998). What is structural realism, *Studies in History and Philosophy of Science* **29**: 409–424.

Laudan, L. (1982). A confutation of convergent realism, *The Journal of Philosophy* **48**: 19–48. Reprinted in J. Leplin (ed.), Scientific Realism (1984), 218–49. References are to the latter edition.

Mitchell, S. (1988). Constructive empiricism and anti-realism, *in* A. Fine and J. Leplin (eds), *PSA 1988*, Vol. 1, Phil. Sci. Assoc., pp. 174–180.

Nagel, T. (1974). What is it like to be a bat?, *Philosophical Review* **83**: 435–450.

Nagel, T. (1986). *The View From Nowhere*, Oxford University Press, New York.

Newman, M. (2004). Ramsey-sentence realism as an answer to the pessimistic meta-induction, *Proc. Phil. Sci. Assoc. 19th Biennial Meeting – PSA2004: Contributed Papers*, Phil. Sci. Assoc.

Newton-Smith, W. (1978). The underdetermination of theories by data, *Aristotelian Society* **suppl. 52**: 71–91.

Newton-Smith, W. (1981). In defence of truth, *in* U. J. Jensen and R. Harré (eds), *The Philosophy of Evolution*, Vol. vol. 26 of *Harvester Studies in Philosophy*, The Harvester Press, Bristol, pp. 269–289.

Niiniluoto, I. (1977). On the truthlikeness of generalizations, *in* R. Butts and J. Hintikka (eds), *Basic Problems in Methodology and Linguistics*, Reidel, Dordrecht, pp. 121–147.

Pooley, O. (2005). Points, particles, and structural realism, *in* S. French, D. Rickles and J. Saatsi (eds), *Structural Foundations of Quantum Gravity*. (OUP, in preparation).

Popper, K. R. (1972). *Objective Knowledge: An Evolutionary Approach*, Oxford Universtity Press, Oxford.

Psillos, S. (1995). Is structural realism the best of both worlds?, *Dialectica* **49**: 15–46.

Psillos, S. (1996). Scientific realism and the 'pessimistic induction', *Philosophy of Science* **63**: S306–S314.

Putnam, H. (1978). *Meaning and the Moral Sciences*, Routledge and Kegan Paul, London.

Putnam, H. (1981). *Reason, Truth and History*, Cambridge University Press, Cambridge.

Quine, W. V. O. (1969). *Ontological Relativity and Other Essays*, Columbia University Press, New York.

Ramsey, F. P. (1930/1990). *Philosophical Papers*, Cambridge University Press, Cambridge. Edited by D. H. Mellor.

Salmon, W. C. (1984). *Scientific Explanation and the Causal Structure of the World*, Princeton University Press, Princeton.

Smart, J. C. C. (1963). *Philosophy and Scientific Realism*, Routledge & Kegan Paul, London.

van Fraassen, B. C. (1980). *The Scientific Image*, Oxford University Press, Oxford.

Worrall, J. (1989). Structural realism: The best of both worlds?, *Dialectica* **43**: 99–124.

To be or not to be: An ancient Danish dialogue concerning appearance and reality

Kevin T. Kelly

Department of Philosophy, Carnegie Mellon University

Andurtes, a gruff Viking with unusually enlightened ideas for the time, and Kevo, a callow foreign youth, are strolling on the beach below the celebrated white chalk cliffs on the southern shore of the isle of Møn, a long, long time ago.

ANDURTES: How are things back in the Empire?

KEVO: The troops are returning from our recent, glorious war against the ocean, Andurtes. At first, the citizens were skeptical about the Emperor's declaration of victory, but when he had huge sacks of shells dumped in the Senate, all opposition was silenced. Had he not abolished the Republic and rescinded civil liberties soon enough, our very baths would be terrorized by renegade flounders and monkfish by now!

So how are things in these barbarous regions? We never hear anything about Denmark back home. Are Henning and Vincent and Arne still away raiding the English coast?

ANDURTES: Heh, heh. You foreigners have such quaint ideas about modern Denmark. We are raiding Normandy these days.

KEVO: It was quite a surprise to me, the first time I visited here, that the same people who pillage the whole Northern Sea maintain such humane social and medical plans at home.

ANDURTES: Ha! How do you think we pay for them?

KEVO: I see. Appearance can be quite different from reality, can't it? Say, I brought a Bear Brew along, but I don't have an opener. The Empire confiscates them when you board a ship nowadays.

ANDURTES: Heh, heh, let me do it. What do you think Danes put these horns on their helmets for? By the way, surely, you must have learned something about the nature of reality back in the great philosophical polis of Pittsburgamon.

KEVO: Oh, you mean the great debate over realism—what is and what is not? Yes, of course, Andurtes. Reality is what the true scientific theory says. So the problem of reality is the problem of determining which theory is true. If several theories all agree with current experience, you are justified in believing the best *confirmed* one. So the whole question comes down to confirmation.

ANDURTES: I see, but I asked about reality, not confirmation. Why are you justified in believing that the confirmed theory is true, rather than just confirmed?

KEVO: That's so easy, Andurtes. Belief that *P* is belief that *P* is true, so confirmation justifies belief that *P* is true because it justifies belief that *P*.

ANDURTES: That's it? That's all there is to your account of reality?

KEVO: Oh, no Andurtes. It remains to *explicate* the concept of confirmation by looking at usage, historical case studies and, most importantly, scientific practice. The explication is then *analytically* true, so it is analytic that confirmation justifies belief that the confirmed theory is true.

ANDURTES: So how does one explicate the concept of confirmation?

KEVO: Just like any other concept, Andurtes. One looks at lots of examples or historical case studies, consults one's intuitions about whether the theory is confirmed or disconfirmed, and then constructs a general definition or rule that accords with all the intuitions. Somehow, I have the feeling that you should be telling *me* this.

ANDURTES: So what is confirmation, then?

KEVO: There is a cult from the East called Bayesianism. The high priest Bayesius discovered that confirmation is an increase in the conditional probability of a theory in light of the current evidence. At first, his devotees were used as footballs in the Colosseum, but then the new emperor Expectine defeated his rival Minimax and declared Bayesianism the Empire's new State Religion. Now you can't do anything in Philosophy, Economics, Sociology, Statistics, or Machine learning without first lighting a candle in the Temple of Bayes, or Expectine's censors will crack down.

ANDURTES: What do these followers of Bayesius preach?

KEVO: Well, Andurtes, there have been many schisms and heretics since the time of Bayesius, but the basic picture is this. You start out with an arbitrary probability measure p over everything you might possibly see and over every theory you might possibly think of in the future, before you think of most of them. Each time you acquire new evidence E, you update your degrees of belief in hypotheses according to $p(H,E)$. Then E confirms H to the extent that $p(H,E) > p(H)$. It's very clever, for there is a mystical formula

$$p(H,E) = \frac{p(H)p(E,H)}{p(E)}$$

that explains all sorts of intuitions about confirmation. For example, confirmation is greater insofar as $p(E)$ is lower (i.e., the observation is surprising) and insofar as $p(E,H)$ is higher (i.e., insofar as the hypothesis explains the data better). One can argue that there are diminishing returns for repeating similar experiments and for other intuitive principles.

ANDURTES: Hmm.

KEVO: What is it, Andurtes?

ANDURTES: Well, I used to be Tycho Brahe's lab assistant.[1] He threw great parties at Uraniborg, heh, heh, but I couldn't go to them because I was never done with those blasted Mars measurements. Anyway, I remember quite distinctly that the big debate in those days was between Ptolemaic astronomers, who held that the planets, including the Sun, revolve about the Earth, and the Copernicans, who said that the planets, including the Earth, revolve around the Sun. Experts thought that the Copernican theory was *simpler* than the alternative and that was widely taken to be evidence or confirmation. They justified their conclusion by citing "Ockham's razor".

KEVO: What do you mean by simplicity, Andurtes?

ANDURTES: I couldn't say in general, Kevo. But at that time it came down to this. The planets occasionally seem to reverse course against the fixed stars. Ptolemy explained this with epicycles, or circles that rotate about a center that is, itself, carried in a circle centered on the Earth. The composition of the two circular motions occasionally produces an apparent, backwards motion from an Earth-bound viewpoint. The issue is that the motions of the circles can be tuned to produce the backwards motion at any time whatever but, in fact, it is observed in Mars, Jupiter and Saturn only when these planets are $180°$ opposite from the sun and in Mercury and Venus only when they are in conjunction with the Sun. In Copernicus' theory, this phenomenon is unavoidably entailed by the fact that retrograde motion is just the Earth passing or being passed by the other planets, because inner planets are in conjunction with the Sun when they lap the Earth and external planets are in opposition to the sun when the Earth laps them.

KEVO: Oh, now I see. Just put the same prior probability on Copernicus' theory C that you put on Ptolemy's theory P. Let E be the fact that retrograde motions have always been viewed at conjunction or opposition. Then $p(E,C) \approx 1$, but that is not true of P. Let θ denote the vector of parameters determining initial position and velocity for each epicycle explaining retrograde motion. Then Ptolemy's theory is

---

[1] Translator's note: the original, runic manuscript, engraved on the bottom of a wassailing bowl preserved in the crypt of the Roskilde cathedral, seems to have been corrupted at the time of Christian IV, the much-beloved Danish king who sadly confused downsizing government with downsizing the nation.

actually the existential statement $(\exists\theta)\, P[\theta]$, where $P[\theta]$ is an open formula with free variables for the components of θ. Then

$$p(E,P) = \int p(E,P[\theta])p(P[\theta],P)\, d\theta.$$

Then $p(E,P[\theta])$ is unity for only a small range of values of θ, so the integral evaluates to a very low number, given that $p(\theta)$ is fairly uniform over the range of θ and, hence, $p(E,P)$ is very near to zero. But then by Bayesius' golden formula:

$$\frac{p(P,E)}{p(C,E)} = \frac{p(P)}{p(C)} \cdot \frac{p(E,P)}{p(E,C)} \approx \frac{.5}{.5} \cdot \frac{0}{1} = 0.$$

So P gets clobbered by C if C is compatible with E. The argument is so strong that the same would be true even if the prior probability of P were quite a bit greater than that of C. That is what we mean when we say that it would be a *miracle* if the parameters in P were set just so as to duplicate the appearances predicted by C.

ANDURTES: Very nice, Kevo. But something still bothers me.

KEVO: What, Andurtes?

ANDURTES: I was fishing once and my father asked me if I thought I would catch a cod. I had no idea what I would catch, so I said the odds were 50/50. Then he asked me if I thought I would catch a cod or a halibut or something else, and I realized that it was I who was caught, because 50/50 on the former question entailed 50/25/25 on the latter, which suggested some wisdom I lacked to the effect that I would catch a cod. I realized that it's impossible to be "fair" with probabilities, because fairness in one partition implies bias in another. Isn't that essentially what you just did, except on a far grander scale? Let G consist of the values of θ such that $p(E,P[\theta])$ is near unity and let B consist of the remaining values of θ. Then "Fairness" over partition C vs. P implies a strong bias against G in the partition of C vs. G vs. B. Then by the mystical formula, you get

$$\frac{p(G,E)}{p(C,E)} = \frac{p(G)}{p(C)} \cdot \frac{p(G,E)}{p(C,E)} = \frac{p(G)}{p(C)} \approx 0.$$

But $p(P,E) \geq p(G,E)$, so if you had started out with the contrary, prior bias $p(G) > p(C)$, then you would have ended up with $p(P,E) > p(C,E)$. So the decisive premise in your argument for Ockham's razor is your strong, prior bias for the simple worlds in C over the complex worlds in G.

KEVO: So what? Bayesius is very liberal about what sort of bias you start with, so there is nothing wrong about starting out with a strong bias toward simple worlds.

ANDURTES: The question is why one is justified in believing that the world is simple. So you start out with a strong bias toward simple worlds and, lo, this bias gets passed along through Bayesius' formula. What do they call that sort of argument in the Empire, Kevo?

KEVO: Why, that would be called a *circular* argument, Andurtes.

ANDURTES: An edifying, broad circle that links many different sorts of ideas and considerations, or a vicious, narrow, uninformative circle that just assumes a bias toward simplicity in order to justify a bias toward simplicity?

KEVO: I suppose the latter, Andurtes.

ANDURTES: Do you know what Vikings do to sophists who purvey narrow, viciously circular arguments, Kevo?

KEVO: I have heard about it! They are tied to a small chariot wheel and rolled around and around on the ramparts of Kronborg as an example to all young philosophers.

ANDURTES: The narrower the circle, the smaller the wheel. And that is far too lenient but, as you say, we Vikings are gentle at home to a fault.

KEVO: Be careful, Andurtes. Expectine's censors are everywhere. And anyway, the bias in question could be part of the explication of confirmation, in which case it is analytically true that simpler theories are better confirmed by the data they explain.

ANDURTES: Suppose that King Knut were to send you to investigate a piracy suspect named Henning. Would you just ask Henning if he is a pirate and then pass along his response to the king?

KEVO: No, Andurtes. Surely, some independent investigation would be required!

ANDURTES: Indeed, it would be a terrible miscarriage of justice to reward Henning for shirking his piracy duties. But isn't that pretty similar to your approach to philosophy?

KEVO: Please explain what you mean, Andurtes.

ANDURTES: As a human, you have some wired-in learning dispositions. They might be the best possible. But then, again, they might not be... after all, our Nordic gods live in a town that is destined to collapse and burn down with them in it, eventually. Surely that raises some questions about their engineering abilities! Now, when you consult intuition to "explicate the meaning of confirmation", you are, in effect, passing along reports from your learning dispositions without independent investigation as to the soundness of their engineering. But even a puppet with a head of wood can be constructed to report "my opinions are highly justified" whenever a button on the back of its neck is pushed.

KEVO: My head is not wooden!

ANDURTES: Of course, not, Kevo. It suffices to suppose that your learning dispositions are O.K., but not nearly so great as some basic engineering improvements might have made them. Then which dispositions would be better justified?

KEVO: The better dispositions, presumably.

ANDURTES: Then what would the explication based upon your dispositions be an explication of, "justification" or "human foible"?

KEVO: "Human foible", I suppose. I think I see what you are getting at. You want some independent reason why we *should* prefer simpler theories, other than the mere fact that it feels good to do so.

ANDURTES: Exactly. I think you finally get it.

KEVO: No problem, Andurtes. There are powerful, independent arguments for favoring the simpler theory. For example, it's better *testable*, for if it is wrong, it is crisply refuted (you'll see retrograde motion at a time other than opposition or conjunction), whereas falsehood of the alternative theory might be followed only by a long run of null experiments (failure to detect such an "effect"). Again, the simpler theory is more *uniform* (such an "effect" will never be detected), and more *unified or explanatory* (no unexplained, independent postulates concerning the time of the retrograde motion are required).

ANDURTES: What do you call it, Kevo, when someone believes that the world is a certain way simply because he wishes it to be that way?

KEVO: That is called *wishful thinking*, Andurtes, a most grievous fallacy. Violators are thrown into a deep dungeon and left there until they can wish their way out of it.

ANDURTES: Isn't that what you have just done, Kevo? Didn't you just argue from some nice methodological features of a theory to its truth? If a complex theory were true, then the truth wouldn't be uniform, explanatory, or unified, so why are you entitled to assume that it does, other than by wishful thinking?

KEVO: You asked me to explain why simple theories are better. Then whatever I tell you, I am blamed for wishful thinking!

ANDURTES: That is going too far. I said that citing the advantages you mentioned amounts to wishful thinking. Why?

KEVO: Because they don't have any clear connection with truth?

ANDURTES: Just so.

KEVO: So you want me to explain how a learning disposition biased toward simplicity is more likely to produce true answers?

ANDURTES: Indeed, I would be satisfied if you were merely to explain how a prior bias toward simplicity helps us find the true theory in *any* clear sense that doesn't beg the question.

KEVO: Oh, why didn't you say so earlier! Bayesians have a story connecting simplicity to truth. If you start out with a simplicity bias and update according to the mystical formula forever, then you are guaranteed, in a sense, to converge to the true theory in the limit of inquiry. Of course, if the Ptolemaic parameters are set to duplicate the Copernican theory exactly, no possible data would tell the difference, but in that case one may as well excuse every method for failure. The interesting question is convergence to the truth when the data do eventually differ, and that is guaranteed. Is that more satisfactory, Andurtes?

ANDURTES: That's the spirit, Kevo, but it falls short of the mark, I am afraid, because the convergence argument would also work with other prior biases, as long as no possibilities are ruled out a priori, so it doesn't single out the simplicity bias at issue.

KEVO: Why isn't it good enough to show that the simplicity bias suffices for finding the truth, Andurtes?

ANDURTES: Suppose your sail rips on a voyage to Greenland, Kevo. Does that prevent you from getting there?

KEVO: No, Andurtes, if the ship is equipped with needles and thread.

ANDURTES: Would you view the big rip as help or as a hindrance?

KEVO: Surely, a great hindrance, Andurtes.

ANDURTES: A hindrance that slows you down even though it can be overcome eventually, with time, toil, and trouble?

KEVO: Truly, Andurtes.

ANDURTES: Isn't that what the prior bias is like in the Bayesian convergence argument? Future evidence and inquiry overcomes the bias even if it is false. And if you started with the contrary bias, future evidence and inquiry would also overcome that bias even if it is false.

KEVO: I suppose so, Andurtes. Followers of Bayesius speak of the prior bias "washing out" with further evidence. That turn of phrase suggests the very picture you are painting.

ANDURTES: So have you explained how a learning disposition equipped with such a bias—the very disposition you repeatedly query in your explication of confirmation—helps you find the truth?

KEVO: Not at all, Andurtes. It seems that I must show that our disposition is better at finding the truth than competing dispositions, not merely that it overcomes its own limitations eventually.

ANDURTES: Just so. Any other ideas?

KEVO: Now I remember an argument I once heard from a very sober philosopher during the reign of Minimax. It can be shown that the expected squared error of a prediction can be greater if the prediction is made with a complex theory than if it is made with a simple one. The basic idea is that the best fitting curve in a complex family may depend crucially upon each datum. The prediction produced by means of the best fitting curve in a simple family will depend less upon each individual datum than the best fitting curve in a complex family (think of the extreme case in which the curve allows for one inflection point per datum), so the variance or probability spread of the prediction around the true value will be less for the simple family since, as it were, more data are relevant to the setting of each free parameter. An overly simple theory might do worse as well, because although the variance of the prediction is reduced, the mean value of the prediction is far off of the true curve. So the theory used for predictive purposes should neither be too simple nor too complex.

ANDURTES: Very nice, Kevo. Now you have an argument that uniquely singles out the right bias to have in order to find the truth. But there is still one problem.

KEVO: What now, Andurtes? Isn't that what you wanted?

ANDURTES: What if Odin were to tell you that the true theory is very complex when the sample size is small?

KEVO: Then there would be no problem. I would know that the truth is complex.

ANDURTES: But if the sample were small, what sort of theory would the preceding argument direct you to choose?

KEVO: Why, a much simpler one, Andurtes.

ANDURTES: Simpler than the true one?

KEVO: Yes, Andurtes.

ANDURTES: So you have not explained how it is that choosing a simpler theory helps us find the true theory; you have explained, instead, why it would be better to use a simple theory (perhaps simpler than the known, true theory) for predictive purposes. That hardly sounds like a reason to believe that the simpler theory is true Kevo, so we have strayed from the problem of reality with which we began.

KEVO: Hmm, I suppose so, Andurtes. But come on. Western science is so successful. Surely that testifies to the reliability of Ockham's razor!

ANDURTES: Do you know that it is always successful in every application or only that it was sort of successful until now in the applications it was applied to?

KEVO: The latter, strictly speaking.

ANDURTES: Then you are assuming that the future will be like the past and that other applications will be like past applications?

KEVO: Yes.

ANDURTES: And what is the assumption that the future will be like the past?

KEVO: Uniformity of nature.

ANDURTES: And is the most uniform world compatible with experience complex or simple?

KEVO: A uniform future is presumably more simple than an irregular or surprising future, Andurtes.

ANDURTES: So you have just appealed to Ockham's razor in an argument for Ockham's razor. What sort of argument is that?

KEVO: That would be another circular argument, Andurtes. Will you have me sent to Kronborg?

ANDURTES: Heh, heh. I won't tell, as long as you don't make the same mistake again.

KEVO: Thank you! But something had to engender our penchant for simplicity. Unless you want to say that Odin or some other god put it in us, it must have evolved through trial and error, along the lines Empedocles suggested. Potential ancestors with other biases must have been stepped upon by Titans or something like that (though it's hard to say exactly what happened). So it must be that we live in a world in which the simplicity assumption isn't far from the mark.

ANDURTES: Does the argument from selection imply that the simplicity bias is selected for always and in every application, or only that it has been selected for in applications and survival situations encountered up to now?

KEVO: I see. I have just repeated my last circular argument, right after you so kindly overlooked it.

ANDURTES: Hmm, I'm afraid your performance so far merits the smallest possible wheel.

KEVO: Now I understand what they say about you, Andurtes.

ANDURTES: What's that?

KEVO: They say your scalp is so bright that it dazzles the mind until people forget everything they once knew.

ANDURTES: Heh, heh, that's why I always wear this old helmet. Don't get sore about it, Kevo. A fundamental question so thoroughly botched through the ages is a philosopher's windfall.

KEVO: You are making sport of me.

ANDURTES: I would never make sport of such a buff, handsome fellow. Ahem. Why do you say so?

KEVO: The problem you posed for me is set up to be hopeless. For how *could* Ockham's razor help us find the truth better than other biases? If we already know that the truth is simple, we don't need Ockham's razor. And if we don't already know that the truth is simple, how could a frozen bias toward simplicity help us find it? It is no more possible for a fixed bias of any kind to indicate the unknown truth than for a compass with a fixed needle to indicate direction.

ANDURTES: Since neither of us knows the solution, I propose that we consider the question together, afresh. For although I don't know many things, one thing I would gladly fight for ... aside from aquavit and French girls and plunder... and especially coffee... is that it is better to seek the truth scientifically than by other means, so some story explaining how scientific method is a better route to reality than witchcraft is better than none at all.

KEVO: Fight all you want, but that won't explain how a fixed bias insensitive to what it indicates can be an indicator—unless you beg the question by saying that you already know it is pointing in the right direction, as the Bayesians and I have been doing.

ANDURTES: By Odin, Kevo, I think you are onto something!

KEVO: Eh? I just said the problem is hopeless.

ANDURTES: But the key to your argument is a tacit assumption—that the only way simplicity could help us find the truth is by pointing at it.

KEVO: Of course. What else would or could such help consist of?

ANDURTES: Perhaps we should stop to consider how we are routinely helped to find things. Suppose that you are lost in an obscure little town that nobody would ever think of going to.

KEVO: Like the one you used to live in before you moved to Roskilde?

ANDURTES: Exactly. Now suppose you were to need help finding your way back to Copenhagen.

KEVO: From there? Anyone would, Andurtes!

ANDURTES: Suppose you pass a local resident and ask for directions. What is she likely to say?

KEVO: Well, funny you should mention it. I did get lost in a small town on my way see you in Roskilde and when I asked for help from an old woman, she gave me directions to the high road, which was just a kilometer or so from town, beyond some trees.

ANDURTES: Was that helpful?

KEVO: Of course! I would still be lost in a maze of hedge-rows if it weren't for her.

ANDURTES: Did she know where you were headed?

KEVO: She didn't need to. She could tell I was a foreigner from the toga I was wearing. The high roads are the best way to get anywhere in Denmark that a foreigner might be headed.

ANDURTES: So she helped you find your destination without pointing at it or even knowing where you were headed?

KEVO: It was hardly a magical feat. She gives the same directions to every foreigner she sees, without even thinking about it, I'm sure.

ANDURTES: Her directions to the high road might have taken you for a short distance directly away from your destination.

KEVO: As a matter of fact, that's what happened. I was headed northeast and the path to the high road took me nearly southwest for a mile. But I was happy to jog a bit in the wrong direction to avoid that maze of fences and hedges in the fields.

ANDURTES: So there is no paradox in helping one to find something without pointing at it. Indeed, there is no paradox in advice that helps you find something by pointing you directly away from what you are looking for. So much for the Kevo paradox.

KEVO: Ha. In that homely case, I suppose so. But what does that have to do with simplicity and truth?

ANDURTES: Well, if the woman's advice skirts the paradox, perhaps Ockham's advice somehow does so in a similar way. Why don't we see if we can tell a similar story in that case?

KEVO: Let's see. Then the destination of the scientist is the true theory.

ANDURTES: So it would seem.

KEVO: So what is the high road to the truth?

ANDURTES: Perhaps we should ask ourselves, first, what distinguishes the high roads from mere footpaths?

KEVO: King Knut burned all the villages in the way to make the high roads as straight as possible. The footpaths shuttle back and forth through gaps in the hedges and fences that partition every square mile of the country.

ANDURTES: So what would correspond to straightness or reversals in the course of inqiry?

KEVO: Reversals in the course of inquiry would seem to correspond to reversals of opinion. We even use the same phrase "changes in attitude" to refer both to belief and direction.

ANDURTES: Then what does that make the high road to the truth?

KEVO: … I suppose that would be the path to the truth with the fewest possible reversals of opinion. Hmm, that is something we didn't think of earlier. Lots of prior biases suffice for mere convergence to the truth, but perhaps there is some special connection between simplicity and converging with the fewest possible number of reversals of opinion! In other words, the simplicity bias may keep one on the straightest possible path to the truth even though it does not necessarily point at the truth at any particular stage.

ANDURTES: We seem to have made some progress.

KEVO: I'm not so sure. Why doesn't the straightest path to the truth point straight at it, like a perfectly straight highway?

ANDURTES: That reminds me of my cousin Leif.

KEVO: How so, Andurtes?

ANDURTES: He had a crazy scheme to sail straight into the great western ocean to discover new islands. He hasn't returned yet.

KEVO: Crazy, indeed. I have had quite enough of sailing, to tell the truth. Always leaning over and wearing a rubber jacket and getting soaked. And you have to clean the barnacles off every season.

ANDURTES: Suppose you are in his shoes and you want to determine exactly how many islands you will ever see. You have no idea how big the ocean is, or even whether it will ever end. Nor do you know how many islands you will see, but we will assume that you will see at most finitely many. What would Ockham's razor say?

KEVO: Hmm, given sufficient ignorance of geology and geography....

ANDURTES: Leif is a great sailor, but is ignorant as an ox, as you can tell from his scheme!

KEVO: Ockham's original formulation was to presume no entities without necessity. I guess that means not to guess more islands than one has seen. Or fewer, for that matter, since it is necessary to posit at least as many as one has seen, if one posits at all. It isn't so clear to me what this has to do with minimizing free parameters in scientific theories, though.

ANDURTES: That sounds about right. Now, Kevo, picture this: no islands appear. Just bleak, grey swells on all points. What do you say?

KEVO: I conclude nothing. An island might appear at any time. See? You didn't trick me, for a change.

ANDURTES: Still no islands.

KEVO: Nothing has changed, so I still reserve judgment.

ANDURTES: Are you going to withhold judgment forever?

KEVO: Yes. I'll never be a sucker again.

ANDURTES: Fine, but before you said the aim was to converge to the true answer. If you never give in, you'll never converge to the true answer, will you?

KEVO: Hmmm. So after a sufficiently long span without islands, I have to cave in and plump for the view that the count is zero.

ANDURTES: Evidently.

KEVO: Strange. Before our discussion, I would have said that eventually sufficient evidence accumulates to justify the conclusion that no more islands are coming, but it seems that the mere ambition to converge to the truth implies the same, overall, diachronic behavior with no appeal to the simile of "support". And, if anything, the convergence argument seems to carry more normative force than merely reporting my intuitions and calling the result "confirmation theory".

ANDURTES: So now you think you will see no islands. Lo, look at the hideous crag that has just loomed out of the fog, covered with ill-starred albatrosses! Now what do you say?

KEVO: Drat! I don't know what to say now. I could be wrong again.

ANDURTES: Again, no land is in sight. Swells and foam everywhere. A vast, empty waste. A whale snorts in the distance. (Odin, I wish I were fishing!) More swells. The mast creaks in its step. . . . You're in the same tight spot as before, Kevo.

KEVO: So just as before, I have to cave in, eventually, and say "one".

ANDURTES: Does the argument continue in this manner?

KEVO: It seems so. Given that I converge to the truth, I can be presented experience that forces me to reverse course at least once per island, regardless of the total number of islands.

ANDURTES: So for each convergent solution to the overall problem, the solution retracts at least n times in some world in which answer n is true?

KEVO: Exactly. So these unavoidable, worst-case retractions correspond to kinks in the high road to the truth.

ANDURTES: An instructive simile, indeed, Kevo.

KEVO: So advice that keeps you on the high road to the truth needn't point at the truth until it is finally reached. That is how it is possible to explain how a fixed bias could help one find the truth without presupposing that the truth is what the fixed bias points at currently.

ANDURTES: Interesting. So how does Ockham's razor keep one on the high road to the truth?

KEVO: That's easy. If I simply count the islands as I see them, I retract at most n times in answer n, which is the best performance possible in light of the preceding argument.

ANDURTES: Excellent. But our argument for Ockham's razor is not yet complete, since other biases would keep you on the high road to the truth as well—in which case the argument would be moot.

KEVO: O.K., suppose that one island has been seen by stage t, but I boldly conclude "two", just to spite Ockham. So what?

ANDURTES: Seagulls, grey swells, whale snorts, creaking mast. . . .

KEVO: You devil, Andurtes! You are forcing me back to the conclusion that there is just one island, on pain of not converging to the truth.

ANDURTES: So you have retracted back to "one". Lo! A nice, flat, grassy island full of golfing Scots passes to larboard, for a total of two.

KEVO: I see your game. You can force $n + 1$ retractions out of me in each answer $n \geq 1$. I see an analogy to the high road example. Had I disregarded the advice of the old woman, I would have gotten nowhere in the maze of fields and would have had to reverse course and take her advice anyway, earning an extra U-turn for my efforts (not to mention the humiliation of aving to walk right past her again). Thereafter, the extra U-turn would be added to every other unavoidable curve in the high road. So I should have taken her advice.

ANDURTES: Could you have done better in the island example?

KEVO: Well, suppose that you agree with me prior to t, but then you return the current island count from t onward.

ANDURTES: In other words, I agree with you prior to t and I become a born-again Ockhamite from t onward.

KEVO: Exactly. So you do just what I do along prior to t, but at t you say "one" while I say "three".

ANDURTES: What if the truth is "one"?

KEVO: That's easy. You never retract from t onward, but I have to retract at least once to converge to the truth, since I say "three" at t. So I retract at least one more time than you do. But you agree with me prior to t, so I could still switch to your strategy and do better from now on than I am destined to do if I say "three".

ANDURTES: What if the truth is "two"?

KEVO: Well, nature can present no marbles until I retract from "three" to "one", and can then present a marble until I converge to "two", for a total of three retractions in answer "two". And you don't have to play the game again to convince me that I will retract $n + 1$ times in answer $n \geq 1$. So you do better than me in each answer compatible with experience gathered up to time t.

ANDURTES: Why don't we say, then, that I *strictly beat* you at the moment t at which you violate Ockham's razor by guessing too many islands?

KEVO: That manner of speaking sounds quiet from the authorities natural, Andurtes. So violating Ockham's razor earned me a strict beating in terms of retractions. Say, that's swell Andurtes. And the counting method does as well as an arbitrary, convergent method in each answer. Let's say that such a method is *efficient*. That's far stronger than never being strictly beaten.

ANDURTES: So there exist efficient Ockham methods and each violation of Ockham's razor is rewarded by a strict beating in terms of overall retractions. Does that sound like a non-circular explanation that one ought to abide by Ockham's razor?

KEVO: It is a compelling and unexpected explanation, Andurtes.

ANDURTES: Would you prefer to consult intuitions and report that you feel like producing the simple hypothesis?

KEVO: That seems like a very empty and unexplanatory approach to me now, Andurtes.

ANDURTES: Or would you prefer the circular argument that the world is probably simple so you should probably say that it is?

KEVO: Not at all, Andurtes. I appreciate that you are keeping my former indiscretions quiet.

ANDURTES: Or would you prefer to cite some nice properties of simple theories and then wishfully assume that the truth has them?

KEVO: That would be quite wrong, Andurtes. It's amazing what passes for wisdom when nobody has found a genuine explanation.

ANDURTES: That isn't unusual, Kevo. When nobody understood chemistry, Aristotelians explained the soporific powers of opium by attributing to it a "dormative virtue". Your attempt to sweep the mysteries of scientific method under the blanket of "confirmation" is quite similar. If we hadn't come up with this explanation of Ockham's razor just now, philosophers might still be repeating the usual nonsense right into the twenty-first century!

KEVO: Oh, the philosophical world would never be that stubbornly foolish, Andurtes! But I admit that we are on to something. And it seems that there is more to be gleaned from the argument.

ANDURTES: What?

KEVO: Nothing in our argument assumes that one must produce the simplest answer right away.

ANDURTES: Right, for if I retract, say, from "zero" to "I don't know", then no further retraction is incurred by continuing to say "I don't know" for any finite amount of time.

KEVO: And then nature can withhold further islands until you say "one", on pain of not converging to "one" if no further islands ever appear. Now what if you were to retract "one", after having selected it, even though no further islands appear? Would that seem strange?

ANDURTES: My red Viking blood would never tolerate abandoning the best explanation for no concrete reason! Anyone caught doing so on my ship is tied to the mast to bolster his resolution, heh, heh.

KEVO: Notice that after you take back answer "one", nature can coax you back to "one", on pain of not converging to the truth, and then can coax you into every subsequent answer, for a total of $n + 1$ retractions in each answer $n \geq 1$. But refusing to retract an answer until the next island appears results in just n retractions in answer n.

ANDURTES: That is the same sort of explanation we gave for Ockham's razor, Kevo. So taking our conclusions together, retraction efficiency implies that one should never produce an answer violating Ockham's razor and that, having adopted such an answer, one should never drop it until it violates Ockham's razor.

KEVO: That allows for producing a count that is too low.

ANDURTES: An under-count couldn't possibly be true, so how could the interests of finding the truth be furthered thereby?

KEVO: Indeed, Andurtes. So it seems that just about every intuitive, diachronic feature of inquiry is captured by the idea of efficient convergence to the truth. The only room left for individual variation is how long to wait before leaping to the simplest answer compatible with experience. When one drops it is uniquely determined and you can't choose any answer other than the simplest answer compatible with experience.

ANDURTES: So it seems, Kevo.

KEVO: But what does the island example have to do with scientific practice?

ANDURTES: Scientific models explain empirical *effects*. For example, if you are looking for a polynomial law, each monomial term is an effect. To find the true form of the law, one must determine which effects obtain and which do not. Now, suppose that the data are imprecise, so that they determine only open intervals around the points on the curve. Then any finite number of such intervals compatible with a constant function is also compatible with a tilted line (since the open intervals always leave a tiny bit of "wiggle room"). For the same reason, any finite number of open intervals around points in the line is compatible with a parabola and any finite number of open intervals around the points in the parabola is compatible with a cubic curve, and so forth. But once the lower-order family is refuted, it is verified that the degree of the equation must be at least one unit higher. So the effects that refute lower polynomial degrees are analogous to the appearance of islands. And polynomial equations with coefficients left as free parameters are analogous to guesses at the total number of islands.

KEVO: Let me see if I have the idea. Do you remember our canoe trip in Sweden?

ANDURTES: On the Vänern? Oh yes. It rained delightfully the whole time! That brought back sweet memories of Viking boot camp. Especially when my air mattress started to float inside the tent.

KEVO: But not every day. On the second day it was sunny and I was in the stern, steering the canoe to the next camp site. At first I noticed nothing. Then I came to suspect a slight tendency for the canoe to veer to the left. Finally, we were thirty compass points off course and the effect was undeniable. That's when I noticed the topless Swedish canoeists dead ahead. So an empirical effect can be sufficiently small to go unnoticed for an arbitrary amount of time, but sufficient data make it unmistakably manifest, eventually.

ANDURTES: I stand convicted, heh, heh. I am a red-blooded Viking warrior, after all.

KEVO: So the idea is that the islands in your example are just a proxy for empirical effects. Then there is some deep connection between minimizing entities and minimizing polynomial degree, after all. What about uniformity of nature?

ANDURTES: Well, the most uniform future is the one in which no more effects appear, so if islands are effects, it is is also the future with the fewest entities.

KEVO: I never thought of it that way before. I have also heard philosophers speak of the simplest hypothesis as the best explanation.

ANDURTES: For each effect, there is the magnitude of the effect, which determines the time at which it is first discovered. In the case of the islands, there is the time at which each island is discovered. None of these magnitudes is explained by the theory itself. Each such parameter setting looks like an unexplained, extra assumption.

KEVO: Excellent, Andurtes. Now I see how the island example relates to a range of familiar symptoms of empirical simplicity. But there is something rather unsettling about our story, after all.

ANDURTES: What do you mean, Kevo?

KEVO: Philosophical skeptics, of a pallid, theoretical sort, are legion back in the Agora, but nobody takes them seriously because, rhetorically speaking, Ockham's razor is more plausible than the skeptical doubts they direct against it.

ANDURTES: By Odin, Kevo, rejecting the best possible route to the truth is like never going home because the king's high road isn't good enough. We make such people eat sour grapes until they reform!

KEVO: But in a way, Andurtes, haven't we just invented a much more virulent form of skepticism? For we don't reject scientific methods as sour grapes; we embrace them as the best possible means to the truth and urge others to follow them. And we have a better argument for that conclusion than card-carrying scientific realists, who bring nothing but circles, wishful thinking, and irrelevancy to the table. But nothing in the argument precludes an arbitrary number of reversals in the future, or that the true theory is not far more complex than our current theory. Nothing in our account adds up to empirical "support" or "confirmation" or any other philosophical safety net. It's all a matter of staying on an arbitrarily long, arbitrarily winding, but nonetheless straightest path to the truth.

ANDURTES: Hmm, that does trouble my hearty Viking nature, Kevo. After all, here we are, walking on a hard, wet beach below the huge white cliffs of Møn. It is quite impossible to entertain that at any moment the sand might suddenly open and swallow us, any number of times, before we reach stable ground.

KEVO: Exactly. And furthermore ... YIPE!

ANDURTES: Kevo! Kevo! Where are you?

KEVO: Down here! Help!

ANDURTES: Heh, heh, heh. The chalk from the cliffs made a quicksand pit on the beach here!

KEVO: One minute I was walking at sea level along a hard, wet beach and the next I was in free-fall in a pool of liquid cement! It was not only the last thing I would have expected—it was an absolute reversal of what I thought I knew would happen.[2]

ANDURTES: Let that be a lesson for us, Kevo. Scientists must feel no less secure in textbook wisdom prior to the discovery of new effects that force its revisiontheory!revision of. So perhaps the relation between science and reality at any given time is no firmer than this quicksand, even though science keeps us on the straightest possible path to the truth, crooked though it may be.

KEVO: That seems deeply correct, Andurtes, although it is hard to keep securely before one's mind in ordinary life.

ANDURTES: Or even in the ordinary course of science, itself, Kevo.

KEVO: Indeed, Andurtes. Shall we get back so I can change out of this wet toga?

ANDURTES: Of course, Kevo. Anyway, I'm sure we won't encounter any more of those quicksand holes. Look, the sand is hard as concrete now.

KEVO: Indubitably, Andurtes. :-D

---

[2] This anecdote is literally true (except that Andurtes insisted upon photographs). It was a most convincing demonstration that full beliefs are indeed revisable.

Part III

Philosophy of medicine and engineering

Shaping engineering knowledge

Arne Jakobsen

Formerly at Technical University of Denmark (now retired)

1 Introduction

The topic of this paper is the process of shaping engineering knowledge. The paper is the result of a study of a line of development of engineering knowledge, namely methods and background knowledge based on theory of plasticity for designing structural elements of concrete. The focus of the study is on how engineering methods are shaped by many different agents with different interests and by different types of intellectual work and rationalities.[1]

Engineering knowledge has been studied from different perspectives. There has been a considerable interest in the relation and the differences between pure science, applied science and engineering knowledge. There has been an interest in the way engineers use knowledge and what types of knowledge they use, and in the last decades there has been an interest in the shaping of technological products. But in this study my interest is in the process of developing engineering knowledge – methods and background knowledge – in how different agents and interests influence the outcome.

---

[1] The article has been discussed by a group of structural design engineers from The Technical University of Denmark, University of Southern Denmark, and from building firms together with a researchers from the Technical University engaged in philosophy of engineering science.

The group of structural design engineers include:
Bent Steen Andreasen (Rambøll, Consulting Engineers),
Bent Feddersen (Rambøll, Consulting Engineers),
Bjarne Chr. Jensen (University of Southern Denmark) and
Mogens Peter Nielsen (Technical University of Denmark).
This article is to a large extent based on interviews with this group.

In some cases development of engineering knowledge follows a breakthrough in general scientific theory or it has relation to new (types of) products. In this case however the development has no relation to new general theoretical developments, the starting point being a rather intuitively understandable insight in the process of a failure. And the reason for developing the methods is not to make new types of constructions possible. They are intended to give advantages in design work. And, by addressing the problems of collapse in structural elements more directly than in methods formerly used, the constructions can be more optimal and materials saved. That holds especially for statically indeterminate constructions.

2 A plasticity based theory of concrete elements

The case is the development of methods, based on theory of plasticity, for designing structural elements of concrete. There are many types of structural elements: beams, columns, shells, etc., and they are exposed to different types of loads. A structural element must meet demands given by the *use* of the element. For instance, a ceiling should not deform too much. This is called the serviceability limit state. And they must be safe against failure and collapse. This is called the ultimate limit state.

Design of a concrete element – the calculation of strength with respect to both serviceability and ultimate limit states – has ordinarily been based on the *elastic* properties of the element. That is, for a given load the task is to decide how large stresses the element should be allowed to reach, and then calculate the matching strength and thus the dimensions of the element. That gives directly the serviceability limit state. Knowing from experiments the relation between deformation and failure, a figure for the ultimate limit state can be calculated. That is an indirect way of finding the ultimate limit state but it has been, and is still being, widely used.

In this case, however, the methods for calculating the ultimate limit states are developed on the assumption that concrete is *plastic* (even ideally plastic). As the problem about serviceability limit state only concerns minor deformations, the calculation of these is based on the elasticity of the material. The direct way of calculating the ultimate limit state, the state where the material collapses however, is in the plastic state of the material.

An ideally plastic material will (in a given interval of loading) yield continuously under a constant load. A certain deformation thus corresponds to a given amount of work – a certain force working on a certain distance – and the force corresponding to a permissible extent of deformation can be calculated. In fact, concrete is rather far from being ideally plastic, but it turns out that with simple corrections the results from the calculations are very close to what can be measured to be correct.

To calculate the work needed for the element to collapse, one must know the yield line – the way in which the concrete will yield. There are often no clear rules by which the position of yield lines can be deduced – one has to rely on intuition and make the best possible estimate. Because any estimate will result in an amount of work equal to or bigger than the actual work by failure the calculation will necessarily be either exactly correct or overestimating the permissible load – an upper bound solution. By repeated estimates one can approach the correct value.

Often a lower bound solution, i.e. a solution on the safe side, will be sought by calculating that dimension of the element, which gives permissible stress distributions in the material by the load in question. By repeating the calculations it is possible to get the results to converge and thus obtain a solution, which is both safe and optimal.

3 The research and development programme

The development of methods for design of structural elements based on the assumption of concrete being plastic is a predominantly Danish phenomenon. The origin of the line of development can be traced back to an article written by the Danish engineer, later professor at the Technical University of Denmark, Aage Ingerslev, in 1921 (Ingerslev, 1921) in which simple methods for determining moments in plates with two way slabs was described. The methods were developed by observations of yield lines in experimental settings and it was concluded that further development of methods should be possible through further experiments and observations.

The method and the way of reasoning proposed in Aage Ingerslev's article was followed up by K.W. Johansen, later his successor as professor. In an article from 1931 (Johansen, 1931) he extended the theory to the description of reinforced slabs, and described more precisely how the theory could become of practical use. In his dissertation K.W. Johansen (1943) developed a more general theory of yield lines.

Both the articles and the dissertation were met with a considerable interest, and especially after Johansen's article in 1931 the methods were used in practice for calculating plates – especially plates with geometrically complex support.

Until around 1960 the use of the method was limited to the calculation of slabs. The strength of other structural elements was decided by the reinforcement. But around 1960 the strength of concrete started to be decisive also in beams and columns, and that gave rise to a broad development process at the Department of Structural Engineering at the Technical University of Denmark. Professor Mogens Peter Nielsen, who succeeded K.W. Johansen, was the central agent in this development. Methods of calculation for different loads and different structural elements were developed, and empirical evidence was established by extensive systematic laboratory tests.[2] Around 1990 methods for calculation of many different constructional elements for different loads – and details such as anchorage of reinforcement, and the impact of casting joints – had been developed.

As mentioned, concrete is rather far from being ideally plastic, but corrected by a so-called "effectiveness factor", the results obtained with plasticity based methods are reasonable near to what is correct. By making the rather "wild" assumption of plasticity perform rationally, effectiveness factors has a role as a deus ex machina. There are rational arguments for using the effectiveness factors, but only in the case of pure bending can they be explained analytically. In the period from 1967 to about 1990 effectiveness factors were determined empirically for different types of load and different types of constructional elements (Andreasen, 1994).

So, in a period covering most of the 20th century – but in the first decades with very scattered activity – a series of developments of structural design methods for dimensioning concrete elements took place, and a sort of engineering-scientific R&D programme was formed.

Apart from the extensive use in Denmark, methods based on plasticity theory have been used in Switzerland since the mid-sixties, but there has been great reluctance towards using such methods in other countries.[3]

In the line of development, different agents with different interests – research institutes, code commissions, private firms – have taken part. It has comprised several phases and different parallel development processes dominated by different academic, administrative, economic and practical interests. The activities have resulted in different types of products – a considerable amount of academic and non-academic publications, codes of practice, computer programs for calculation and, of course, solutions to practical problems.

An important part of the introduction of new methods is connected with convincing other academic and engineering communities about their merits. For a long time there had been elasticity based methods for calculating concrete elements, and the supporters of the plasticity based methods have had to struggle to

---

[2] In the 50's there was a general increased interest in theories of plasticity. It still remains to be investigated what influence this might have had on the Danish development in question.

[3] In cooperation with Danish engineers an early development of the methods took place in Switzerland. An interesting point is that a development somewhat similar to the Danish development took place in the Soviet Union in the 30's. This work was not known in the West until the translation of A. A. Gvozdev's article in 1960 (Gvozdev, 1960), appeared in International Journal of Mechanical Sciences.

convince academic groups, engineers and firms engaged in designing of the comparative advantages of plasticity based methods.

When one reads scientific papers it is easy to get the impression that obtaining agreement about what seems to be purely technical questions is just a matter of rational arguments and empirical evidence – that the scientific methodology ensures rational technical solutions. In many instances, however, that is not the case. In the process of developing and introducing plasticity based methods for concrete constructions, there have been many manifestations of personal disagreement of a nature which seem very far from exchanges of technical rational arguments.

The opposition has been of different natures and stemmed from different types of agents: academics opposed to changes in paradigms, engineers and engineering firms. And a very important battleground has been the development of the *codes of practice*.

In the early phases, convincing skeptics about the merits of the methods was often related to the practicality of the plasticity based methods – particularly where the use implied economic advantages over other methods.

In addition to a growing number of economically feasible applications, especially two types of codification of the methods have influenced the acceptance of the methods: Implementation of the plasticity based methods in the codes of practice, and implementation of the methods in elaborated computer programs.

Some of the methods became part of the Danish code of practice in 1973. Since around 1985 there have been negotiations about the position and prevalence of the methods in the common European codes (Eurocodes), and since 1992 the methods have been incorporated into the Eurocodes, but discussions and very different types of arguments and interpretations of already negotiated principles are still going on.[4]

Professional programs containing methods for calculating different constructions have existed since the late 80's and extensive construction and extension of programs is still going on.

Apart from different opinions about the comparative advantages based on technical and practical considerations this comprehensive shift in methods (used in such an extensive field of engineering as calculations of structural components) brings with it considerable repercussions. A great number of engineers and teachers, members of code committees, insurance departments etc. have to get familiar with the new methods or even be retrained. Teaching materials have to be changed

---

[4] An interesting aspect of development of engineering knowledge is the diplomatic, and sometimes less diplomatic maneuvers which intertwine with rational, scientific arguments about for example introduction to codes. An analysis of the negotiations – based on written material and interviews – about the incorporations into the Eurocode would give a most valuable insight into the process of developing engineering codes and norms.

etc. The considerable costs implied make up a hindrance for a broader acceptance of new methods.

4 Truth and utility

As mentioned methods based on theory of plasticity have been developed in a period where other methods (for most part based on elastic property of materials) were already in use. The introduction thus appeared in competition with methods which had already shown their merits. So what are the convincing advantages of the method?

In many cases the advantages of new methods are due to a superior theoretical foundation – new theoretical insights or better utilization of existing theoretical knowledge. That is not the case for this program. The advantages are apparently not related to questions about the theoretical foundation or truth.[5] The advantages are obviously more related to questions about utility.

One decisive type of criteria for utility concerns safety – the demand that results should be reliable. For instance, that we can be certain (to a reasonable degree) that the carrying capacity of a structural element calculated by use of a particular method is sufficient.

Other types of demands are about optimality and effectiveness – when using the methods the results should be economically competitive. For example, that not too much work or material should be used because the carrying capacity is overstated. And it means that not too much of the structural designer's time should be spent on design when using the methods. A convincing argument for the programme has been that economic advantages, not least savings of materials, in some cases are considerable.

But of course truth and utility are related. If the methods for design were true in a scientific sense and used perfectly correct, then both the criteria about safety and the criteria about optimality would be fulfilled.

Engineering knowledge will, however, most often not meet the scientific demands for truth. In many areas of engineering it is more often the rule than the exception that the validity of the used knowledge is uncertain and the theoretical understanding behind it is only partial. The case illustrates this nicely: The central

---

[5] Although truth in a scientific sense is not decisive in engineering science, it is regarded as a value probably rather analogous to the meaning of "values" in T. S. Kuhn's theory of paradigms (Kuhn, 1970). An illustration of this point is that several attempts have been made to find satisfactory theoretical explanations of effectiveness factors, but only in the case of pure bending has an analytical explanation been given (Andreasen, 1989, p. 94). This also illustrates the point that the lack of theoretical understanding has not impeded the practical use.

assumption that concrete is plastic (or even ideally plastic) is far from being true. The effectiveness factor is only empirically validated. And the assumption that if corrected with an effectiveness factor, concrete can, inside given limits, be treated as being perfectly plastic, is rather obscure.

But even apart from this, the theoretical foundation for the methods is defective. Thus, it is not possible to give an adequate theoretical explanation of fundamental parts of the theoretical background: the aforementioned extremal principles or the relation between the deformation of a body and the inner forces – the so called constitutive equations.

Additionally, methods for solving practical problems often address only the most decisive factors and parameters in the problem situation. Practical problem situations are normally so complex that it would be practically impossible to deal with all aspects. That means that less decisive factors are neglected and that complex, real objects and features are approximated to more simple ones. In this case phenomena such as creep and hysteresis are neglected. Also, a lot of approximations about ideal geometry, homogeneity of materials, linearity of relations etc. are made. What are the decisive and what are the less important factors? And what approximations are permitted? These questions are often decided by valuation resting on experience or tradition.

So, both the theoretical indeterminateness and the uncertainty of valuations influence predictions about the validity of a method. Say, for designing structural elements. That means that validity of the methods is confined to reliability for well-known cases and only inside the limits of empirical trial. As is a common condition for use of engineering knowledge and methods, it means that also safety and optimality are only known within these limits.

There are advantages by using methods based on the assumption of plasticity. According to the proponents the calculations give reliable results. They claim that their methods to a higher degree than the elasticity based methods attack what is really the problem, and thus give more optimal results. They are easy to use and what is important, the way the calculations proceed is intuitively understandable. Additionally, you can to a certain degree decide how much accuracy you want, how many repeated calculations you make to reach a certain degree of convergence – and thus how much effort you have to put into the calculation.

Furthermore, the proponents claim that the methods are shown to be reliable when used inside well known limits. The use of both upper and lower limit calculations (based on different theoretical assumptions) of course adds to the reliability. And, as the calculations can be repeated, it is possible to get upper and lower bound solutions to converge. To some extent this will permit the user to choose the degree of accuracy and thus the degree of both reliability and optimality. Finally, the methods are for many cases relatively easy to use, giving calculations which are normally rather simple.

One further quality of methods based on plasticity is that the calculations are open to the designer's intuition. When calculating the strength of a structural element using methods based on plasticity, the engineer has to imagine the most probable line for a possible rupture. Knowing this and knowing the strength of the material, it is possible to calculate the amount of the force necessary for failure along the line. The structural engineer is thus compelled to imagine and understand the mechanism of collapse of the construction. "Extremal principles possess, besides their relative simplicity the fascinating quality that the result depends on the designer's imagination and ability to familiarize himself with the mode of operation of the construction" (Sandbye, 1977, my translation).

Another quality in connection with the use of methods based on plasticity theory is that it is possible – and necessary – intuitively to understand the methods and follow how a calculation progresses. We have argued that it is a common condition for engineering work that the reliability of methods and knowledge used is known only inside the limits of empirical trials. As engineering work – apart from trivial cases – is most often outside the limits of empirical trials this intuitive understanding is a great advantage. As one of the researchers in developing the programme wrote: "by the use of the theory of plasticity it is possible to avoid gross blunders, because the theoretical considerations automatically show prospective upper limits and the like" (Andreasen, 1989, p. 207).

Throughout the development of the programme of plasticity based methods, there has been an outspoken appraisal of the intuitively understandable – good common sense – against the too complicated and unduly theoretical approach. It is probable that the future will turn out more scientifically correct methods and perhaps also methods with wider limits for their reliability. It is very likely that such methods will be more complicated and less intuitively understandable. The above mentioned researchers praise the simplicity of the plasticity based methods: "in the future in the case of concrete structures, the calculations based on the theory of plasticity will probably be replaced by other methods, where the concrete is described more accurately. These calculations must undoubtedly be carried out numerically by means of a computer. Nevertheless, the solutions from the plastic calculations *and other simple methods* [my italics] will still be indispensable as a control for the results from the numerical calculations" (Andreasen, 1989). And, it is very likely that methods based on plasticity will be used as an easy, reliable second pencil and paper-check to more sophisticated and nontransparent methods.

References

Andreasen, B. S. (1989). Anchorage of Ribbed Reinforced Bars, *Serie R, 238*, Afdelingen for Bærende Konstruktioner, Danmarks Tekniske Universitet.

Andreasen, B. S. (1994). The effectiveness factors for concrete strengths, *Bygningsstatiske Meddelelser* **65**(2–3–4).

Gvozdev, A. A. (1960). The determination of the value of the collapse load for statically indeterminate systems undergoing plastic deformation, *International Journal of Mechanical Sciences* **1**.

Ingerslev, A. (1921).   Om en elementær Beregningsmaade af krydsarmerede Plader. [About an elementary method for designing crosswise reinforced slabs], *Ingeniøren* **69**.

Johansen, K. W. (1931). Beregning af krydsarmerede Jernbetonpladers Brudmoment. [Calculating Failure Moment for crosswise reinforced Concrete Slabs], *Technical report*, Bygningsstatiske Meddelelser, Dansk Selskab for Bygningsstatik, Copenhagen.

Johansen, K. W. (1943). *Brudlinieteorier. [Theories of yield lines]*, Polyteknisk Forening, Copenhagen.

Kuhn, T. S. (1970). *Structure of Scientific Revolutions*, second edn, University of Chicago Press, Chicago.

Sandbye, P. (1977). Plasticitetsteori 1. Kontinuumsmekanik for begyndere. [Continuum mechanics for beginners], *Technical report*, Danmarks Ingeniørakademi.

Nomological machines in science and engineering

Frederik V. Christiansen[1] and Camilla Rump[2]

[1] Institute for Medicinal Chemistry, Danish University for the Pharmaceutical Sciences
[2] Centre for Science Education, Faculty of Science, Copenhagen University

> *Det er i det hele taget sundt at konstruere sådan,*
> *at ens forudsætninger opfyldes så godt som muligt.*
>
> K. W. Johansen[3]

1 Introduction

What is engineering? What is technology? What is the role of science in engineering and technology?[4]

Engineering comprises extremely heterogenous activities, objects, theories, methods, values and knowledge forms. Attempts to give short, clear cut answers to the above questions are therefore almost bound to be problematic. At most one can hope to come up with more or less appropriate *metaphorical* answers to such questions. That means that answers will be only partial – they will never convey the full complexity that reality provides.

What a metaphorical answer *can* do is to highlight some aspects while downplaying others. This is both the strength and the danger in the use of metaphors. In so far as the features highlighted are "real", the metaphor may enable us to see things clearly that were previously obscure. But metaphors can also be misleading. Their apparent clarity shape our understanding and, in turn, lead our actions. Metaphors may make us disregard important differences in the world.

---

[3] "In general it is healthy to construct in such a manner that ones conditions are fulfilled as closely as possible." K. W. Johansen was professor in Statics at the Technical University of Denmark (1950–1972). The quote is from (Johansen, 1979, p. 4).

[4] We shall use the term *science* as denoting primarily *physics*. This is certainly objectionable, but in accordance with the general tradition in philosophy of science and technology where "the laws of nature" provided by physics has stood as the prototype of science.

The best way to avoid this is perhaps to possess an arsenal of metaphors, so that one metaphor may be countered with another. In the present essay we shall contribute to the establishment of such an arsenal by describing the conception of science provided by Nancy Cartwright, and discuss implications for our understanding of engineering and technology.

We will start by outlining two other (more frequently encountered) metaphors of engineering, and then move on to present Cartwright's conception of science with appropriate generalizations. We will discuss engineering examples in the light of this view, and discuss the implications for the understanding of the use of science in engineering.

2 Engineering as applying science

In the extensively used textbook on university physics by Alonso & Finn, originally from 1969 with a second edition from 1980, the authors discuss the relation of Physics to the other sciences:

> We indicated . . . that the objective of physics is to enable us to understand the basic components of matter and their mutual interactions, and, thus, to explain natural phenomena, including the properties of matter in bulk. From this statement we can see that physics is the most fundamental of all natural sciences. Chemistry deals with one particular aspect of this ambitious program: the application of the laws of physics to the formation of molecules and the different practical means of transforming certain molecules into others. . . . The application of the principles of physics and chemistry to practical problems, in research and development as well as in professional practice, has given rise to the different branches of engineering. Modern engineering practice and research would be impossible without a sound understanding of the fundamental ideas of natural sciences. (Alonso and Finn, 1980, p. 10)

The view presented above, and especially the special status afforded to physics, represents a special kind of physicalism which has been very influential in, at least, the 20th century. It is a view where the laws of physics are taken to be universal statements from which it is possible to deduce a host of specific consequences. Such views of physics were held by, among others, the early logical positivists and also, to a degree, by Karl Popper. The related view that engineering is mainly concerned with "applying science" in the production of technology has been widespread in the positivist era and has also been advocated philosophically, for instance by Mario Bunge (1966). Many relevant reasons can and have been given for rejecting this position (see for instance Mitcham, 1994, pp. 199–204). We may label this metaphor as "Engineering as applying science".

3 Engineering as designing

In practical problem solving (such as engineering work) science often plays a rather limited role, as the problems dealt with are not scientific in nature, but must be recognized and delimited with respect to a host of different types of considerations. Therefore engineering and technology cannot be grasped when conceived of as bare application of scientific methods and knowledge. Donald Schön formulates this sharply, as a dilemma between "rigor or relevance" – the rigorous methods provided by science are often unable to deal with the problems encountered in professional practice:

> In the varied topography of professional practice, there is a high, hard ground where practitioners can make effective use of research-based theory and technique, and there is a swampy lowland where situations are confusing "messes" incapable of technical solution. The difficulty is that the problems of the high ground, however great their technical interest, are often relatively unimportant to clients or to the larger society, while in the swamp are the problems of greatest human concern. Shall the practitioner stay on the high, hard ground where he can practice rigorously, as he understands rigor, but where he is constrained to deal with problems of relatively little social importance? Or shall he descend to the swamp where he can engage the most important and challenging problems if he is willing to forsake technical rigor? (Schön, 1983, p. 42)

Schön advocates what he calls an "epistemology of practice", wherein the process of *designing* becomes the crucial element, and the role of science and the scientific method is downplayed. Schön is certainly right that the role of scientific methods and science with respect to practical subjects such as engineering has often been overemphasized. However, there can be no denying that scientific knowledge increasingly plays a role in much engineering work – for instance in such fields as construction, biotechnology, nanotechnology and computer science.

4 Nomological machines

We think there is a need for new ways of conceiving of engineering where the engineering activity is, on the one hand, not reduced to the application of scientific principles, but where, on the other hand, the crucial role of science in engineering is acknowledged. We shall argue that it may be possible to arrive at such a position by generalizing Nancy Cartwright's view of science as presented in her book *The Dappled World*. This will provide us with a metaphor which could be formulated as "Engineering as the design and innovation of nomological machines".[5]

---

[5] Admittedly, this does not sound as catchy as the above-mentioned metaphors.

Cartwright's notion of nomological machines pertains in particular to science, only partially to engineering.[6] What Cartwright opposes is the view of science as a system of universal laws connected in a deductive system, normally ascribed to the logical positivists (and Carnap in particular). On Cartwright's view the laws of physics have no claim to universality, in the sense that they are applicable to nature "in itself". Rather, they describe the world through particular models, and such models are always limited in scope. That is what is aimed at by the metaphor of the world being *dappled*: there are small areas of light where things are well understood and a lot of darkness in between.

Cartwright argues that our most basic scientific knowledge is not about "detached" laws, but about *capacities* and *natures* of things, that is, how things behave (or tend to behave) in specific contexts.[7] Roughly, Cartwright argues that physical laws are only valid for, and should be understood with respect to, the behaviour of *nomological machines*. Cartwright describes such machines as follows:

> What is a nomological machine? It is a fixed (enough) arrangement of components, or factors, with stable (enough) capacities that in the right sort of stable (enough) environment will, with repeated operation, give rise to the kind of regular behavior that we represent in our scientific laws. (Cartwright, 1999, p. 50)

The archetypal nomological machine is the physical experiment in the laboratory. In such experiments measurements are made upon samples that are quite unlike our normal objects of experience: They are specially prepared samples which as closely as possible meet the demands of our theoretical models.[8] A specific example could be Galileo's description of the inclined plane experiment in "Two New Sciences":

> A piece of wooden moulding or scantling, about 12 cubits long, half a cubit wide, and three finger-breaths thick, was taken; on its edge was cut a channel a little more than one finger in breath; having made this groove *very straight, smooth, and polished*, and having *lined it with parchment, also as smooth and polished as possible*, we rolled along it a *hard, smooth and very round* bronze ball. Having placed this board in a sloping position ... we rolled the ball ..., noting ... the time required to make the descent. We repeated this experiment more than once in order to measure the time

---

[6] Although many examples discussed by Cartwright are engineering examples rather than scientific examples.

[7] To give justice to this claim would require a lengthy argument. We will not provide this here, but refer the reader to Cartwright's main text on capacities: *Nature's Capacities and Their Measurement* (Cartwright, 1994).

[8] See also (Pedersen and Jakobsen, 2003) for a discussion of this.

with an accuracy such that the deviation between two observations never exceeded one tenth of a pulse-beat. Having performed this operation and having assured ourselves of its reliability, we now rolled the ball only one quarter of the length of the channel; and having measured the time of its descent, we found it precisely one-half of the former. [Our italics] (Galilei, 1638, p. 179)

Indeed, it is *not* a universal feature of the world that falling objects obey Galileo's law of falling bodies ($s \propto t^2$). Rather, the motion described by Galileo's law is something that can be brought to occur in the inclined plane experiment, or even more accurately in a vacuum.

The quote is interesting because is displays the importance of the *shielding conditions* for the function of the nomological machine. Such shielding mechanisms are absolutely crucial for the working of the machine; without them the laws of physics "won't work". The experimental setup serve, among other things, to eliminate external factors of disturbance so it is possible to *reproduce* the causality in question.

In many ways Cartwright's view can be paralleled with Aristotle's. That is true not only for her discussion of capacities, but also with respect to Aristotle's distinction between the sublunar and superlunar motions. According to Aristotelian dynamics, entities below the lunar sphere are subject to such effects as "generation", "destruction", "increase", "diminution" and "violent motions" in addition to their rectilinear natural motion (Aristotle, 1971). In short, the sublunar entities are subjected to a constant flux. The celestial bodies on the other hand experience no such flux, as there is no contrast to their perfect circular motions. For this reason, it is possible to describe accurately (i.e. mathematically) the motions of the celestial bodies, whereas this – because of the constant flux – is much more difficult with our earthly affairs. This account has been scorned quite a bit over the last several hundred years, but it has to be admitted that there's some truth to it. The solar system actually provides a system of dynamics which is extremely stable and unusually undisturbed. It is a nomological machine, peculiar in the respect that it is not man-made. It is perhaps for this reason that astronomy, unlike "terrestrial" physics, has been paradigmatic (in the sense that there has been a basis for "normal science"), for nearly 2000 years. Ian Hacking makes the same point in *Representing and Intervening*:

Outside of the planets and stars and tides there are few enough phenomena [i.e. observable regularities] in nature, waiting to be observed. ... Every time I say there are only so many phenomena out there in nature to be observed – 60, say – someone wisely reminds me that there are some more. But even those who construct the longest lists will agree that most of the phenomena of modern physics are manufactured [in experimental setups] (Hacking, 1983, pp. 227–28)

It is because regularities are generally hard to come by that nomological machines become so crucial to science; by means of nomological machines regularities come into the world:

> Ian Hacking is famous for the remark, 'So far as I'm concerned, if you can spray them then they are real.' I have always agreed with that. But I would be more cautious: '*When* you can spray them, they are real'. . . . Hacking's point is not only that when we can use theoretical entities in just the way we want to produce precise and subtle effects, they must exist; but also that it must be the case that we understand their behavior very well if we are able to get them to do what we want. That argues, I believe, for the truth of some very concrete, context-constrained claims, the claims we use to describe their behavior and control them. But in all these cases of precise control, we build our circumstances to fit our models. I repeat: that does not show that it must be possible to build a model to fit every circumstance. Perhaps we feel that there could be no real difference between the one kind of circumstance and the other, and hence no principled reason for stopping our inductions at the walls of our laboratories. But there is a difference: some circumstances resemble the models we have; others do not. And it is just the point of scientific activity to build models that get in, under the cover of the laws in question, all and only those circumstances that the laws govern. Fundamentalists want more. They want laws; they want true laws; but most of all, they want their favorite laws to be in force everywhere. I urge us to resist fundamentalism. Reality may well be just a patchwork of laws. (Cartwright, 1999, p. 30)

5 Objects, nomological machines and model objects

Even in as simple an experiment as Galileo's inclined plane, surprisingly much idealization is needed in order for the experiment to be in accordance with the theoretical predictions. In fact, no matter how much effort is put into the experiment, the plane will never be completely smooth nor the ball completely round. Therefore there will always be differences between the ideal situation (that is, the theoretical model) and the actual situation. Ernst Cassirer puts this elegantly in his criticism of Ernst Mach's nominalism:

> . . . any attempt to interpret the concepts of natural science as mere aggregates of facts of perception must necessarily fail. No scientific theory

is directly related to these facts, but is related to the ideal limits, which we substitute for them intellectually. We investigate the impact of bodies by regarding the masses, which affect each other, as perfectly elastic or inelastic; we establish the law of the propagation of pressure in fluids by grasping the concept of a condition of perfect fluidity; we investigate the relations between the pressure, temperature and volume of gas by proceeding from an "ideal" gas and comparing a hypothetically evolved model to the direct data of experience. (Cassirer, 1910/1953, p. 143)

Thus, scientific knowledge does not describe nature as it is or appears to us. Indeed, knowledge as described by the laws of nature goes beyond anything which may be "given in experience". The laws of science are not metaphysical entities governing and constituting reality, they are rather *regulative ideas* in Kant's sense, that is, ideas which *transcend* what is given in experience, but play a crucial part in our making order of phenomena. Kant argued that the fundamental laws of mechanics were *a priori* and constitutive of our conception of phenomena (in virtue of the close connection between the laws of mechanics and the categories of substance, causality and interaction). Thus, these laws had precisely the universal and metaphysical grounding that Cartwright objects to, and it is not unfair to say that history has proven Kant wrong on this point. But Kant also argued that many other sciences did not have and could not have such a "secure foundation", for instance biology and chemistry. In these sciences we have to rely upon ideas of reason that, like the Platonic ideas, transcend the phenomena. For instance, many judgments in biology rely upon the idea of teleology. But this idea, Kant claims, has no constitutive function with respect to objects of experience. That does not make the idea useless, indeed it is a necessary idea if we are to make sense of living beings. But it does not describe the world "as it is". As Kant puts it, such transcendental (regulative) ideas

have an excellent, and indeed indispensably necessary, regulatory employment, namely that of directing the understanding towards a certain goal upon which the routes marked out by all its rules converge, as upon their point of intersection. This point is indeed a mere idea, a *focus imaginarius*, from which, since it lies quite outside the bounds of possible experience, the concepts of the understanding do not in reality proceed. (Kant, 1781, A645/B673)[9]

---

[9] Kant gives a relevant example and elaboration: "These [regulative ideas] are not derived from nature; on the contrary we interrogate nature in accordance with these ideas, and consider our knowledge as defective so long as it is not adequate to them. By general admission, *pure earth, pure water, pure air* etc., are not to be found. We require, however, the concepts of them (though, in so far as their complete purity is concerned they have their origin solely in reason) in order properly to determine the share which each of these

Figure 1 shows how the Kantian/Cartwrightian view might be illustrated. The figure distinguishes two "levels of being": the empirical and the transcendental level. The empirical level consists (mainly) of the objects we encounter, distinguish and relate to in our daily lives (ordinary objects). Nomological machines are likewise empirical objects, but they are of a special kind constructed so as to resemble our theoretical constructs as closely as possible. At the transcendental level we have our theoretical constructs and models (regulative ideas). Pedersen and Jakobsen (2003, p. 4) suggest the name "Model objects" for these abstract conceptual entities and argue that these are "the proper tagets for scientific theories". Nomological machines do not "reflect" precisely objects of experience, nor do model objects "reflect" precisely the functioning of nomological machines (as the Galileo example shows).

Cartwright argues that what we here term model objects function as "blueprints for nomological machines", but that these blueprints do not tell us in any precise way how to build a nomological machine (Cartwright, 1999, p. 58). The term blueprint is perhaps somewhat misleading, because the very function of a blueprint is to facilitate and regulate the construction of the artifact in question. That is not always the case with model objects, as can be seen from the fact that often model objects describe situations which cannot be realized physically (e.g. the Carnot engine). So model objects are more than blueprints, but may of course have a regulative function with respect to the construction of nomological machines. The important point is that model objects do not correspond directly to nomological machines (and even less to "ordinary" objects of experience).

The relationship between model object and nomological machine is therefore more ambiguous than the relation between blueprint and artifact. Still, model objects are in fact useful in the understanding of empirical objects and processes and in the process of construction of artifacts. In general we may say, that there is not identity, only analogy between model object and nomological machine.

6 Engineering use of nomological machines

In the light of the described conception of science, how should we make sense of the role of science in engineering and technology? Let us discuss a couple of examples.

---

natural causes has in producing appearances." (Kant, 1781, A646/B674). Pure earth etc. refers to the antiquated chemical concepts of Kant's time, but that doesn't change the basic idea: The regulative ideas of substances in their pure form help us to determine their share in compounds. But such pure forms do not correspond to what is found in nature.

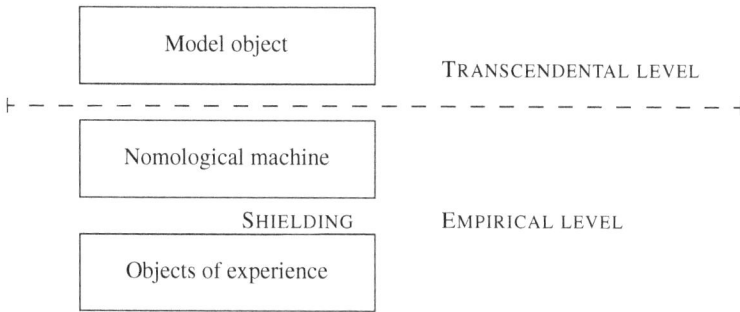

Fig. 1. Illustration of the "levels of being" described by the ideal model objects and the empirical nomological machines, objects of experience. Relations of analogy exist among the transcendental and empirical levels. The figure is an adaptation from (Christiansen and Thingstrup, 1999).

6.1 Examples of engineering nomological machines

In 1976 the building of the 59 story Citibank corporation headquarter in New York was commenced. It was designed by William LeMessurier, a renowned structural engineer with extensive experience with skyscrapers. The design was unusual due to a peculiar constraint on the building site: The building should allow for a free-standing building (a church) on the one corner of the block. Thus, the Citicorp building rests on nine-story stilts, with the church underneath. Moreover, as the church had to be on the corner of the block, the stilts are placed in the middle of each of the buildings four sides, rather than on its corners. In order for this to work statically, the building is equipped with diagonal girders which transfer the weight of the building to vertical columns which run throughout the building (see figure 2). Now, this is the basic design. However, the lightweight design provided by the system of girders would, on its own, make the building likely to sway in the wind. Therefore, the top of the building is supplied with a mechanical damper – a 400 ton concrete block with springs attached to it, floating on pressurized oil bearings (Whitbeck, 1998, p. 148).

Certainly, there is a lot of physics – specifically mechanics – involved in this construction. The basic design with the stilts and diagonal girders could be said to constitute the basic nomological machine. The overall function of the machine is to provide a static construction, given the peculiar constraints on the site. Had there been no wind (or other uncontrolled forces) there would be no need for mechanical damping as there would be no oscillations to dampen. The mechanical damping of the building, therefore, can be understood as a shielding mechanism – a mechanism that serves to ensure that the basic function of the nomological machine is not

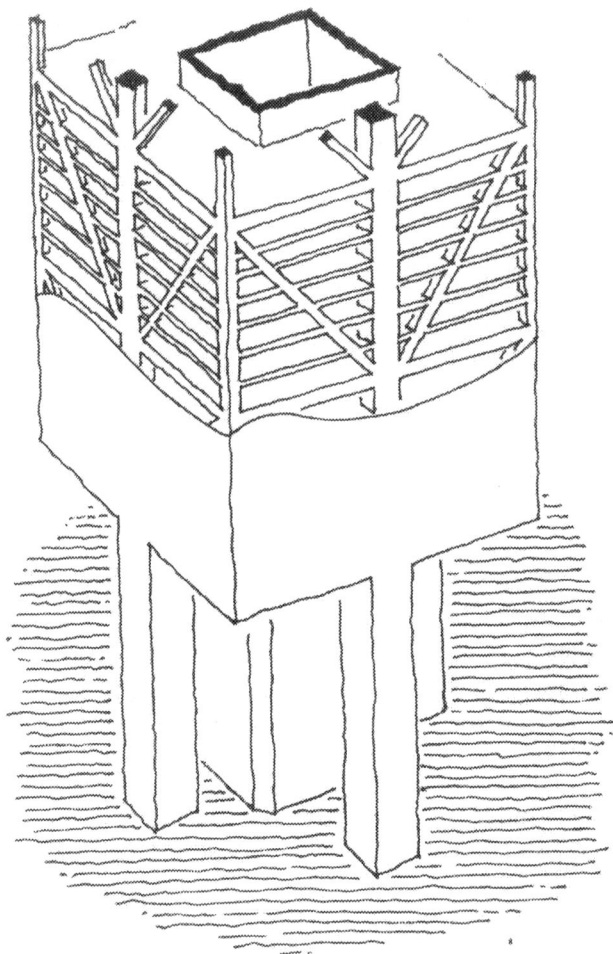

Fig. 2. Sketch of the Citicorp structure. Illustration by Per Galle.

disturbed by the wind – a factor which is not well described by our physical theory. The wind's influence on the building was estimated by scale model/wind tunnel experiments and statistical wind data (Morgenstern, 1995, p. 47).

Similarly, the steel lattice that made up the facades of World Trade Center 1 & 2, were, in effect, wind braces, designed so that the respective central cores of the Towers took only the gravity loads of the building (Heyer, 1993, pp. 194–195).

To take a less spectacular example, let us consider microprocessors. Any modern computer has a microprocessor chip (or several). It is safe to say, that without extensive scientific knowledge such chips could never have been developed. Fabrication of such chips require extremely controlled conditions – cleanroom facilities. Once inside the computer and brought into operation, a different kind of shielding becomes needed. Modern microprocessors produce *a lot* of heat (typically in excess of 50 Watt). Without a CPU cooler a modern microprocessor will burn out in a matter of seconds.

These examples show, that shielding mechanisms play an even more central role in engineering design than it does in science, because the nomological machines devised by engineers do not have the "natural" shielding provided by the laboratory. "Engineering nomological machines" have to withstand not only the competent use by careful physicists, but also relatively careless use by lay persons and uncontrolled forces of nature not shielded by the confines of the laboratory.

Engineering design often starts with the development of a prototype. A prototype of a technological artifact is often a nomological machine which is not sufficiently shielded to be used in more general contexts of use. Bruno Latour vividly describes the development of the Diesel engine, from its conception as an idea of a machine with constant temperature ignition based on Carnot's principles, to the cumbersome development of a prototype working with constant pressure ignition (an entirely different physical principle). The former is a model object, the latter a nomological machine. This machine was highly unstable, however, and never really came to work outside the laboratory facilities. It was not until more than ten years later that a stable product had been created (Latour, 1987, pp. 104–108). In many cases, a crucial part of the *process of innovation* consists in getting from a nomological machine that works in the laboratory (a prototype) to a nomological machine that works outside of the laboratory (a product).

7 Conclusion

We have argued, that at least some parts of engineering are involved in the production of nomological machines, that is, technology which relies upon relevant shielding in order for scientific knowledge to come into play. Engineering and science activities are alike in this respect. What remains to be discussed is the specific role of nomological machines in science and engineering, respectively.

As is rightly stressed by Bunge (1966), there is a crucial difference in the overall goal of science and engineering. Science has a cognitive goal; that of enhancing and expanding our scientific knowledge. The nomological machines created by physicists in the laboratories serve a cognitive function, that of enhancing our (theoretical) knowledge. Thus, for instance, the Aspect experiment can be considered a nomological machine which served the function of physically testing the completeness of the theory of quantum mechanics (Aspect et al., 1982).[10] Roughly we may say that *in science, nomological machines serve as a means to gain insight into the nature of model objects* (see figure 1).

The primary goal of engineering is to solve practical problems. This often involves creating nomological machines considered useful with respect to fulfilling this goal. As nomological machines are often to be used under extreme conditions, by diverse agents with diverse purposes, the question of devising appropriate shielding mechanisms comes far more into focus in engineering than in science. The process of innovation, crucial to much engineering work, often consists in making nomological machines less "lab-like", that is, *making nomological machines robust enough to withstand encounters with the world outside the laboratory*. This is sometimes done by securing "lab-like" conditions in relevant spatial areas of the technology.

Unlike in science, the question of how model objects may be turned into stable nomological machines is fundamental to much engineering. In engineering, unlike in science, the nomological machines are ends in themselves, not means for understanding model objects.

References

Alonso, M. and Finn, E. J. (1980). *Fundamental University Physics*, Vol. 1, 2 edn, Addison-Wesley Publishing Company, USA.

Aristotle (1971). On the heavens, *in* J. Barnes (ed.), *The Complete Works of Aristotle*, Vol. 1, Bollinger, Princeton, pp. 447–511.

Aspect, A., Grangier, P. and Roger, G. (1982). Experimental realization of Einstein-Podolsky-Rosen-Bohm Gedankenexperiment: A new violation of Bell's inequalities, *Physical Review Letters* **49**: 91–94.

Bunge, M. (1966). Technology as applied science, *Technology and Culture* **7**: 329–347.

---

[10] The experiment was based on a thought experiment by Einstein, Podolsky & Rosen (Einstein et al., 1935).

Cartwright, N. (1994). *Nature's Capacities and Their Measurement*, reprint edn, Oxford University Press.

Cartwright, N. (1999). *The Dappled World*, Cambridge University Press, United Kingdom.

Cassirer, E. (1910/1953). *Substance and Function*, Dover Publications, Inc. English translation and publication 1923 (with Einsteins Theory of Relativity), Dover republication 1953.

Christiansen, F. V. and Thingstrup, T. (1999). *Erfaringsobjekter i kvantemekanik*, Master's thesis, Roskilde University, Dept. for Philosophy and Science Studies, Denmark.

Einstein, A., Podolsky, B. and Rosen, N. (1935). Can quantum-mechanical description of physical reality be considered complete?, *Physical Review* **47**: 777–780.

Galilei, G. (1638). *Two New Sciences*, Dover Publications, Inc. English translation by Henry Crew & Alfonso de Salvio, first published 1914 by The MacMillan Company.

Hacking, I. (1983). *Representing and Intervening - Introductory Topics in the Philosophy of Natural Science*, Cambridge University Press, Cambridge, UK.

Heyer, P. (1993). *Architects on Architecture: New Directions in America*, Van Nostrand Reinhold.

Johansen, K. W. (1979). *Hvad laver vi ingeniører egentlig?*, Dansk Selskab for Bygningsstatik, Lyngby. Special publication.

Kant, I. (1781). *Critique of Pure Reason*, Macmillan Press, London. English translation by Norman Kemp Smith, first published 1929.

Latour, B. (1987). *Science in Action*, Harvard University Press.

Mitcham, C. (1994). *Thinking Through Technology: The Path Between Engineering and Philosophy*, University of Chicago Press, USA.

Morgenstern, J. (1995). The fifty-nine-story crisis, *The New Yorker* **May 29**: 45–53.

Pedersen, S. A. and Jakobsen, A. (2003). Engineering science and reality, *Preprints and reprints* **3**. Published by Department for Philosophy and Science studies, Roskilde University, Denmark.

Schön, D. (1983). *The Reflective Practitioner*, Basic Books, New York.

Whitbeck, C. (1998). *Ethics in Engineering Practice and Research*, Cambridge University Press, Cambridge, UK.

Philosophy of medicine in the late 20th century – an evolution from order to chaos?

Søren Holm

Cardiff Law School, Cardiff University
Section for Medical Ethics, University of Oslo

Philosophy of medicine emerged as an international field of academic activity in the late 1960s and early 1970s when international academic societies and journals specifically focusing on this area were established.

During the history of philosophy and medicine we can identify prior periods where there has been significant interaction between these two areas. Plato used many medical examples in his writings, Hippocratic and Galenic medicine were based on philosophical ideas about nature and the human body, the ideas of Paracelsus[1] influenced both medicine and philosophy (Fisher, 2003; Ebbesen and Koch, 2003), many famous doctors have written works on the philosophical underpinnings of medicine, and there was a thriving Polish school of philosophy of medicine in the early part of the 20th century (Brorson, 2000; Lowy, 1990).

The most recent resurgence of philosophy of medicine is thus part of a larger history, but in this paper I will focus on the developments since approximately 1970. I have decided to concentrate on this period primarily because this is the period where Stig Andur Pedersen has been active in the field, but also because the rise of modern philosophy of medicine was to a large extent unrelated to the earlier interactions between philosophy and medicine in the sense that no unbroken history can be traced between the periods and places where philosophy of medicine became prominent.

The analysis will focus on 1) the reasons for the rise and subsequent fall of philosophy of medicine in relation to clinical medicine, medical research methodology and medical ethics or bioethics, and 2) some major topic areas in philosophy of medicine.[2]

---

[1] Teophrastus von Hohenheim.

[2] Because of my own linguistic limitations the analysis will focus on developments in the Nordic countries and the English and German speaking parts of the world.

1 The rise and fall of philosophy of medicine

The rise of philosophy of medicine began in the late 1960s and lasted until the mid-1980s.

The topics covered in this early phase were very varied, and there was no clear demarcation between philosophy of medicine and medical ethics. If we look at the first years of the journals "Journal of Medicine and Philosophy" and "Theoretical Medicine" (later renamed "Theoretical Medicine and Bioethics") we find articles on "The image of man in medicine" (Lieb, 1976), "A meliorist view of disease and dying" (Lewis, 1976), "The concept of disease" (Margolis, 1976), "Causality and medicine" (Agassi, 1976), "The outcome as cause: Predestination and Human Cloning" (Eisenberg, 1976), "Rational justification for therapeutic decisions" (Card, 1980), "A systems-tensorial interpretation of psychomedical concepts" (Houghton, 1980), "Medicine under capitalism" (Kröner, 1980), and "Omissions and other negative actions" (Walton, 1980).

Some of these topics have disappeared from the agenda today, e.g., "Medicine under capitalism", but some are still discussed.

At the 2005 congress of the European Society for Philosophy of Medicine and Health Care (ESPMH) there was, for instance, presentations on the concept of disease, on epistemology and research methodology and on causality in medicine, alongside many, many papers on ethical issues.

The main reason for the rise of philosophy of medicine was probably that medicine itself was in flux. There was a rapid rise in number of medical doctors, in effective treatments and in research output. All of these developments in medicine called for a response, and philosophy of medicine was just one of a number of areas that benefited from this (another example is medical sociology).

An important publication which marked the coming of age of philosophy of medicine was the book "Philosophy of Medicine" published by Blackwell in 1986 and in Danish as "Medicinsk filosofi" in 1990. This book written by Henrik R. Wulff, Raben Rosenberg and Stig Andur Pedersen (a gastroenterologist, a psychiatrist and a philosopher/mathematician) is still one of the only textbooks covering a broad range of issues in the philosophy of medicine. At the same time as this book marks the coming of age of philosophy of medicine, its publication can in retrospect be seen to coincide with the beginning of the decline of the subject as well.

The early phase of the Philosophy of medicine was characterised by a fruitful interaction between philosophers, clinicians and medical researchers. Both sides thought that they were interested in the same issues, and found the contributions of the other side relevant and interesting. As a result there was often direct engagement in common analysis of the problems.

Early developments of the randomised, controlled clinical trial were, for instance to a significant degree underpinned by primarily Popperian ideas concerning the nature of scientific activities, and philosophical analysis of rational choice played a role in developing medical decision theory.

2 The decline

From the mid 1980s philosophy of medicine has been in decline. In terms of research activity and impact on medical practice and in terms of the perceived importance of the topic in medical education.

The reasons for the decline of Philosophy of medicine are manifold. One reason is, paradoxically, the relative success of medical ethics or bioethics.[3] Both activities seem to occupy the same niche – medical humanities – when seen from the outside, and the outside view is clearly important when it comes to things like research funding or time on the medical curriculum. There are many reasons for the growth in interest in bioethics. One is the great expansion in research ethics regulation and in clinical ethics consultation, first in the USA and later in Europe. Another is, probably, that issues that were previously overtly political or ideological are now discussed in terms of ethics (e.g., resource allocation in the health care system). Whatever the reasons for the increase in interest in ethics, the effect has been that from the early 1980s in the USA, and the mid- to late 1980s in Europe bioethics has gradually crowded out philosophy of medicine in the medical humanities niche,[4] and this has also lead to a gradual reorientation of the main journals and main academic societies.[5] At the annual conferences of the European Society of the Philosophy of Medicine and Health Care there are often discussions of how to maintain the balance between philosophy of medicine and ethics in the conferences, and these discussions always focus on the possibility that the ESPMH may become an almost purely ethics society.

Another reason is probably that medicine and medical research as such is no longer conceptualised as a "hot" and philosophically interesting research area by philosophers. The focus has moved to molecular biology, genetics and genomics,

---

[3] As professor of bioethics in Cardiff and professor of Medical Ethics in Oslo I am hardly in a position to complain about this factor.

[4] There are now attempts to fight back from medical humanities, but in this fight back philosophy of medicine is not highly valued (being perceived as too much like bioethics) and at the front we instead find literature and medicine and similar areas of study.

[5] The journal "Theoretical Medicine" has, as mentioned above been renamed "Theoretical Medicine and Bioethics" and the original American academic society in the field the "Society for Health and Human Values" has merged with a number of other American societies to form the "American Society for Bioethics and Humanities".

and a philosopher who in the 1970s might have pondered the concept of disease or the randomised controlled clinical trial, will today be busy considering the nature of the gene or the relation between heritability and causality.

Finally, the areas of medicine and medical research that had the most fruitful engagement with philosophy have become highly specialised areas in their own right with relatively fixed research paradigms of their own, and therefore less perceived need for philosophical analysis and justification. This is, for instance the case for medical decision making (e.g., see the journal "Medical Decision Making" with highly technical papers on Markov chain modelling and the best measures for health related quality of life), and for clinical research methodology (e.g., see the journal "Clinical Trials").

3 The "concept of disease" debate

A central issue in philosophy of medicine has been a discussion about the proper analysis and understanding of the concept of disease or illness and the concept of health. What do we mean when we say that "XX is ill", that "YY has breast cancer" or "Z is a disease"?

Bjørn Hofmann has recently published an exhaustive review of this discussion and identified more than 80 different analyses (Hofmann, 2001). They can be grouped into three positions:

1. Biological / Realist
2. Action theoretical
3. Non-realist / Nominalist

All of these positions were fully developed at the end of the 1970s, and later contributions have mainly been small revisions of each of the main positions. It is even the same diseases that are used as the main examples in the discussions in the 1970s and today. The discussion thus reached an impasse early on, with none of the positions having a compelling, knock down argument against the other positions.[6]

If one of the positions had won by the compelling force of its justifying arguments, or if medical doctors had noticed one of the few areas of agreement in the discussion, namely that it is not a simple matter to define whether a specific condition should count as a disease or not, the discussion about the concept of disease might conceivably have had some impact on medical practice. But the impasse identified above has made it impossible to wrest medical practitioners from

---

[6] One of the only areas of agreement between the positions is that the World Health Organisation definition of health as "Complete physical, psychological and social wellbeing" is absurd as a definition of health.

their, often rather naïve and unreflective, biological realism. This may not be very important in the core areas of medical activity, but it may well be a problem in "fringe areas" such as public health where it does matter whether we, for instance, conceptualise obesity as a disease or a result of lifestyle choices, and it may also be a problem in discussions about whether the pharmaceutical industry invents new diseases to enlarge the markets for their products (e.g., "attention deficit disorder" for rowdy boys or "social anxiety disorder" for shy people).

This is not to say that philosophical analysis of the concept of disease has had absolutely no impact in clinical medicine. There is one area where a specific analysis of disease has had a major impact. This is in psychiatric diagnosis where the American DSM (Diagnostic and Statistic Manual) diagnostic system in its latest revision, DSM-IV-TR, builds on a set of ontological commitments identified by Sadler as:

> ... six ontological assumptions (assumptions about the way things are) underlying the DSM effort (empiricism, hyponarrativity, individualism, naturalism, pragmatism, traditionalism). (Sadler, 2004, p. 13)

and thereby on a realist analysis of psychiatric disease (Sadler, 2005).

However, the philosophy of psychiatry has, at least partly, split from philosophy of medicine as a separate field with its own leading journal "Philosophy, Psychiatry, & Psychology" that focuses on the area of overlap among philosophy, psychiatry, and abnormal psychology and is not much read by the average person interested in philosophy of medicine.

4 Philosophy of science and philosophy of medicine

One major contribution of philosophy of medicine to medicine has been to introduce concepts from philosophy of science into medicine. Popperian ideas about falsifiability are now part of the common understanding of medical researchers, whether or not their research practice actually aims at falsification (Senn, 1991), and Kuhnian ideas concerning the historical development of science and the occurrence of paradigm shifts have been important in discussions concerning the relation between orthodox medicine and complementary medicine. In the latter discussions a standard move by proponents of complementary medicine is to claim that orthodox and complementary medicine each has its own paradigm, that these paradigms are incommensurable and therefore equally good, and that the paradigm of complementary medicine will transform itself into a new holistic paradigm that will in the end replace the paradigm of orthodox medicine. All of these claims are problematic, but showing that is beyond the scope of this paper.

5 Paradigm shifts in medicine?

One side-effect of the import of Kuhnian ideas in the philosophy of medicine is that people have been avidly looking for paradigm shifts in medicine. This search for new paradigms has either been a purely academic activity, or has been undertaken in order to be able to claim to belong to the new paradigm.

These analyses have often looked for truly fundamental paradigm shifts (like the shift from a geocentric to a heliocentric worldview), and many instances of what could more reasonably be analysed as speciation events have become heralded as paradigm shifts. This has occurred partly because of the mis-perception that anything that can be called a scientific revolution must involve a paradigm shift. But this is clearly a mistake. The invention and introduction of the so-called "gene chip" which allows for the simultaneous analyses of many genes was clearly a scientific revolution, but it did not involve any paradigm shift in the Kuhnian sense. And even the more general turn towards genetics and genomics during and after the decoding of the human genome in the Human Genome Project does not qualify as a Kuhnian revolution.

A better way of understanding the geneticisation of medicine is probably as a fairly fundamental shift in focus, but still constituting an incremental and not a revolutionary change. Many of the research questions are, for instance, still the same, but they are now being investigated using the new genetic tools that have become available.

If we really want nominate a modern development as a candidate of a Kuhnian paradigm shift in medicine the very long process towards a radically evidence based medicine from early uses of statistical methods in the 1800s to the evidence based medicine and meta-analysis movements of today seems to be a better candidate. The basic ideas about what counts as medical knowledge has changed fundamentally in the process, and is still changing (Sweeney, 2006). But because the process has taken so long, and because it has been intertwined with fundamental changes in the way health care is delivered and organised, it does not fit the intuitive idea that revolutions have to be quick, and that may make it more difficult to realise how fundamental the change has really been. In this we have to remember that one of the "classic" Kuhnian paradigm shifts, the shift from a geocentric to a heliocentric cosmology as a matter of historical fact also took a very long time.

6 Conclusion

In this paper I have traced the rise and fall of modern philosophy of medicine. The fall has not been complete, the field has not disappeared, but the decline in interest has nevertheless been quite marked. Is there hope for a resurgence?

Perhaps. We can clearly identify areas of medicine that are in need of philosophical analysis.

One of these is the concept of causality in modern geneticised medicine. What do we actually mean when we say that a specific gene is a risk factor for developing a specific disease? Is this just a way of summarising the results of a statistical analysis of gene-epidemiological data, or does it imply a direct causal role for the gene? It is clear that the justification for statements of this kind are often only statistical, but that they are often understood or misunderstood as causal statements, both by lay people and by experts, partly because of the very simple notions of causation that are still prevalent in medicine.

Another area ripe for philosophical analysis is the concept of evidence in evidence-based medicine (EBM). What are the epistemological commitments of EBM, and to what degree does these commitments lead to the often rather constrained understanding of what counts as evidence that are put forward by EBM proponents?

In both cases the main obstacle to a fruitful engagement between philosophy and medicine will be to convince medical practitioners that what the philosophers have to say is worth listening to, and relevant to medical practice. Physicians are in general rather practical people, who want practical solutions to their problems, not further philosophical complications. But a philosopher who begins an engagement with medicine on these terms will often find that medicine actually opens up questions that are of considerable philosophical interest and, moreover, that even doctors can become interested in these over time.

References

Agassi, J. (1976). Causality and medicine, *Journal of Medicine and Philosophy* pp. 301–317.

Brorson, S. (2000). Ludwik Fleck on proto-ideas in medicine, *Medicine Health Care and Philosophy* **3**(2): 147–52.

Card, W. I. (1980). Rational justification for therapeutic decisions, *Theoretical Medicine* **1**(1): 11–28.

Ebbesen, S. and Koch, C. H. (2003). *Dansk filosofi i renæssancen*, Gyldendal, København.

Eisenberg, L. (1976). The outcome as cause: Predestination and human cloning, *Journal of Medicine and Philosophy* pp. 318–331.

Fisher, S. (2003). Early modern philosophy and biological thought, *Perspectives on Science* **11**(4): 373–377.

Hofmann, B. (2001). Complexity of the concept of disease as shown through rival theoretical frameworks, *Theor. Med. Bioeth.* **22**(3): 211–36.

Houghton, G. A. (1980). A systems-tensorial interpretation of psychomedical concepts, *Theoretical Medicine* **1**(2): 225–247.

Kröner, P. (1980). Medicine under capitalism, *Theoretical Medicine* **1**(2): 249–
250.

Lewis, T. (1976). A meliorist view of disease and dying, *Journal of Medicine and
Philosophy* pp. 212–221.

Lieb, I. C. (1976). The image of man in medicine, *Journal of Medicine and Philosophy* pp. 162–176.

Lowy, I. (1990). Medical Critique [Krytyka Lekarska]: a journal of medicine and
philosophy-1897-1907, *Journal of Medicine and Philosophy* **15**(6): 653–73.

Margolis, J. (1976). The concept of disease, *Journal of Medicine and Philosophy*
pp. 238–255.

Sadler, J. Z. (2004). *Values and psychiatric diagnosis*, Oxford University Press,
Oxford.

Sadler, J. Z. (2005). Social context and stakeholders' values in building diagnostic
systems, *Psychopathology* **38**(4): 197–200.

Senn, S. J. (1991). Falsificationism and clinical trials, *Stat. Med.* **10**(11): 1679–92.
Review.

Sweeney, K. (2006). Personal knowledge, *BMJ* **332**: 129–130.

Walton, D. N. (1980). Omissions and other negative actions, *Theoretical Medicine*
1(3): 305–324.

Wulff, H. R., Pedersen, S. A. and Rosenberg, R. (1986). *Philosophy in medicine:
An introduction*, Blackwell Scientific Publications, Oxford.

Wulff, H. R., Pedersen, S. A. and Rosenberg, R. (1990). *Medicinsk filosofi*, Munksgaard, København.

Does modern neuropsychiatry threaten human values?

Raben Rosenberg[1] and Jakob Hohwy[2]

[1] Centre for Basic Psychiatric Research, University of Aarhus
[2] Department of Philosophy, University of Aarhus

Bit by experimental bit, neuroscience is morphing our conception of what we are. The weight of evidence now implies that it is the brain, rather than some non-physical stuff, that feels, thinks, and decides. That means there is no soul to fall in love. We do still fall in love, certainly, and passion is as real as it ever was. The difference is that now we understand those important feelings to be events happening in the physical brain (Churchland, 2002)

1 Introduction

The provoking question in our title is adapted from the title of an editorial in the prestigious journal *Nature Neuroscience* (1998). The editorial was based on a conference that had been held in Washington D.C. with the bold title "Neuroscience and Human Spirits" and organised by a conservative Washington think-tank with the agenda to reinforce link between Judeo-Christian moral tradition and public policy. The worry expressed at the conference was based on the exponentially growing research in neuroscience during the previous two to three decades that had questioned deeply seated concepts of human nature including belief in free will and moral choice as the basis for moral responsibility, that is, questioned essential human values.

Many of the concerns expressed by the conference participants about the materialistic account of human nature found in modern neuroscience probably express worry and concern in a much broader context. How is it possible to apply the concept of free will in the explanation of human action when a deeper understanding of the functions of the brain convincingly shows the importance of biological factors? As stated in the editorial "it is now clear beyond doubt that many behaviours are to some extent 'determined' by biology" (1998). Thus, people who inherit specific genotypes and who are exposed to certain environments are more likely to engage in criminal or anti-social conduct. Hence, if behavioural choices are constrained by biology, how is it possible to regard them as truly "free"? Related to this: how can it be justified to punish people who are "victims of their biology"?

The editorial seems specifically worried by the fact that a wealth of studies in behavioural genetics have consistently demonstrated the influence of genes on human behaviour, including antisocial behaviour such as crime. It also notes that even religious behaviour may have a significant genetic component, and according to the editorial it is reasonable to ask whether religious belief and ethical precepts just represent evolutionary adaptations.

The purpose of this paper is to discuss some of the concerns expressed in the editorial with a focus on neuropsychiatry, a discipline within psychiatry that applies neuroscientific theories and research methodology. It is argued that modern neuropsychiatry does not threaten human dignity and values. Instead, it helps us understand and accept the limitation of human freedom due to disease processes in the brain. These points emerge when we let neuroscience and philosophy interact, and, accordingly, it is observed that neuroscience should continue its recent close collaboration with philosophy.

2 Neuroscience and psychiatry

Neuroscience developed as a distinct discipline in the late 1950s and early 1960s and has led to a convergence into a unified discipline of the traditional fields of neurophysiology, neuroanatomy, neurochemistry, and behaviour studies (Cowan et al., 2000). In fact, today its scope is much broader as it also includes *cognitive science,* a discipline that attempts to understand the mind in terms of information processing, that is, representational structures in the brain and computational procedures that operate on them (Gazzaniga et al., 1998).

Psychiatry is a branch of medicine that traditionally has employed a wide range of theories and methods from biomedicine, psychology, sociology and the humanities. Neuropsychiatry is not yet a firmly established subdiscipline of psychiatry. It has been considered an amalgam of neurology and psychiatry, dealing with disorders that cross the boundary between the two disciplines, but this is a narrow and unsatisfactory view. A broader definition considers neuropsychiatry to be the application of neuroscientific principles to all psychiatric disorders (Sachdev, 2002).

To many neuroscientists, the worries and concerns expressed in the *Nature Neuroscience* editorial will appear exaggerated and extremely pessimistic considering the potential benefit of neuroscientific research for humankind. In fact, modern neuroscience is characterized by great expectations for improvement in our understanding of the function of the brain in the near future, and thereby expectations for better possibilities for preventing and treating those severely incapacitating conditions that are contained under the umbrella of neurological and psychiatric diseases.

3 Expectations and results

We have just ended *the Decade of the Brain (1990–1999)* that was proclaimed by the President of the USA as a major political manifestation of the needs for understanding the functions of the brain concerning normal psychology, social behaviour as well as the neurobiological mechanisms of neurological and psychiatric disorders. There are good reasons for our great expectations to modern neuroscience. Within the last 50-100 years new significant understanding of basic mechanisms of brain function from the clinical to the genetic level has been obtained, and technological advances, for instance within neuro-imaging (i.e., PET-, MRI-scanning etc) has made it possible to study mental phenomena and brain function in humans.

A variety of books have been published that introduces modern neuroscience in a broader context. In 1985, Jean-Pierre Changeux's comprehensive presentation of major results from neuroscience was presented to an English-speaking audience. In his impressive book *Neuronal Man, the Biology of the Mind* (Changeux, 1985), Changeux summarized the long history of our efforts to understand brain function and elegantly presented the neuroscientific view of complex human behaviour including consciousness. Higher mental phenomena viewed in a neuroscientific perspective have been a major topic in several other books, as evidenced by the provoking subtitle of a book written by the Nobel laureate Francis Crick: *The Astonishing Hypothesis, the scientific search for the soul* (Crick, 1994).

Within a few decades, neuroscience has changed both the theoretical conceptions of many psychiatric disorders and the treatments options. Today, mental disorders are conceived as the results of a complex interaction between genes and biological, psychological and social environmental factors. The rich clinical phenomenology (i.e., hallucinations, delusions, anxiety etc) are seen as indicators of dysfunctional brain processing. A variety of genetic, biochemical, physiological and anatomical aberrations have been demonstrated for almost all major psychiatric disorders (Charney et al., 1999). However, due to the complexity of the brain with its billions of neurons and synapses we are just at the starting point of a long process that hopefully will lead to elucidation of the causes of incapacitating diseases such as of dementia, schizophrenia and affective disorders. Nevertheless evidence-based treatments are available for psychiatric disorders by means of psychotropic drugs and behavioural cognitive psychotherapy. The rise of modern neu-

roscience has changed psychiatry from a psychoanalytically oriented discipline to modern neuropsychiatry firmly rooted in a biomedical framework with its insistence on evidence-based treatment. To Eric Kandel, neuroscience represents a new intellectual framework for psychiatry (Kandel, 1998), and, to many psychiatrists, a neuroscientific perspective in the future will guard against the predominance of ideology over science, and the disciplines' demedicalization into community and social service psychiatry (Yudofsky and Hales, 2002).

This might indicate that there is little concern about a future psychiatry based on neuroscience but this would be a premature conclusion. In fact, a biologically oriented psychiatry that focus on brain mechanisms has consistently been criticized in the broader setting of public debate during the last couple of decades. It was the core programme of the anti-psychiatric movement in the 1960s and 1970s (Ross, 1979) and it has been the underlying theme in the heavy criticism of the widespread use of newer antidepressants ("happiness pills") for a variety of complaints (anxiety, shyness, lowered mood etc). Therefore there is good reason to believe that the *Nature Neuroscience* editorial focuses on essential problems for neuroscience in modern societies.

4 Nothing but a pack of neurons…

The editorial raised the question how it is possible to apply the concept of free will to explain human action when the understanding of the functions of the brain convincingly has shown the importance of biological factors. How can human beings be ethically responsible if neuroscience can reduce consciousness to biochemical, physiological processes or genetic mechanisms?

As stated by the Crick:

> The Astonishing Hypothesis is that "You", your joys and your sorrows, your memories and your ambitions, your sense of personal identity and free will, are in fact no more than the behavior of a vast assembly of nerve cells and their associated molecules. As Lewis Carroll's Alice might have phrased it: "You're nothing but a pack of neurones" (Crick, 1994, p. 3)

It may be reassuring that the astonishing statement is claimed to be just a hypothesis, but there has been a long tradition in neuroscience where textbook chapters on consciousness have been absent while the main chapters have focused on cellular and molecular processes in the neurons. Recently chapters on attention, memory and perception do appear, but it may be fair to claim that consciousness as a specific human feature linked to personal identity and integrity has had only a small role to play in neuroscience, as evidence also in the citation from Crick.

To paraphrase Monod's view on molecular biological theories of the evolution of man (Monod, 1997): "How can the astonishing hypothesis be accepted? – It is due to its impressive empirical support". There is firm empirical evidence demonstrating the close correlation between mental phenomena and neurobiological processes. There is a wealth of experiences available for the sceptic, the effects of anaesthetics or alcohol perhaps being the most impressive ones. Thus, there is close dose and time dependence between unconsciousness and the concentration of anaesthetics applied. Similarly, a bottle of good wine can change almost all aspects of those psychological functions that are essential for the experience of personal identity. Finally overt changes of behaviour, for instance by external frightening stimuli, are consistently accompanied by a variety of neurophysiological changes.

Very convincing is also the history of modern neuroscience (Crick, 1994). Brain processes underlying mental phenomena have across time been explained with reference to biological processes of ever smaller magnitudes: i.e., from actions of cells to metabolic processes in the subcellular synapses and recently to genes. A variety of technical advances – from the microscope to modern biochemical methods – has made such scientific progress possible. Brain function at the clinical level has been reduced to the function of billions of nerve cells with even more impressive numbers of synapses, and the chemical machinery of cells has been elucidated at the molecular-atomic level. On top of this, the Human Genome Project have unveiled our genome.

In psychiatry, drugs are used for a variety of purposes in treating psychotic, anxiety and depressive symptoms. The appearance of modern psychopharmacological drugs in the 1950s was due to serendipity but was followed by a period were several new drugs were marketed based on their action on neurotransmitters in the synapses of the brain. The success of the drug industry has culminated in the development of newer antidepressant drugs that are efficacious towards a variety of psychiatric symptoms that are of a milder severity than those qualifying for a diagnosis of a severe psychiatric disorder. To many people with significant social anxiety or depressive symptoms newer antidepressants have significantly improved their quality of life. Most embarrassing to our common sense conception of the role of consciousness is probably that improvement of severe psychiatric disorders can be obtained by drugs via their interaction with neurotransmitters in the brain. Thus psychological understanding does not seem necessary for improvement and is bypassed by chemical remedies.

5 Mind-body issues in neuroscience

There is little doubt that a materialistic ontology prevails in modern neuroscience. To excellent scientists such as Francis Crick the identity theory (that mental states are identical to brain states) is probably the natural answer to questions on mind-body viewpoints. To many neuroscientists it is rather trivial to claim that the mind is identical with biological processes in the brain. Any kind of dualistic thinking would demonstrate ignorance to modern neuroscience.

Eliminative materialism is a well known body-mind position in philosophy (for references, see Kim, 1998) and reading a few modern textbooks in psychiatry may easily convince you that this position has its advocates in psychiatry. Thus, there is little room for *common sense* or *folk psychological* explanation of abnormal mental symptoms. In fact, such explanations are often bypassed in clinical settings as our common-sense psychological framework is considered a misleading conception of the causes of abnormal human behaviour in a broad sense. As an example, for psychotic disorders, it seems that purely psychological explanations have little to offer for the development of hallucinations and delusions. Severe depressive disorders are often characterized by such misinterpretation of events that psychological explanations, that involve some degree of rationality, seem out of place. Thus, people may claim themselves guilty or responsible for wars in Africa or consider earthquakes as a natural consequence of some bad behaviour of theirs 60 years previously.

A consequence of neuroscientific research that focus on "low level" scientific disciplines is that causes of abnormal behavior and psychiatric diseases are explained solely at such basic level. The most influential theories in psychiatry of schizophrenia and depression have for several decades been theories concerning dysfunctional neurotransmitters, i.e., dopaminergic neurotransmission in schizophrenia and serotonergic in depression or anxiety disorders. In the setting of public debate it has often been stated so simply that depression is due to lack of serotonin in the brain. The rich clinical phenomenology of lowered mood, lack of initiative, feeling of guilt, memory problems and sleep disturbances are attributed to a single chemical molecule, and the effort needed is solely the ability to swallow a tablet regularly.

The strength of the "low-level" approach in biomedicine can be further illustrated by the great expectation to the near future. In fact a revolution in biomedicine is expected due to our insights into disease phenomena at the level of the function of the genes and their proteins This might lead to completely new systems of classifications, treatment and prevention. Interestingly, the molecular biological reduction of diseases individualizes diseases in the sense that diseases are conceived as aberrations of normal processes specific for the individual person. As an example, it is due to a specific genetic build-up that cancer will develop in a person who smoke heavily or never smoke. Similarly, scientific treatment at the nano-level ("Nano-science") is expected to revolutionize future drug treatment allowing for a specific treatment of those deficient genes in cells that cause mental disorders.

To many prominent neuroscientists and psychiatrists, progress of scientific explanation has been at the level of characterization of complex biological phenomena in the basic biological disciplines, that is, at the molecular-genetic level. They will probably object against such progress being seen a threats to human dignity, and they will claim that the benefit to mankind will be much greater than the losses due to elimination of classical concepts of consciousness and free will. Such is the myth concerning the explanatory power of neuroscience that underlies the ethical concerns expressed in the editorial.

6 Consciousness revived

Cognitive science is a discipline that attempts to understand the mind in terms of information processing, that is, representational structures in the brain and computational procedures that operate on them. However for many years, neurobiology failed to play a central role in cognitive science. Many researchers influenced by mind-body functionalism (for discussion, see Churchland, 2002) believed that the level of physical implementation of the information processing functions was irrelevant to the understanding of the mental states they were studying. This is because the same functions can in principle be implemented by different physical properties (e.g., organic brains or silicon computer chips). Similarly, the humanities have paid little attention to neuroscience as their focus of research have been the *products* of consciousness in broad sense.

With the development over the last 30 years of sophisticated neuroimaging and neurophysiological methods for studying the living brain, this picture has changed. Now the dominating science of the mind is the amalgamation of neuroscience and cognitive science in *cognitive neuroscience* (Gazzaniga et al., 1998; Ilardi and Feldman, 2001).

Among the basic tenets of cognitive neuroscience are:

- The human brain is an organ to represent and process salient information about the environment
- All human mental events occur as the results of neural information processing
- All human (nonreflexive) overt behavior occur as the results of neural information processing
- There is a close correspondence between mental events (e.g., thoughts, emotions, perceptions, and so on) and brain events
- Genes and protein products are important determinants of the pattern of interconnection between neurons in the brain and the details of their function (modified from Ilardi and Feldman, 2001)

Similarly Eric Kandel has suggested a new intellectual framework for psychiatry based on the following principles:

- Principle 1. All mental processes, even the most complex psychological processes, derive from operations of the brain.
- Principle 2. Genes and their protein products are important determinants of the pattern of interconnections between neurons in the brain and the details of their functioning
- Principle 3. Altered genes do not, by themselves, explain all of the variance of a given major mental illness. Social or developmental factors also contribute very importantly. Just as combinations of genes contribute to behaviour, including social behaviour, so can behaviour and social factors exert actions on the brain by feeding back upon it to modify the expression of genes and thus the function of nerve cells. Learning, including learning that results in dysfunctional behaviour, produces alterations in gene expression. Thus all of "nurture" is ultimately expressed as "nature."
- Principle 4. Alterations in gene expression induced by learning give rise to changes in patterns of neuronal connections. These changes not only contribute to the biological basis of individuality but presumably are responsible for initiating and maintaining abnormalities of behaviour that are induced by social contingencies (Kandel, 1998).

Finally, when the development of consciousness in humans is seen in an evolutionary perspective, its role is not a passive accidentally developed human function but rather an essential mechanism for the survival of the individual in its interaction with the surroundings. It is not an epiphenomenon to brain function similar to the smoke of the fire. Consciousness, in this functional sense (Churchland, 2002), has developed into an integrating system with high-level self-representational capacities. The complexity of the human brain enables the evaluative, planning and predictive structure of consciousness, that is, the production of those wonderful fruits that are found in the garden of Karl Popper's 'World III' (Popper and Eccles, 1981).

The last couple of decades have seen a surge in the study of consciousness and, whereas there has been some success in explaining the psychological function served by consciousness (Dehaene and Naccache, 2001), there is somewhat less optimism concerning the prospects for an explanation of the subjective side of consciousness (Block, 2001), that is, of *what it is like* to have conscious states (Nagel, 1974). A core question is whether and how it is possible to give reductive neuroscientific explanations of consciousness without simply eliminating or reducing away the conscious states themselves: how can conscious states *fit* into the natural world. Though there has been significant progress in identifying the neural correlates of consciousness, we are nowhere near a comprehensive theory that explains consciousness.

The problem is that one can grasp a complete functional, structural account of human brain activity without having any idea that this is an explanation of a conscious creature, rather than an explanation of a purely information-processing but non-conscious creature: it will always seem a legitimate *further* question whether a creature with this pattern of activity is also conscious. The debate is conducted under the assumption that very few people would deny that the mind in some sense *is* the brain (as we said above, this at least is what lesion studies, psychotropic drugs etc give us strong reason to believe) (see also Churchland, 2002). The problem is to understand *how* this could be true, and it is especially pertinent to psychiatric illnesses that often essentially involve disturbances to consciousness, rather than disturbances to mere information processing.

Cognitive neuropsychiatry is one of the newest branches of psychiatry (Halligan and David, 2001; Hohwy and Rosenberg, 2005a). It is a multidisciplinary approach trying to escape the classical Scylla-Charybdis situation: either a biologically oriented ("mindless") psychiatry or a mentalistic psychiatry ("brainless"). The scientific discipline of cognitive neuropsychiatry attempts to understand psychiatric disorders as disturbances to the normal function of human cognitive organisation, and it attempts to link this functional framework to relevant brain structures and their pathology. This new discipline is the natural extension of cognitive neuroscience into the domain of psychiatry (Hohwy and Rosenberg, 2005a,b). It presents an advance over traditional biological psychiatry, which tends to go directly from clinically defined disease entities to molecular genetics, neurochemistry and neurophysiology, bypassing the cognitive level. In addition, philosophy has an important role to play in cognitive neuropsychiatry since it analyses its core concepts of consciousness concerning motivation, rationality, free will and moral values. Consciousness, in its functional as well as its subjective manifestations, has re-entered the scene as a legitimate and necessary object of research.

7 Concluding discussion

The fear that modern neuroscience is a threat to human dignity is based either on misconceptions of the nature of neuroscience and the implications of its results, or on mistaken conceptualisations of the phenomena in question. Or both. Thus, on the one hand, modern neuroscience cannot any longer be identified with a scientific discipline that reduces consciousness to cellular and molecular mechanisms of nerve cells in the brain. Its scope of interests comprises studies of brain function at different levels, from the phenomenological to the molecular-atomic level. On the other hand, a little conceptual analysis shows, for example, that there could be no such thing as completely undetermined free will worth caring about: wholly undetermined decisions are arbitrary, and arbitrary decisions are not 'free' in any important sense. Rather, we say of ourselves that we are free when we act according to the most important of our desires, and that is compatible with the fact that all our desires are determined by nature, and potentially explorable by science (see, e.g., Frankfurt, 1988). The worry about free will discussed in the *Nature Neuroscience* editorial is thus based on an inadequate analysis of 'free will'.

In terms of metaphysics, modern neuroscientific results seem consistent with some kind of emergentism (for discussion, see Kim, 1998) where brain function is seen as multi-layered and in need of theoretical explanations at different levels, that is, at the levels of genes, biochemistry, physiology, cell function, neural network, attention, memory, mood, emotions, decision etc. According to this conception, consciousness is an emergent property of the brain that is irreducible to lower-level phenomena from which they arise. This is a multi-layered naturalistic model.

This would be a systems-theory oriented approach where properties and processes at a given level must be analysed with level specific concepts to account for level processes but where they at the same time can be correlated with lower levels concepts and processes. If emergentism is true, it would be impossible to explain why mental-physical correlations hold. As stated by Kim:

> According to the emergentist, why pain emerges when C-fibres are excited will forever remain a mystery: we have no choice but to accept it as an unexplainable brute act (Kim, 1998, p. 229)

To science, this is no fatal objection as we have plenty of good work to do by correlating between levels. That being said, the hope is of course that we will be able to go beyond emergence and develop a firmly stated theoretical framework for understanding how consciousness arises from brain activity. Until that happens, scientists will continue their research and try to build models that transgresses levels. Some such efforts have already demonstrated their value. Black-box receptors for drug interaction with cells have been replaced by three-dimensional models of protein receptors. Their amino acid structure and coding genes have been identified.

A cautious, and relatively agnostic, strategy would be to consider the problem of consciousness as a problem with a potentially *empirical* solution. Apart from purely philosophical speculation, our only rational option is to engage in further research. And future research may allow us to decide between a reductive body-mind identity thesis and a non-reductive emergence thesis. On this conception, we should not expect the truth of an identity thesis to become apparent in one revolutionary step. Rather, the hope for the identity thesis will be small, empirical steps that each serve as inductive evidence for that thesis (see, e.g., Hohwy and Frith, 2004).

As mentioned, on the right analysis of the concept of free will, there is no loss of human dignity by the demonstration that neurobiological, psychological or social factors constrain consciousness' apparent free will to form those plans and decisions. In her analysis of free will, Patricia Churchland (2002) addresses the neurobiology of decision making and free choice, and concludes that we should see free choice in terms of a multidimensional notion of a parameter space, where the dimensions of the parameter space reflects the primary determinants of in-control behaviour (p. 212). Decisions in daily life are made as the result of the influence of neurotransmitters, hormones, and neuronal networks analysing input from the environment.

Specific gene variation may constrain the spectrum of options for decisions. Some mutations or biological dysfunctions may be very influential and impose severe restrictions on the available behavioural repertoire. In fact, there is a long tradition in Denmark for offering treatment instead of punishment for crime committed by psychotic patients, with reference to their grossly disturbed reality testing performance. Similarly, we have little scientific evidence to conclude that ordinary criminals have so few degrees of freedom when making their choices for criminal acts that they must considered "determined by biology", although biological factors for evident reasons underlie their behaviour. It is through our neuroscientific knowledge that we can give good reasons for deciding under which circumstances people should be excused from responsibility or be granted diminished responsibility. Put differently, responsibility can be diminished when the 'parameter space' is so distorted by underlying pathology that we cease to see subjects as able to act according to their most important desires, or as able to arrive at important desires in a normal, rational way in the first place.

Paradoxically, neuropsychiatry based on modern neuroscience has had an enormous impact on our view of the nature of mental disorders, as well as on the creation of a respectful and humanistic attitude towards psychiatric patients. Severe psychiatric diseases, such as the psychoses, are not due to the influence of evil spirits or supernatural powers, unbalanced fluids of the body, unethical life style or a dehumanizing society. They are diseases with symptoms that mirror aberrant neurobiological function – primarily of genetic origin.

References

Block, N. (2001). Paradox and cross purposes in recent work on consciousness, *Cognition* **79**: 197–219.

Changeux, J.-P. (1985). *Neuronal Man: The Biology of Mind*, Princeton University Press, Princeton, NJ.

Charney, D. S., Nestler, E. J. and Bunney, B. S. (1999). *Neurobiology of Mental Illness*, Oxford University Press, New York.

Churchland, P. S. (2002). *Brain-wise: Studies in neurophilosophy*, MIT Press, Cambridge, Mass.

Cowan, W. M., Harter, D. and Kandel, E. R. (2000). The emergence of modern neuroscience: Some implications for neurology and psychiatry, *Annu Rev Neurosci* **23**: 343–391.

Crick, F. (1994). *The astonishing hypothesis: The scientific search for the soul*, Simon & Schuster, New York.

Dehaene, S. and Naccache, L. (2001). Towards a cognitive neuroscience of consciousness: Basic evidence and a workspace framework, *Cognition* **79**: 1–37.

Editorial (1998). Does neuroscience threaten human values?, *Nat Neurosci* **1**: 535.

Frankfurt, H. G. (1988). *The Importance of What We Care About*, Cambridge University Press.

Gazzaniga, M. S., Ivry, R. B. and Mangun, G. R. (1998). *Cognitive Neuroscience: the Biology of the Mind*, W.W.Norton, New York.

Halligan, P. W. and David, A. S. (2001). Cognitive neuropsychiatry: Towards a scientific psychopathology, *Nat Rev Neurosci* **2**: 209–215.

Hohwy, J. and Frith, C. D. (2004). Can neuroscience explain consciousness?, *Journal of Consciousness Studies* **11**: 180–198.

Hohwy, J. and Rosenberg, R. (2005a). Cognitive neuropsychiatry: Conceptual and methodological issues, *World Journal of Biological Psychiatry* **6**: 192–197.

Hohwy, J. and Rosenberg, R. (2005b). Unusual experiences, reality testing, and delusions of control, *Mind & Language* **20**(2): 141–162.

Ilardi, S. and Feldman, D. (2001). The cognitive neuroscience paradigm: A unifying metatheoretical framework for the science and practice of clinical psychology, *J Clin Psychol* **57**: 1067–1088.

Kandel, E. R. (1998). A new intellectual framework for psychiatry, *Am J Psychiatry* **155**: 457–469.

Kim, J. (1998). *Philosophy of Mind*, Westview Press, Boulder, Colorado.

Monod, J. (1997). *Tilfældigheden og nødvendigheden. Et essay om den moderne biologis naturfilosofi*, Fremad, Copenhagen.

Nagel, T. (1974). What is it like to be a bat?, *Philosophical Review* **83**: 435–450.

Popper, K. R. and Eccles, J. C. (1981). *The self and its brain*, Springer International, Berlin.

Ross, A. (1979). Opgør med antipsykiatrien, *Bibl Laeger* **171**: 241–268.

Sachdev, P. S. (2002). Neuropsychiatry – a discipline for the future, *J Psychosom Res* **53**: 625–627.

Yudofsky, S. C. and Hales, R. E. (2002). Neuropsychiatry and the future of psychiatry and neurology, *Am J Psychiatry* **159**: 1261–1264.

Part IV

Metaphysics and epistemology

Good news for prisoner A

Jeff Paris

School of Mathematics, University of Manchester

1 When we were small

The well known *three prisoners puzzle* was apparently first stated by Martin Gardner in *The Second Scientific American Book of Mathematical Puzzles and Diversions* in 1961. A somewhat simplified version of it goes as follows.

> Three prisoners, A, B, C are languishing in gaol. The gaoler tells prisoner A that out of the three of them two are to be executed. At this point then it seems not unreasonable (by an appeal to some sort of symmetry or indifference presumably) to suppose that A gives each of them probability $1/3$ of being the lucky one who is reprieved. A now says to the gaoler, "Look, I know you are not allowed to give me any information about whether or not I'm going to be executed but since at least one of B and C are for sure going to be executed it won't give me any information as far as my own fate is concerned if you were to tell me the name of one B or C who is condemned". Well, that sounds reasonable to the gaoler, so s/he gives A a name, B say. "Aha!" says A, "So now it is either C or I who will be spared, which means that my chances have gone up to $1/2$!"

Many authors have written about this puzzle and as far as I am aware the generally accepted view[1] is that unfortunately (for him) A is fooling himself. The formal explanation of A's fallacious reasoning is as follows. Initially A's probability function, P_0, is specified by

$$P_0(a \wedge \neg b \wedge \neg c) = P_0(\neg a \wedge b \wedge \neg c) = P_0(\neg a \wedge \neg b \wedge c) = 1/3,$$

---

[1] For a state of the art exposition of this interpretation see (Grünwald and Halpern, 2003).

with all other atoms getting probability O, where the language consists of three propositional variables, a, b, c, standing to 'A reprieved', 'B reprieved', 'C reprieved' respectively. So according to P_0,

$$P_0(b) = P_0(a) = P_0(a \wedge \neg b) = 1/3.$$

Then, following the gaoler's revelation, prisoner A applies standard Bayesian conditioning on the event $\neg b$ (i.e. B is for it) to give a new probability function P_1 with

$$P_1(a) = \frac{P_0(a \wedge \neg b)}{P_0(\neg b)} = \frac{1/3}{1 - 1/3} = 1/2,$$

The fallacy here, so generally accepted wisdom goes, is rooted in A's naive assumption that all s/he[2] has learnt from the gaoler is that B will be executed, i.e. $\neg b$.

The 'problem' here is that the language with just a, b, c is inadequate to fully express the conditioning information that A has learnt from the gaoler. A suitable language contains not only a, b, c but also g_b, g_c standing for 'gaoler says B', 'gaoler says C'. Let us suppose for simplification that A believes for certain that the gaoler will not be able to resist answering his question. In this case it seems not unreasonable that in this extended language prisoner A's probability function, Q_0 say, prior to questioning the gaoler should[3] be specified by

$$Q_0(a \wedge \neg b \wedge \neg c \wedge g_b \wedge \neg g_c) = Q_0(a \wedge \neg b \wedge \neg c \wedge \neg g_b \wedge g_c) = 1/6$$

and

$$Q_0(\neg a \wedge b \wedge \neg c \wedge \neg g_b \wedge g_c) = Q_0(\neg a \wedge \neg b \wedge c \wedge g_b \wedge \neg g_c) = 1/3$$

with all other atoms of the language getting zero probability. As, in the first case,

$$Q_0(b) = Q_0(a) = Q_0(a \wedge \neg b).$$

However, what is relevant here now are the figures

---

[2] No, I can't do it, it has to be 'he', I do not want to entertain the idea of any woman having to suffer the appalling conditions that I am sure must prevail in that house of *correction* (well, for sure two of them won't be repeat offenders). To even things up I shall make the gaoler female.

[3] Though it is not clear how much this obligation is not derived from a desire to end up with the right answer, i.e. that the gaoler's information should not have any effect on the probability A assigns to his beating the rap.

$$Q_0(g_b) = 1/2, \quad Q_0(a \wedge g_b) = 1/6,$$

and if we, correctly it seems, condition on the actual evident, g_b, we obtain

$$Q_1(a) = \frac{Q_0(a \wedge g_b)}{Q_0(g_b)} = \frac{1/6}{1/2} = 1/3.$$

In other words, in his naive approach prisoner A has been fooling himself. In reality he has no reason to pop the champagne, except of course the fear that otherwise it could be going to one of the beneficiaries of his will.

Clearly this argument relies on some possibly questionable appeals to symmetry (or insufficient reason) but basically I suppose most of us feel that it is somehow 'right', well at least 'sensible'. So, no joy for prisoner A it seems.

2 After we got big

But all this happened a long time ago. It turns out that this same situation has been repeated many times over since then. A whole stream of smart alec prisoners A's have gone through the system and our gaoler, patient and forbearing as she may be, is getting pretty fed up with answering the same question over and over again. So one day, because she's got a lot on, she doesn't bother to wait for prisoner A to ask the question, she just blurts out, "And before you ask your smart question, prisoner B is going to be executed". Now as it happens, on this occasion prisoner A hadn't even thought of asking that question. In fact, although it is hard to believe, he hadn't even got round to working out that as things stood he sensibly had only a 1 in 3 chance of coming out of this mess in one piece. But now his brain suddenly leaps into action. What he knows amounts to saying that B will die and that exactly one of himself or C will be spared. He applies the old 'symmetry' and comes up with the answer $1/2$. The moral is that sometimes it pays not to think about things too much.

At this point the gaoler was pretty annoyed with herself for inadvertently improving A's prospects and vows not to do it again. Nevertheless not long afterwards she unfortunately made another mistake, with similar consequences. This time she says to A, "Two of you, B,C will be executed", A asks his question right on cue, she replies, "B", but then, glancing at her watch, realizes that it's already happened and adds, "in fact he is already dead". In consequence A now reorders his information, putting B's demise *before* he learnt that one of himself, B,C would be spared and by 'symmetry' again comes up with $1/2$.

At this point there are two possible futures. The first is where the prisoner A of the previous paragraph is indeed the lucky one reprieved. Back on the outside, and thinking clearly again he realizes that actually the information that B was already dead was immaterial. After all, built into this whole puzzle is the implicit assumption that there are not going to be any 11th hour pardons from the governor, so saying that B *will* be executed means he is already *as good as dead*. The trick of it then, which he subsequently makes public in (Prisoner A, n.d.), is simply to reverse the order of the information and read 'dead' for 'good as dead'. Good news for prisoner A's.[4]

The second future I am afraid, is not so good for prisoner A's, at least not the studious ones amongst them. In this case the prisoner A of (now) two paragraphs ago does land up against the wall (or else is reprieved but fails to get (Prisoner A, n.d.) accepted for publication). In this case the gaoler realizes that she has the power to reduce future, clever, prisoner A's to abject hopelessness. Namely she will recall to memory numerous past executees, $C_1, C_2, C_3, \ldots, C_m$ and start the conversation with prisoner A with the information "Only one out of you, $B, C, C_1, C_2, C_3, \ldots, C_m$ has not been condemned to death". Then the clever prisoner A goes through the motions by asking the innocuous question "tell me $m+1$ of these ...", but he knows it's really to no avail, being so clever he's read the articles about the 'correct answer' and will still only give himself probability $1/(m+1)$ of being the lucky one. On the other hand, if he fortuitously didn't read the articles he can push it right back up there to $1/2$ again!

3 What does it all mean?

One possible initial conclusion we might draw from this discussion is that the 'answer' seems terribly susceptible to the temporal order of the information, despite the fact that at the end of the day what A has learnt amounts to essentially the same thing in each case. That does not seem quite right, time did not seem to be so much of the essence when the problem was originally formulated, so what has gone wrong?

I would suggest that the 'mistake', which is widespread in the area, even amongst people who should know better[5] concerns the status of the probabilities involved, in particular those assigned by P_0. I believe that the justification in most reasonable people's minds for P_0 being chosen as it is is on grounds of symmetry, or indifference, or insufficient reason. In short the initial information is invariant under any permutation of A, B, C and P_0 should respect, or preserve, this invariance. Given the constraints then this forces

---

[4] Game show addicts should note however that none of this helps with the superficially similar Monty Hall puzzle (vos Savant, 1990; Wikipedia contributors, 2006). The element of choice involved in that puzzle renders it essentially different.

[5] Because I've already spent some time explaining it to them, unfortunately to no avail.

$$P_0(a \wedge \neg b \wedge \neg c) = P_0(\neg a \wedge b \wedge \neg c) = P_0(\neg a \wedge \neg b \wedge c) = 1/3$$

with all other atoms getting zero probability. That is fine, one can hardly complain about this assignment, if we *have* to make an assignment then it seems the only common sensical one to make. However, despite it's inevitability that should not fool us into giving these figures an authority they do not really posses. They are, after all, just personal assignments, subjective probabilities, manifestations of willingness to bet. In short, we, the subjects, have made them up.

And it is the failure to properly appreciate this which causes the problems. For once we forget it and set P_0 in stone as 'established knowledge' then the arrival of gaoler's second piece of information leads to the idea that we must somehow 'update' our 'established knowledge' to encompass this new facto. But the truth of the matter is that the new information is no different in quality than the gaoler's initial revelation. The decent thing to do is to start again from scratch and evaluate *all* the evidence on a level playing field.[6]

So what is the answer? Well, that would seem to depend not just on the two blatant pieces of evidence but also on whether or not *A* thought of, and asked, the question before the gaoler blurted out the answer, the timing (real or imagined) of *B*'s execution, … . Personally, I think we should leave it to old *A* to come up with the answer he feels happiest with.

References

Gardner, M. (1961). *The Second Scientific American Book of Mathematical Puzzles and Diversions*, Simon and Schuster, New York.

Grünwald, P. D. and Halpern, J. Y. (2003). Updating Probabilities, *Journal of Artificial Intelligence Research* **19**: 243–278.

Prisoner A (n.d.). Think your way out of clink, *Studia Penitentia*, **CXLVII**: 1238-1433, ed. A. Pedersen.

Voorbraak, F. (1996). Probabilistic belief expansion and conditioning, *Technical Report LP-96-07*, Institute for Logic, Language and Computation, University of Amsterdam.

Voorbraak, F. (1999). Probabilistic belief change: expansion, conditioning, and constraining, *in* K. B. Laskey and H. Prade (eds), *Proc. 15th Conference on Uncertainty in Artificial Intelligence*, Stockholm, Morgan-Kaufman, San Francisco CA, pp. 665–672.

vos Savant, M. (1990). Ask Marilyn, *Parade Magazine*, p. 12, Feb. 17.

---

[6] This is not the first paper to make such a point, see for example the much more detailed arguments presented by Frans Voorbraak in (Voorbraak, 1996, 1999).

Wikipedia contributors (2006).   Monty Hall problem.   URL: *http://en.wikipedia.org/wiki/Monty_Hall*. Date of last revision: March 6th, 2006. Accessed March 8th, 2006.

State consciousness – two defective arguments

Oliver Kauffmann

Department of Media, Cognition and Communication, University of Copenhagen

1 Introduction

In (Lycan, 2001) a simple deductive argument for the higher-order representation theory of state consciousness (HOR) is presented. Lycan takes the argument to be valid for his own and similar 'inner sense' accounts of HOR (see Armstrong, 1981; Lycan, 1996, 2004) as well as for Rosenthal's higher-order thought version (see e.g. Rosenthal, 1997, 2004, 2005). From a so-called first-order representational perspective on state consciousness (FOR) Lurz (2001) presents an argument against Lycan's account. I demonstrate that Lurz's argument gives a misinterpretation of HOR and that his FOR-compatible account does not capture the self-presentational structure of state consciousness, a structure implied by the stipulative definition of state consciousness shared by him and Lycan. Although Lurz's argumentation fails, Lycan's argument can be shown to be non-valid for another simple reason. Hence both accounts of state consciousness are defective.

In the next section I briefly clarify what the notion 'state consciousness' is taken to mean and present two much discussed features of conscious states: Phenomenal character and the self-presentational structure of such states. In section 3 the general agenda for FOR and HOR – what unites them and what divides them – is spelled out with emphasis on the issue about the self-presentational structure of conscious states. This feature plays an axiomatic role for Lurz's and Lycan's arguments and is the touchstone in my critique. Section 4 presents Lycan's argument. In section 5 I give an exposition of the premises in Lurz's argument against Lycan. Section 6 proceeds with a demonstration that Lurz's critique of Lycan's argument is defective. Section 7 is a critique of Lurz's version of FOR. Lycan's argument is demonstrated non-valid in section 8. Section 9 sums up.

2 State consciousness – the notion and the explanandum

What does the term 'state consciousness' denote? In philosophy of mind there is widespread agreement that there is a difference between ascribing the property 'consciousness' to creatures and to mental states (see e.g. Bayne and Chalmers, 2003; Block, 1995; Carruthers, 2005; Dretske, 1995; Gennaro, 1995; Jacob, 1999; Kriegel, 2004; Lurz, 2004; Lycan, 1996; Manson, 2000; Rosenthal, 2005; Van Gulick, 1995). First, an animal or a human being can be described as being conscious, by which we mean, roughly, that the subject considered as a type or token is awake. Second, some subjects are also routinely characterized as being conscious *of* things and state of affairs: By having a perceptual mental state, a creature can be said to be conscious *of* some parcel of its environment. A creature's being conscious and a creature's being conscious *of* something are different characterizations of the subject as such. But third, mental states themselves can also have the property of being conscious:

> When we describe desires, fears, and experiences as being conscious or unconscious we attribute or deny consciousness, not to a being, but to some state, condition, or process in that being. States (processes, etc.), unlike the creatures in which they occur, are not conscious of anything or that anything is so although their occurrence in a creature may make that creature conscious of something or that something is so. (Dretske, 1995, p. 98).

Put somewhat differently, if mental states are properties of creatures, properties of those states are second-order properties of the creatures harboring such states. Therefore consciousness as a property of a mental state is different from those properties of a creature we ascribe to it when we say it is conscious (period) or conscious *of* something. So, even if there exists a relation of interdependency between the ascription of, say, a creature's being conscious of something and its being in a conscious state, these properties are distinct. Our matter of concern can now be raised: What is, then, this property of consciousness attributed to mental states (for short: 'state consciousness')? What are the characteristic features of a conscious state, to be more precise about the explanandum?

In recent discussions of the prospects for representational reductionism two features have been in focus. These features are the so called phenomenal character and a certain self-presentational structure both of which have been claimed to characterize conscious states. The phenomenal character of a conscious state is that there is something it is like for the subject to have that experience, pace (Nagel, 1974). When you are looking at a banana there is something it is like for you to have that particular experience of yellow, the claim goes.[1] The claim about the self-presentational structure of consciousness is that a subject's having a conscious state implies that the subject is aware of being in that state.[2] This claim is essential for HOR's explanation of state consciousness, but is generally denied by FOR. But before turning to that issue, what is representationalism, generally spoken?

3 Representationalism – the explanans

What does a representational reduction of state consciousness amount to? FOR and HOR claim that all the features of a conscious state should and can be exhaustively accounted for in terms of the representational properties of that state. This perspective is different from a straight reduction of consciousness to physical or neurological properties; instead, consciousness is claimed to be reducible to intentional and functional relations. A second reductive move is then to explain such notions in informational and/or teleological relations between brain states and physical states of the environment.

In accordance with the representational approach, when I look at the banana, what it is like for me to have that experience is merely a matter of the state's representing the yellow banana. The phenomenal character of the conscious state is nothing above the content of that state. Some FOR proponents have hypothesized that a content of a conscious state must be of a non-conceptual kind and have a non-symbolic, fine-grained, picture-like format (see e.g. Dretske, 1995; Kirk, 1994; Tye, 1995, 2000). In addition the content is typically claimed to have the functional role property that it is available for further cognitive and conative states in the subject. See (Kirk, 1994; Tye, 1995, 2000).

But HOR proponents argue that such kinds of representational relations are not sufficient to explain state consciousness. Another type of representational relation is required to explain what it is for mental states to be conscious, with their characteristic features of phenomenal character and self-presentational structure. And HOR's hypothesis is that necessarily, for a mental state to be conscious, the presence of that state must be represented by another state. A meta-representation

---

[1] While phenomenal character is an acclaimed mark of conscious states of a sensory kind, many philosophers deny that pure cognitive conscious states like thoughts and beliefs exhibit this feature.

[2] In this paper the expressions 'aware of' and 'conscious of' are used synonymously in formulation of the self-presentational structure of conscious states. This is also the case when it comes to HOR's turning this formulation into an explicit explanatory account of state consciousness. A practice licensed by representationalists of both sorts; see e.g. (Rosenthal, 2004, p. 41, note 4), (Dretske, 1995, p. 98), (Lycan, 1996, p. 3).

is needed, which is the rationale behind the name 'higher-order representation theory'. According to HOR a subject S is in a conscious mental state M_1 iff S by having a mental state M_2 in a suitable way W is aware of being in M_1, where 'aware of' indicates the representational relation between M_1 and M_2 and 'W' is placeholder for the different specific HOR suggestions concerning the precise nature of this relation. Thus the divide between HOR and FOR can be seen to turn on whether or not the existence of a meta-representation is deemed necessary in order for a mental state to be conscious.

Why should we wish to 'go HOR'? Defenders of HOR point out that their explanatory sketch simply reflects a pre-theoretically acceptable mark of state consciousness – namely what I have called the self-presentational structure of a conscious state. Clearly HOR proponents take this feature as pre-theoretically vindicated. Lycan states that he can't hear a natural sense of the phrase 'conscious state' other than as meaning 'state one is conscious of being in' (Lycan, 1996, p. 25), and similarly Rosenthal declares that:

> whatever else we may discover about consciousness, it's clear that, if one is totally unaware of some mental state, that state is not a conscious state. (Rosenthal, 2002, p. 408)

Now, one thing is whether or not HOR's explanation of state consciousness is correct. Another is whether the structure of self-presentation is correct as an initial characterization of the explanandum. Certainly not all FOR proponents assent to this pretheoretical 'mark of state consciousness'. As Lycan points out, Dretske does not (Lycan, 2001, p. 3, note 1). Neither does Tye (1995, pp. 5–6) or Kirk (2005, p. 216).

The question is whether it is possible to defend a FOR account without denying the HOR's axiomatic claim about the self-presentational structure of a conscious state. This is what Lurz (2001) claims and Lycan (2001) denies.

4 Lycan's argument for HOR

Lycan presents his argument for HOR in the following way:

(1) A conscious state is a mental state whose subject is aware of being in it. [Stipulative definition]

This premise lends expression to Lycan's acceptance of the claim about the self-presentational structure of a conscious state.

(2) The 'of' in (1) is the 'of' of intentionality; what one is aware of is an intentional object of the awareness.

(3) Intentionality is representation; a state has a thing as its intentional object only if it represents that thing.

Therefore,

(4) Awareness of a mental state is a representation of that state. [2,3]

And therefore,

(5) A conscious state is a state that is itself represented by another of the subject's mental states. [1,4]

Lycan's challenge to the opponent of HOR is: If he accepts (1), but wishes to reject (5), which one of the premises (2) and (3) will he abandon?

5 Lurz's argument against Lycan

Lurz (2001) argues against Lycan by pointing out two different readings of (1): A higher-order reading of 'aware of' which makes Lycan's argument valid, and a 'lower-order reading', according to which S's awareness of his mental state M simply is S's awareness of what his mental state M represents, and as such lends no support to HOR. Let me expose what I take Lurz's premises to be before I state my case against his argument:

Lurz's first three premises – let me denote them (L1), (L2) and (L3) – are identical with (1), (2) and (3) respectively.

(L4) 'A subject's state of *awareness of* X' (Where X has representational content) can be read as 'A subject's state of *awareness of what* X represents'.

(L5) A subject's state of *awareness of what* X represents is not equivalent to the subject's state of *awareness that* X represents such-and-such or equivalent to a state of *awareness as of* X's representing such-and-such.[3]

(L6) A subject's state of *awareness of* his mental state M is not necessarily a state of *awareness that* M represents such-and-such. [L4, L5]

Therefore,

---

[3] Lurz's disjunction between '*awareness that* something represents such-and-such' and '*awareness as of* something representing such-and-such' is borrowed from McGinn (1982) and is introduced to capture the difference between thought- and perceptual versions of HOR respectively (Lurz, 2001, p. 315). The distinction neither plays a role for Lurz's argument against HOR nor for my case against Lurz. It will therefore be ignored in what follows.

(L7) A subject's state of *awareness of* his mental state M is not necessarily
       a state that is a representation of M. [L2, L3, L6]

Thus, since Lurz's lower-order reading of (1) does not affect the supposed truth of
Lycan's premises (2) or (3) – occurring as (L2) and (L3) – Lycan's sentence (4)
does not logically follow from his premises (1)-(3). According to Lurz a subject's
state of *awareness of* his mental state M might be read as a state of *awareness of
what* his mental state M represents. And for a subject to be in a state of awareness
that represents the *content* of his mental state M, it is not the case that the subject
is in a further representational state that represents M as such:

> Hence, the subject is *not* in a higher-order representational state when
> he is simply aware of *what* his mental state M represents [...] it is *not*
> a representation of the mental state M as such – no more so than one's
> awareness *of what a trompe l'oeil* painting represents is a representation
> of the *trompe l'oeil* painting as such. (Lurz, 2001, p. 316)

From this Lurz's conclusion immediately follows:

(L8) A subject's having a conscious state is not necessarily to have a mental
       state M that is itself represented by another of the subject's mental
       states; that is: HOR as expressed in (5) does not logically follow from
       the premises (1-4). [L1,L7]

Thus, Lycan's deductive argument for HOR fails.

6 A first reply to Lurz

Lurz's argument against Lycan is invalid since his higher-order reading is a misinterpretation of HOR. Lurz's higher-order reading can be rejected without rejecting
HOR as such. This is so because HOR does not claim that *awareness of* M is
to be explained in terms of *awareness that* M represents such-and-such. Thus the
proponent of HOR might embrace the general distinction stated in (L5) between
awareness *of what* something represents and awareness *that* it represents such-
and-such, as well as this distinction applied to mental states, that is (L6), without
this implying (L7) and (L8). It is not part of Lycan's argument that awareness of
M – in terms of having a representation with the intentional object M – is to be
understood as awareness that M represents such-and-such.

   To read HOR's notion of 'awareness of M' in terms of 'awareness that M
represents-such-and-such' is to read HOR as putting forward an explanation of
a subject's state consciousness in terms of the subject's having explicit awareness *of M as representing such-and-such*. There is a word for this specific kind

of awareness of one's ongoing mental states as representing such-and-such, which is *introspection*. A state of ongoing introspection is a conscious mental state. But not all conscious states are states of ongoing introspection. The claim that a conscious state is identical with an ongoing state of introspection is untenable. If true, all mental states except those where the subject has explicit awareness of the fact that it's having a mental state should be deemed unconscious, which makes state consciousness a rather rarely occurring phenomenon. On the contrary, when we introspect our mental states, they are conscious to begin with. Lurz's reading of HOR's notion 'awareness of M' in terms of 'awareness that M represents such-and-such' ((L6)) implicitly presents HOR as an explanation of state consciousness by way of the subject's having introspective consciousness. But this reading puts HOR's card before the horse: such an explanation would be circular. HOR is an attempt to explain state consciousness in terms of a second-order representation of the state in question, wherefore state consciousness itself should not be introduced at the second-order representational level as a part of the *explanans* of the first-order state consciousness.

Hence, while Lurz is right in pointing out that 'S's awareness of what M represents' is not equivalent to 'S's awareness that M represents such-and-such' (L6), he is wrong when he takes HOR to explain state consciousness in terms of the latter.

The higher-order thought version of HOR clearly acknowledges a distinction between a conscious state and an introspective state of consciousness, where the latter denotes a state in which a subject is aware that his mental state is representing such-and-such; see (Rosenthal, 1997, pp. 730, 745–746), (Rosenthal, 2000), (Gennaro, 1995, pp. 16–21). But the inner sense version of HOR acknowledges this distinction as well. Unfortunately, however, Lycan and David Armstrong have both used the locution 'introspection' (as well as terms like 'scanning', 'inner perception' and 'monitoring') to describe the activity by which a subject's mental states are rendered conscious for it according to the perceptual version of HOR. This is unfortunate, since it might lead someone to misrepresent HOR as a thesis where introspective consciousness holds a position either as *explanandum* (which is false) or as *explanans* (which is viciously circular). The last option is an implicit part of Lurz's argument against Lycan. In fact, though, Armstrong makes a clarifying distinction between 'reflex' introspective awareness and 'introspection proper', a clearing up which Lycan points out too (Lycan, 1996, pp. 162–163, note 2).

> It is a plausible hypothesis that the latter ['introspection proper'] will normally involve not only introspective awareness of mental states and activities but also introspective awareness of that introspective awareness. (Armstrong, 1981, p. 63)

Put in Armstrong's terms: By attributing an 'introspection proper' reading of 'awareness of one's own mental state' to HOR, Lurz's case against HOR is valid: Since a subject's state of *awareness of* his mental state M is not necessarily a state of *awareness that* M represents such-and-such (L6), (L7) and (L8) follow immediately and Lycan's deductive argument for HOR fails.

But if one adopts Armstrong's 'reflex'-introspective reading of 'awareness of one's own mental state' instead, a subject is said to be consciously aware of *what* his mental state M represents by being aware of M – a reading which does not lead to (L7) and (L8). In Armstrong's famous example of the absent-minded driver, who suddenly – after a long time of driving without awareness of what he has been doing – 'comes to' and realizes what he is doing, the lack of awareness of the ongoing perceptual representations is a lack of 'reflex' introspective awareness, see (Armstrong, 1981, pp. 59–60). The driver is absent-minded in a double sense: He is not consciously aware of the road, although he is still having unconscious representations of it while driving. The road is the object represented by his first-order perceptual states.[4] Secondly, he is not aware of these first-order states.

But, according to HOR, it is only because of one's becoming aware of the latter that one is consciously aware of what they represent – and did unconsciously represent all along – namely the road.

Certainly there is a difference between being aware of what a *trompe l'oeil* painting represents and being aware that a painting represents such-and-such (L5). But this is a false analogy to HOR since a subject needs not be aware *that* he is having a representation of a mental state in order for the mental state to be conscious by having a representation of it.[5] To put it metaphorically: Lurz's higher-order reading of HOR is simply too high.

7 A second reply to Lurz

Lurz might object that even if his higher-order reading illustrates a misconception of HOR along the lines indicated above, his argument still has teeth against the 'true HOR account'. This is so because he also argues *via positiva* for FOR:

---

[4] In Armstrong's terminology, the absent-minded driver has 'perceptual consciousness' of the road. "Above all, how is it possible to drive a car for kilometres along a road if one cannot perceive that road?" (Armstrong, 1981, p. 60).

[5] Likewise David Rosenthal's higher-order thought version of HOR does not imply that the second-order thoughts necessary for the first-order states to become conscious need themselves be conscious, although these second-order states might sometimes themselves in turn become conscious by the subject's having third-order thoughts about *them* (Rosenthal, 1997, p. 742).

(L9) All cases supporting a higher-order reading of the expression 'S's awareness of M' in (L1) can be interpreted in the lower-order way as well, that is, as instances of '*awareness of what* M represents'. And this reading is in accordance with a FOR account of state consciousness.

Therefore

(L10) There is no reason for accepting the higher-order reading of the expression 'S's *awareness* of M' in (L1) over the lower-order reading except by begging the question against the FOR account of state consciousness.

Thus, even if Lurz's own higher-order reading of HOR is mistaken, it might be true that all cases supporting 'true readings of HOR' can be interpreted in the suggested lower-order way, a way which conforms with a FOR account. Lurz takes it as a fact that we use the expressions 'aware of' and 'aware of what' interchangeably, a fact which lends abductive support for the hypothesis that these locutions really do express the same idea (Lurz, 2001, p. 317). Accordingly there would be no reason except by begging the question against FOR for accepting *any* higher-order reading of 'state of awareness of his mental state' (Lurz, 2001, pp. 317–318).

Against this I argue that (L9) must be rejected. By adopting Lurz's lower-order interpretation of (L1) his stipulative definition of state consciousness is rendered incomprehensible as a criterion for distinguishing between conscious and unconscious mental states. (L9) does not capture the self-presentational structure of a conscious state expressed by (L1), the feature by which we had the first handle to distinguish conscious from unconscious instances of mental states. Therefore (L10) is false.

Consider what Lurz's presents as the basis for his claim that 'aware of' and 'aware of what' are used interchangeably:

> Whenever we find it natural to say of a subject who is distracted, for instance, that she is not aware *of* perceiving some item or fact (which she is in fact perceiving), we find it just as natural to say that she is not aware *of what* she is perceiving when she is perceiving the item or fact. And vice versa. (Lurz, 2001, p. 316)

Lurz points out that this fact is indeed exemplified by the wording of David Armstrong and Peter Carruthers in their descriptions of the absent-minded driver case (Lurz, 2001, pp. 316–317).[6]

---

[6] See Armstrong *op.cit.* and (Carruthers, 1989, p. 258). See also (Carruthers, 2000, p. 149) for a similar description.

But first, this is not to say that 'aware of' and 'aware of what' are used interchangeably when someone *is* (consciously) perceiving. From the fact that the expression 'S is not aware of perceiving some x' is used interchangeably with 'S is not aware of the x he is perceiving' it does not follow that the expressions 'S is aware of perceiving some x' and 'S is aware of the x he is perceiving' are used interchangeably as well.

Second, there is a reason why Lurz *should* acknowledge this distinction. If S is not, some way or another, aware of his perceiving, he is not consciously perceiving x. To substitute 'S's awareness of what M represents' for S's awareness of M' leaves us with a reading of (L1) by which we no longer have a criterion for distinguishing between instances of conscious and unconscious mental states. Lycan and Lurz agree upon the stipulative definition of state consciousness: (L1)=(1). Both of them believe that this premise states a 'mark of state consciousness' which gives us a non-question begging first handle on the difference between conscious and unconscious states. Now, by adopting Lurz's lower-order reading of (L1) it gets the following form:

(L1*)   A conscious state is a mental state whose subject is aware of what M represents.

But when a subject has an *un*conscious mental state he might also be said to be aware of what the state represents. Certainly the unconscious driver too might be described as being perceptually aware of the road, other cars, curves, etc. Lurz clearly accepts the existence of unconscious mental states. He describes, with HOR, the unconscious driver's visual states in terms of 'perceiving' and seeing' (Lurz, 2001, pp. 315, 317). So, by accepting (L1*), what is the difference between conscious and unconscious mental states? It would be truly unhelpful and circular to modify (L1*) to 'a conscious mental state is a mental state whose subject is consciously aware of what M represents', in order to explain the difference. Thus, where (L1) captures such a difference, (L1*) does not.

When Lurz interprets (L1) – a premise acceptable to both HOR and FOR – in terms of (L1*) he leaves his own stipulatively defined 'mark of state consciousness' behind. A reading of (L1), which implies that S is aware of his mental state M by being aware of what M represents, runs the risk of wiping out the distinction between conscious and unconscious mental states and leave the self-presentational structure of consciousness behind. As exemplified by Lurz's lower-order account. Hence (L9) must be rejected, and, contrary to (L10), Lurz's lower-order reading of (L1) does not pose a threat to a HOR account of state consciousness.

Lurz describes (L1) as reflecting 'our pre-theoretical intuitions about state consciousness' (Lurz, 2001, p. 314). That is, (L1) reflects a non-explanatory criterion by which we distinguish between conscious and unconscious mental states. Since

(L1*) no longer reflects this distinction, (L1*) is not, strictly speaking, a reading of (L1) at all. Lurz's lower-order reading is simply too low.

8 A reply to Lycan

Although I take Lurz's argument against HOR to be defective, I am not satisfied with Lycan's argument myself. His argument is not valid.

Consider again Lycan's premises and the challenge posed: If the opponent of HOR accepts (1), but wishes to reject (5), which one of the premises (2) and (3) will he abandon?

My answer to Lycan's challenge is that neither of the premises (2) and (3) has to be rejected in order to show that the conclusion (5) does not follow. From (1)-(4) it might still be the case that when a subject is in a conscious state – that is, a mental state whose subject is aware of being in it (1) – this very state represents *itself* to the subject as one he is aware of being in. According to such an account of state consciousness, no *further* mental state is needed for this representation.[7] From acceptance of the subject's awareness of the state as an intentional relation (2) and understanding this relation as the subject's having a representation of the state in question (3), it does not follow that the conscious state does not represent *itself* to the subject. Certainly defenders of HOR argue against this possibility for *other* reasons. But (5) cannot be deduced *solely* from (1)-(4). Acceptance of the proposition that awareness of a mental state is a representation of that state (4), does not necessarily imply that the representation in question is carried out by *another* of the subject's mental states, as (5) says. So, yes, Lurz is right: Lycan's argument is defective, but not because (4) does not logically follow from Lycan's premises (1)-(3), as Lurz suggests. The reason is that (5) does not follow from (4).

A deductive shortcut to HOR has not been brought to light yet.

9 Conclusion

Lurz's case against HOR fails for two reasons: First, his higher-order reading misrepresents HOR as a theory about introspective consciousness. Second, his lower-order reading does not account for the stipulative definition of state consciousness. This reading does not reflect a conception of state consciousness by which we can distinguish a conscious from an unconscious mental state. Thus, on the one hand,

---

[7] Such an account can be found in (Brentano, 1874). For recent versions see (Thomasson, 2000) and (Lehrer, 2002). For a discussion of this view in contrast with HOR, see (Kriegel, 2006). Indeed (Kriegel and Williford, 2006) is a collection of papers all of which in different ways address this issue.

Lurz's argument against HOR is defective and his FOR account does not lend expression to the self-presentational structure of a conscious state stipulated in the definition of this term.

On the other hand, though, Lycan's argument is non-valid anyway. A theory of state consciousness in which the self-presentational structure of consciousness is accounted for by the mental state's representing itself is not an instance of HOR.[8] In other words, from the fact that S's awareness of a mental state is a representation of that state it does not follow that this representation is carried out by S's having another mental state: HOR does not follow. Hence both of the considered arguments by Lycan and Lurz are defective.

Perhaps other versions of FOR or HOR can lend us the explanation of state consciousness. Whether this is the case remains to be shown.

I would like to thank Mikkel Gerken, Bill Lycan and David Rosenthal for helpful comments.

References

Armstrong, D. M. (1981). What is consciousness?, *in* D. Armstrong (ed.), *The Nature of Mind and Other Essays*, Cornell University Press, Ithaca, NY, pp. 55–67.

Bayne, T. and Chalmers, D. J. (2003). What is the unity of consciousness?, *in* A. Cleeremans (ed.), *The Unity of Consciousness. Binding, Integration and Dissociation*, Oxford University Press, Oxford, pp. 23–58.

Block, N. (1995). On a confusion about a function of consciousness, *Behavioral and Brain Sciences* **18**: 227–287.

Brentano, F. (1874). *Psychologie vom empirischen Standpunkt*, Felix Meiner, Hamburg.

Carruthers, P. (1989). Brute experience, *Journal of Philosophy* **86**: 258–269.

Carruthers, P. (2000). *Phenomenal Consciousness. A Naturalistic Theory*, Cambridge University Press, Cambridge.

Carruthers, P. (2005). *Consciousness. Essays from a Higher-Order Perspective*, Oxford University Press, Oxford.

Dretske, F. I. (1995). *Naturalizing the Mind*, Oxford University Press, Oxford.

Gennaro, R. (1995). *Consciousness and Self-consciousness. A Defense of the Higher-Order Thought Theory of Consciousness*, John Benjamins Publishing Company, Amsterdam.

---

[8] Lycan (2004, p. 110, note 1) actually mentions such a theoretical possibility, but dismisses it right away without argument.

Jacob, P. (1999). State consciousness revisited, *in* D. Fisette (ed.), *Consciousness and Intentionality: Models and Modalities of Attribution*, Kluwer Academic Publishers, Dordrecht, pp. 9–32.

Kirk, R. (1994). *Raw Feeling*, Oxford University Press, Oxford.

Kirk, R. (2005). *Zombies and Consciousness*, Oxford University Press, Oxford.

Kriegel, U. and Williford, K. (eds) (2006). *Self-Representational Approaches to Consciousness*, MIT Press, Cambridge, Massachusetts.

Kriegel, U. (2004). Consciousness and self-conscicousness, *The Monist* **87**: 185–209.

Kriegel, U. (2006). The same-order monitoring theory of consciousness, *in* U. Kriegel and K. Williford (eds), *Self-Representational Approaches to Consciousness*, MIT Press, Cambridge, Massachusetts, pp. 149–170.

Lehrer, K. (2002). Self-presentation, representation, and the self, *Philosophy and Phenomenological Research* **64**: 412–430.

Lurz, R. (2001). Begging the question: A reply to Lycan, *Analysis* **61**: 313–318.

Lurz, R. (2004). Either FOR or HOR. A false dichotomy, *in* R. J. Gennaro (ed.), *Higher-Order Theories of Consciousness. An Anthology*, John Benjamins Publishing Company, Amsterdam, pp. 227–254.

Lycan, W. (1996). *Consciousness and Experience*, MIT Press, Cambridge, Massachusetts.

Lycan, W. (2001). A simple argument for a higher-order representation theory of consciousness, *Analysis* **61**: 3–4.

Lycan, W. (2004). The superiority of HOP to HOT, *in* R. J. Gennaro (ed.), *Higher-Order Theories of Consciousness. An Anthology*, John Benjamins Publishing Company, Amsterdam, pp. 93–113.

Manson, N. (2000). State consciousness and creature consciousness: A real distinction, *Philosophical Psychology* **13**: 405–410.

McGinn, C. (1982). *The Character of Mind*, Oxford University Press, Oxford.

Nagel, T. (1974). What is it like to be a bat?, *The Philosophical Review* **83**: 435–450.

Rosenthal, D. M. (1997). A theory of consciousness, *in* N. Block, O. Flanagan and G. Güzeldere (eds), *The Nature of Consciousness. Philosophical Debates*, MIT Press, Cambridge, Massachusetts, pp. 729–753.

Rosenthal, D. M. (2000). Introspection and self-interpretation, *Philosophical Topics* **28**(2): 201–233.

Rosenthal, D. M. (2002). Explaining consciousness, *in* D. J. Chalmers (ed.), *Philosophy of Mind: Classical and Contemporary Readings*, Oxford University Press, Oxford, pp. 406–421.

Rosenthal, D. M. (2004). Varieties of higher-order theory, *in* R. J. Gennaro (ed.), *Higher-Order Theories of Consciousness. An Anthology*, John Benjamins Publishing Company, Amsterdam, pp. 17–44.

Rosenthal, D. M. (2005). *Consciousness and Mind*, Oxford University Press, Oxford.

Thomasson, A. L. (2000). After Brentano: A one-level theory of consciousness, *European Journal of Philosophy* **8**: 190–209.

Tye, M. (1995). *Ten Problems of Consciousness. A Representational Theory of the Phenomenal Mind*, MIT Press, Cambridge, Massachusetts.

Tye, M. (2000). *Consciousness, Color, and Content*, MIT Press, Cambridge, Massachusetts.

Van Gulick, R. (1995). What would count as explaining consciousness?, *in* T. Metzinger (ed.), *Conscious Experience*, Schöningh/Imprint Academic, Paderborn, pp. 61–79.

The logic of temporal beginning

Peter Øhrstrøm

Department of Communication and Psychology, Aalborg University

What does it mean that something begins to be – or begins to be the case? For instance, what is the meaning of expressions such as 'the beginning of his/her life', or what does it means that a certain process begins? Such questions were discussed by Aristotle and, later, very often in scholastic philosophy. Also in various modern contexts, we often wish to establish a clear meaning of the notion of 'beginning'. In e.g. biomedical ethics, it is essential to agree on when the life of a human being begins. For instance, is a human fertilised egg a human being? If not, when does the new life begin its existence as a properly classified human being? Such questions are obviously important, and they arise not only in biomedical ethics but also in many other contexts. In scientific language in general we need some semantic clarifications, if we are to reason about beginnings and endings.

According to the Aristotelian view, we may meaningfully speak of the end of a process, whereas "there is no such thing as a beginning of a process of change, and the time occupied by the change does not contain any primary when in which the change began." (Aristotle, n.d., Book VI, Part 5). The need for such an asymmetry has to do with the idea of time conceived as a continuously ordered set of instants without duration. Given this idea of time, processes are normally understood as corresponding to intervals in time. When something ends in time, something else normally begins. A process may have been ongoing until a certain moment in time, after which the process has stopped, i.e., it is not ongoing. Aristotle argued that since a given process cannot be both ongoing and not ongoing at the same time, there must be a limiting moment in time between the period in which the process is ongoing and the period immediately following in which the process has stopped. Aristotle argued that such a moment has to be an indivisible instant, and he also held that it had to belong to the earlier period in which the process is ongoing. It is, however, very much debatable whether such a limiting instant should belong to the earlier or to the later period. It is definitely not obvious why the Aristotelian choice should be more reasonable than the alternative according to which there would be a first instant (i.e., a beginning) of a process but no last instant. As pointed out by C.L. Hamblin "the best anyone seems to be able to do with the problem is to solve it by fiat" and that the Aristotelian solution is "arbitrary and ad hoc" (Hamblin, 1972, p. 327).

However, if we accept the Aristotelian approach, we also have to accept the truth of the implications:

(1) If the process is now ending, then the process is now ongoing.
(2) If the process is now beginning, then the process is not ongoing.

In accordance with the alternative model, we will have to accept the truth of the following implications:

(3) If the process is now ending, then the process is not ongoing.
(4) If the process is now beginning, then the process is ongoing.

As mentioned above, the choice between the two models – and consequently between (1-2) and (3-4) – can hardly be based on anything else but arbitrariness. Such a situation obviously appears to be very unsatisfactory and problematic. However, there are a few possible solutions, which I will introduce briefly in the following.

One possible response to the problem could be to make a distinction between various kinds of processes (or states) letting one kind of processes work according to (1-2) and all other processes work according to (3-4). It is, however, not obvious how such a distinction can be established in an objective manner. In scholasticism, some philosophers in dealing with beginnings and endings ('desinit' and 'incipit') found it useful to distinguish between 'processes' and 'states' (see Øhrstrøm, 1988, p. 161 ff.). But even though such a distinction is interesting, it is hard to see how it can support a certain choice between (1-2) and (3-4) in any convincing manner.

A second possible response to the problem could be to abandon the very notion of a statement being true (or false) at an instant without duration. We could still do this in the context of a continuous time formulating a logic of durations according to which statements are true (or false) over durations. This is in fact the solution suggest by C.L. Hamblin in his paper on 'Instants and Interval' (Hamblin, 1972), and I have elsewhere (with Per Hasle) discussed in detail how logics of duration can be established (Øhrstrøm and Hasle, 1995, p. 303) and also described some of the new complications to which such a logical model may give rise. It is easy to verify that given a logic of duration we might in fact obtain the truth of the pair, (1) and (4).

A third possibility could be to drop the idea of a continuous time and instead assume a discrete time. In the Middle Ages, this solution was studied by William of Sherwood, who established a rather sophisticated logical model by means of which he could deal with the logic of "beginning" and "ceasing". I have elsewhere (Øhrstrøm, 1988, p. 160 ff.) shown that, with this apparatus, Sherwood was able to provide relevant analyses of statements such as "what begins to be ceases not to be" and that he could analyse iterated uses of the terms in question e.g. "ceasing not to be ceasing".

It is still an open question whether any strong and convincing argument can be given in favour of one of these possible solutions or we just have to accept a solution which is arbitrary and ad hoc.

There are other fundamental questions that occur if we want to obtain a logic useful for reasoning about beginnings and endings. For instance, any such logic must provide clear answers to the questions regarding whether implications such as the following are true:

(5) If q now begins to be the case, then $\neg q$ ends to be the case.
(6) If q now begins to be the case, then q will be the case for some time hereafter.
(7) If q now begins to be the case, then q was not the case immediately before now.
(8) If q now ceases to be the case, then q has been the case for some time.
(9) If q now ceases to be the case, then q will not be the case immediately after now.

The strong interest in the logic of beginning and ending during the medieval period appears to be intimately related to metaphysical and cosmological questions conceived in a theological framework. In particular, a number of logicians and philosophers have examined the cosmological, the metaphysical, and the formal aspects of the radical idea of a very first instant of time. In medieval philosophy, this problem was seen as closely related to the notion of creation from nothing (*creatio ex nihilo*). This is of course still relevant in relation to modern theology, and in addition, various Big Bang-models are based on the notion of a sudden creation of the universe i.e., a beginning of the universe. In the following, I will concentrate on the analysis of the ideas regarding the beginning of the universe. Firstly, I intend to focus on the aspects of time and temporal logic following from the assumption that the universe has been created from nothing. Secondly, I shall briefly discuss the very notion of *creatio ex nihilo*. In both parts of the discussion, I will refer to the works of A.N. Prior (1914-69), who made careful investigations into these problems demonstrating their relevance for a deeper understanding of the notions of time and temporal logic.

1 Thomas Aquinas and A.N. Prior on the beginning of the universe

The notion of an absolute beginning of the universe is difficult, but it has been analysed in details for centuries. One of the most important texts dealing with this crucial notion was written by Thomas Aquinas in his *Quaestiones Disputatae de Potentia Dei*, Q. 3, *De Creatione*, Articles 1 to 3. Prior studied this text very carefully, and he also edited the text (in Shapcote's translation from 1932-34) to be used in his teaching. In particular, he considered Thomas' response to Article 1, objection 2, in which Thomas made the following points:

> "Before the world was, it was possible for the world to be: but it does not follow that there was need of matter as a basis of that possibility. For it is stated in Metaph. V,12, that sometimes a thing is said to be possible, not in respect to some potentiality, but because it involves no contradiction of terms, in which sense the possible is opposed to the impossible. Accordingly it is said that before the world was it was possible for the world to be made, because the statement involved no contradiction between subject and predicate. We may also reply that it was possible by reason of the active power of the agent, but not on account of any passive power of matter . . . ." (Prior, n.d.)

Prior used this passage and other related passages in his teaching. He distributed them in student handouts, which contained not only the original texts but also brief comments by Prior himself. In the handout *St. Thomas Aquinas' 'On the Power of God'* Prior wrote the following comment regarding the passage mentioned above: "It would seem that before the world existed, God *was able* to make it, and it *was possible* (not self-contradictory) that it should be made, but it was not (and there was nothing that was) *able to be made*." (Prior, n.d.). Prior went on to question whether Thomas really believed in a time 'before the world existed' during which God was able to act in a certain way, but chose not to act in that possible way. Prior found that a model based on the idea of "first nothing, then something" is very debatable. In fact he maintained that a proper and consistent reading of Thomas Aquinas should lead us to an interpretation of the text according to which the idea of "first nothing, then something" has to be rejected. It is obvious that Prior realized that an interpretation based the idea of "first nothing, then something" would turn out to be logically inconsistent. However, Prior did not explain the underlying argument in details. But what he had in mind seems to be very close to the argument later put forward by N. Malcolm (Malcolm, 1963). This argument can, in a slightly elaborated form, be presented as the following trilemma (see Øhrstrøm, 1988, p. 147):

(M1) If a thing has come into being, then at some earlier time it did not exist.

(M2) The universe at some time is the totality of truth and reality at that time.

(M3) The universe is a thing which has come into being.

That these three statements constitute a trilemma together can be verified in the following way:

It follows from (M1) and (M3) that

(M4) At some earlier time (before the universe came into being) the universe did not exist.

(M2) implies that

(M5) At some earlier time (before the universe came into being), there was no truth or reality.

I have earlier claimed that (M5) is "obviously absurd", since it falsifies itself (see Øhrstrøm, 1988, p. 147). However, Prof. Stig Andur Pedersen has long ago (actually, as an opponent at my doctoral defence in 1988) pointed out to me that (M5) need not be seen as falsifying itself, since it may be read counterfactually in the following manner:

(M5') If there was a time before the universe came into being, then at this time it would be true to say: 'There is no truth (or reality) now'.

Stig Andur Pedersen's point is clearly correct, which means that we need one more step in the argument. This additional step will be based on the observation that the statement

(M6) At some time it is true to say: 'There is no truth (or reality) now'

definitely falsifies itself. The concluding step in the argument would then be to note that the assumption "There was a time before the universe came into being" leads to a statement, which falsifies itself. For this reason, it is obvious that given (M1-3) we have to conclude that the statement "There was a time before the universe came into being" is false. But this statement immediately follows from the combination of (M1) and (M3). For these reasons, it is evident that (M1-3) constitute a trilemma.

However, we have to realise that (M1) is very problematic – at least when used with the universe as the 'thing' in question. In fact, this may lead us back to Thomas Aquinas, who, dealing with his article 2, considered the question 'Is Creation a Change?' As noted by Prior in his comments, Thomas answers "that although the unreflective may regard it as the most radical change conceivable, it is not properly speaking a change at all" (see also Prior, 2003, p. 87). In his handout, Prior wrote the following comment to Thomas' solution:

> There is no change, or needn't to be, if first of all X is red and then Y
> is blue; for change, one same thing, X say, must be first red and then
> blue, say. But when neither X nor Y yet existed, there was no difference
> between X not existing and Y not existing. Can we not say, though, 'Pre-
> viously, it was not the case that X existed, and now it is the case that X
> exists'? But Thomas says no real time before the world, since God is not
> in time and time cannot exist on its own. Perhaps we can say "It is not the
> case that X's existence *was* the case, but it *is* the case that X's existence
> *is* the case"; but would this express a change? (Both main clauses same
> tense – present. The subordinate past in first clause is only the object of a
> negation; as Thomas says an 'imaginary' not a 'real' past.) (Prior, n.d.)

The underlying problem here can be presented in relation to (7), which can be
exemplified by letting q stand for the statement 'the universe exists' i.e.:

(7.1) If the universe now begins to be, then the universe was not immediately before now.

Obviously (7) and (7.1) may be seen as aspects or consequences of the usual understanding of the notion of 'change'. Letting q stand for the statement 'the universe
exists' and P for the usual past operator in tense logic and B for an operator corresponding to 'begin', a straightforward consequence of (7.1) seems to be:

(7.2) $Bq \supset P\neg q$

which means that, at least at the creation of the universe (i.e. at the very first moment of the universe), we will have $q \wedge P\neg q$. But according to standard semantics
for tense logic $P\neg q$ would mean that there was a time before the universe came into
being at which the universe did not exist. This is obviously exactly the conclusion,
which was rejected above as false.

However, what Prior points out in the above quotation is that although the
conjunction $q \wedge P\neg q$ is false at the very first moment of the universe, the statement
$q \wedge \neg Pq$ will in fact be true at the same time. This also means that according to
the Priorean solution, (7.2) will be rejected, whereas

(7.3) $Bq \supset \neg Pq$

will be accepted. Prior has elsewhere (Prior, 2003, p. 105) pointed out that the
axiom $\neg Pq \supset P\neg q$ added to the basic tense logic K_t will give us a logic corresponding to non-beginning time. Given this axiom, (7.3) obviously entails (7.2).

Prior has also considered the construction of a tense logic corresponding to a
notion of time according to which there is an ending time (see Prior, 2003, p. 139
ff.). I have discussed this logic in (Øhrstrøm, 2004). It is easy to see how Prior's
system for ending time can be transformed into a tense logic corresponding to a
notion of time with a first instant. The crucial axiom will be

(B1) $\neg Pq \lor P\neg Pq$

or alternatively

(B2) $Pq \supset PHp$

where H stands for the usual operator in tense logic corresponding to 'It has always been the case that …" i.e. $H \equiv \neg P\neg$. It is easy to demonstrate that the two above formulations are equivalent. Given (B2), we may substitute $\neg q$ with p getting $Pq \supset P\neg P\neg\neg q$, from which we immediately obtain (B1). Obtaining (B2) from (B1) is slightly more complicated:

    (1) $q \supset (q \supset q)$ [by propositional logic]
    (2) $Pq \supset P(q \supset q)$ [by (1) and standard tense logic, K_t]
    (3) $Pq \supset P\neg P(q \supset q)$ [by (2) and (B1)]
    (4) $\neg p \supset (q \supset q)$ [by propositional logic]
    (5) $P\neg p \supset P(q \supset q)$ [by (4) and standard tense logic, K_t]
    (6) $\neg P(q \supset q) \supset \neg P\neg p$ [by (5) and contraposition]
    (7) $P\neg P(q \supset q) \supset PHp$ [by (6) and standard tense logic, K_t]
    (8) $Pq \supset PHp$ [by (3) and (7)]

The tense logic K_t with the addition of (B1) or (B2) corresponds to a notion of time according to which there is a very first instant i.e., a beginning of time itself. In such a model of the universe, there are no instants before the very first instant; i.e., the instant at which the universe comes into being.

2 Creation and the beginning of the universe

But how can the beginning of the universe be explained? Will it provide any meaning to ask for a causal explanation? The answer given by Thomas Aquinas is of course that the universe is made by God out of nothing (creatio ex nihilo). Prior was an admirer of Thomas. Although disagreeing on some points, he pointed out that Thomas "is wonderfully aware of the difficulties that surround the thought of creation out of nothing, and states these difficulties most perspicuously" (Prior, 1959, p. 82). One of the points made very clear by Thomas (and Prior) is that according to this idea of creation, 'nothing' should not be conceived as something material. In Prior's own words:

> One silly thing it's only too easy to do … is to talk as if "nothing" were the name of some kind of stuff out of which the world was made. I've even read a theologian (Barth) who [in his Dogmatics in Outline, 1949] talks as if "nothing" were a sort of hostile power from which God rescued the world in giving it being. (Prior, 1959, p. 89)

According to Prior, the claim that the world was made from nothing simply means that it *wasn't made out of anything*. Obviously, Thomas does not want to describe creation in terms of usual causality. Prior seems to accept this view in his comments in the handout, although he has some reservations when it comes to Thomas' reduction of creation to something like mere existence. In his comments on Thomas' analysis of article 3, Prior says:

> "Creation taken actively, taken passively", i.e., creating and being created. Being created is not a genuine case of the Aristotelian category of 'passion', i.e., being acted on, having something done to one. Thomas says it is just a 'relation' which the creature has to the Creator once it is in being. (It is truer to say that a thing must *be* in order to *be created*, than to say that it must *be created* in order to *be*; though in replay to Objection 3, Thomas says both.) But hasn't Thomas here escaped flattening creation out of nothing to creation out of something, by flattening it to just *being around* while things start to exist? Insistence on God's being outside of time tends this way. (Prior, n.d.)

In Prior's mind, then, it seems that much of this remains mysterious. On the other hand, Prior clearly acknowledges that it does not give rise to any contradiction to assume that there is a completely unique kind of relation between the Creator and the creatures, which cannot be reduced to anything else in any satisfactory manner. The special nature of such a relation probably has to do with the difficult relation between time and eternity. And here Prior also seems to be right in pointing out that Thomas apparently has made things much too easy with his claim of "God's being outside of time". The relation between time and eternity appears to be much more complicated than indicated by a straightforward reading of Thomas' analysis. As William Lane Craig (2001a, p. 67) has convincingly argued a robust understanding of the doctrine of *creatio ex nihilo* must involve more than just a tenseless approach to time. It must also incorporate the idea of coming into being and the notion of tensed facts (in order to state the proper meaning of something existing now). All this can be analysed within the Priorean framework, but it is very hard to see how the temporal aspects of the doctrine of *creatio ex nihilo* can be conceived in terms of a tenseless (i.e., non-Priorean) theory of time. Craig (2001b, p. 153 ff.) seems to be right in maintaining that in order to obtain a proper understanding of the doctrine of *creatio ex nihilo* and of an absolute beginning as such, we need a hybrid view that combines the ideas of timelessness and omnitemporality. Craig's presentation in (Craig, 2001b) can be read as a strong argument for the importance of using the Priorean framework in the analysis of the temporal aspects of the meaning of the doctrine of *creatio ex nihilo* and of an absolute beginning.

3 Conclusion

In this paper, we have seen that the problems of formulating a logic of temporal beginnings and endings are rather complicated, and that there are still important open questions as to how such a logic should be formulated. On the other hand, it should also be emphasized that the analysis indicates that the notions of beginning and ending can in fact be presented in a consistent manner using temporal logic. The possibility of such a logical language is obviously important as an indication of the possibility of a precise and consistent scientific language which can be used in disciplines dealing with beginnings and endings in a systematic manner. In particular, it is possible to establish a logical language based on the assumption that time itself has an absolute beginning. I have argued that the notions of beginning and ending can be analysed through basic ideas from Priorean temporal logic.

In general, the possibility of a consistent logic of temporal beginning and ending is crucial, since it may turn out to be a necessary conceptual foundation for the discussion of a number of important questions within cosmology, metaphysics and ethics.

References

Aristotle (n.d.). Physics. Book VI, Part 5. Translated by R.P. Hardie and R. K. Gaye. URL: *http://classics.mit.edu/Aristotle/physics.6.vi.html*. Accessed March 10th, 2006.

Craig, W. L. (2001a). Response to Paul Helm, *in* G. E. Ganssle (ed.), *God and Time*, Paternoster Press, pp. 63–68.

Craig, W. L. (2001b). Timelessness and omnitemporality, *in* G. E. Ganssle (ed.), *God and Time*, Paternoster Press, pp. 129–160.

Hamblin, C. (1972). Instants and intervals, *in* J. T. Fraser, F. C. Haber and G. H. Müller (eds), *The Study of Time*, Springer-Verlag, pp. 324–331.

Malcolm, N. (1963). Memory and the past, *in* N. Malcolm (ed.), *Knowledge and Certainty*, Englewood Cliffs.

Prior, A. N. (1959). Creation in science and theology, *Southern Stars* **18**: 82–89.

Prior, A. N. (2003). *Papers on Time and Tense*, second revised and extended edn. Per Hasle and Peter Øhrstrøm and Torben Braüner and Jack Copeland (eds), Oxford University Press.

Prior, A. N. (n.d.). Comments by A. N. Prior regarding St. Thomas Aquinas' *On the Power of God* (translation by L. Shapcote), Q.III. Of Creation. Article 1-3., Student Handout. Box 9, The A.N. Prior Archive, Bodleian Library, Oxford. Undated.

Thomas Aquinas (1932–34). *On the Power of God*, Vol. 1–3., Burns, Oates & Washbourne, London. Translated by L. Shapcote.

Øhrstrøm, P. . and Hasle, P. (1995). *Temporal Logic. From Ancient Ideas to Artificial Intelligence*, Kluwer Academic Publishers.

Øhrstrøm, P. (1988). *Nogle aspekter af tidsbegrebets rolle i de eksakte videnskaber - med særligt henblik på logikken*, Aalborg Universitetsforlag.

Øhrstrøm, P. (2004). The uncertainty of the future, *in* P. H. Harris and M. Crawford (eds), *Time and Uncertainty*, Brill, Boston, pp. 229–244.

Becoming through technology

Jan-Kyrre Berg Olsen

Section for Philosophy and Science Studies, Roskilde University

1 Introduction

The "reality" of time is intertwined with cosmological notions. In general, the way we tend to think about the world, including our common sense notions and the scientific ideas we are working on, are full of metaphysical background theory. This "background" indirectly operates upon our cognitions of the world. For instance, we find that these "ideas in the background" manifest themselves as commitments in our interpretations of phenomena. One such phenomenon is time. What is this "background" that plays such a major role in committing the individual, scientists, and philosophers to views that in some way or another become aspects of their theories about the world? It is evident that the further away our commitments about reality are from our experiences of it, the more indebted are our commitments to ideas and theories that claim to disclose reality without the aid of experience. These ideas and theories function not only to convince the protagonist of the legitimacy of his own endeavours, they also make him want to convince others of their truth. These ideas also function in the sense that they serve to give justification to these commitments.

One of the most influential ideas, within philosophy, concerning the nature of real time is the idea of "Becoming". Becoming describes a specific ontology of time. In this specific context, time is the very fabric of reality. To understand how time works, one must have an idea about how the world is working.

In this essay I want to follow up on some thinkers within physics and related philosophies that do not defer to conventional scientific postulations about a physical reality outside time, frozen in its making. The line of argument goes from a critique of deterministic rationality, to a discussion of experienced temporality and entropy. The last part of this essay focuses on ways to extend the local temporal viewpoint to a more global point of view, through the use of simple technology such as water clocks, sandglasses, thermometers, or even nature's own technology such as pulse and heartbeat. These are all real world phenomena in which temporal direction is not hidden, which in fact is the case with mechanical clocks. Anyway, the debate about time always contains more than just postulations about the real nature of time. The debate about time goes to the core of the essential features of reality itself, the nature of time has to do with human cognition; more precisely it has to do with cognitions and experiences in which true reality is disclosed.

2 The deterministic "world-view"

The ideas that proponents of Becoming are struggling against, are the ideas found within the metaphysical doctrine of "determinism". It is the ideas of the Eleatics and of Plato which constitute the ideological core of modern natural science. It is within this metaphysical framework that time escapes everything we know about it from our everyday experience of it. And to escape everyday experience one must apply a quite different sort of reasoning. The ideological "core" of determinism consists of a set of beliefs that is crucial to a style of scientific thinking that we, conventionally, label rational. In fact this belief in the "rational"[1] has been instrumental in the development of modern ideas about a "non-temporal universe" that simply is. It does not evolve or become.[2] We can say that the style of thinking we find within the physical sciences today, as it is presented through Einsteinian physics and quantum physics,[3] has its origins in the thinking of Parmenides and Plato. The style of thinking that obtained its modern features through the scientific ideas of Galileo Galilei. Before we can go ahead with the Eleatic-Galilean styled thinking within modern science and philosophy of time we have to take a look at the very first known tendencies to "freeze" time.

---

[1] "Rational" comes from the Latin word "*ratio*" meaning "thought" or "of the mind". But now, in our present day, the word means the same as "being able to determine" which, however, must include, in order to be rational, that which is being determined as something entirely "independent of mind".

[2] For a more extensive reading about determinism and its rationale consult the writing of Capek. See for instance his "Introduction" to *The Concepts of Space and Time* (Capek, 1976). Interesting is also Edmund Husserl's *The Crisis of European Sciences* (Husserl, 1970). In the following I will go into detail on aspects that I believe are of the utmost importance to our understanding of the development of today's concepts of "objective" time, and which is devoid of any "subjective" content. However, the following is not a question of the intrinsic validity of science; I am not dismissing science. I do not intend to question science itself but the non-dynamic, non-temporal "realistic" interpretation of it.

[3] At least according to Edmund Husserl's *Crisis*: See (Gurwitsch, 1965, p. 292).

The "eternism" of Parmenides is the fundamental idea. He formulates the idea as follows. Whatever that can be said to be, or that which actually is, can have no beginning or end. If it had a beginning and an end, it would not be, and that is excluded according to Parmenides. In addition to "Being" having no beginning, Parmenides formulates the following proof: "What necessity would force it, sooner or later, to come to be, if it started from nothing? ... It neither was nor will be, since it is altogether now."[4] To sum this up in the words of B. Williams: "Here Parmenides gives the first expression to an idea of eternity." (Williams, 1988, p. 220).

Parmenides' denial of becoming was too radical for the atomists. The atomists retained the principle of the immutability of Being in a slightly different way, so as to make the principle of the immutability of Being fit experience. Democritus, Epicurus and Lucretius did not deny change and becoming, they reduced it "to the displacement of the atoms, each of which was the Parmenidean plenum on a microscopic scale: uncreated, indestructible, immutable, impenetrable", as Capek (1976, p. xxvii) has put it. The universe of the Eleatics consisted of matter and void and there was no place for time. Time was, therefore, explained away as "appearance" (Democritus), "accident of accidents" (Epicurus), and that time has "no being by itself" (Lucretius). Capek comments on these sayings by stating that time become "a mere function of the changing configurations of the immutable particles." And thus the relational theory of time was born.

With time as a relation between things, philosophers focused their attention on the regularity and periodicity of the celestial motions, as well as on day and night, the differing seasons throughout the year and all "events" that could be measured. The significance of the metrical aspect of "time" was growing, as was the notion of regularity and homogeneity of the motions. Furthermore, the cleavage between experience, which is fundamentally qualitative, and the mathematical perspective, became gradually more marked by the lack of corresponding properties between the two (Capek, 1976).

The Parmenidean conception of the universe was extreme; however, the influence of Parmenides' ideas has persisted throughout history as ontological background (beliefs about reality) in the theories of other influential thinkers. Actually, no temporalities apply at all in Parmenides' conception. We cannot separate between past, present and future in the realm of perpetual present; of Being which does not become or change. There is only that which is – that which is not is excluded. In Parmenides' point of view, Being is "uniform, unchanging, has no divisions, is the same under any aspect..." (Williams, 1988, p. 221). This concept of eternal, unchanging and uncreated being has, through Plato's thinking, had a particularly significant impact upon the course of modern philosophical and scientific

---

[4] Here Parmenides is quoted from (Williams, 1988, p. 220).

development. Plato applied the idea of "uncreated Being" in his characterization of the Forms. The Forms are to be understood as the fundamental forms of reality, which exist beyond the apprehension of our senses and experience.

Here we find what can be termed as "Platonic-Parmenidean Reason", that is, "Greek Reason" (Marcuse, 1965). There are some very specific implications of this concept of reason. First of all, and with particular reference to the above notion of a static and uncreated world, we find that in the context of Plato's thinking the true Being becomes ideal Being. This means simply that this is not the kind of being "we experience immediately in the flux of our empirical, practical world", as Marcuse puts it (Marcuse, 1965, p. 281). This is, according to Capek, the same as asserting that we have a "coeternity of truth and fact." (Capek, 1965, p. 443). This implies, as Marcuse points out, that the validity of reason is "supra factual" and "supra temporal". The fundamental and real nature of reality can only be discovered, disclosed and defined by this kind of rational reason, and thus, as rational, it has the mandate to overrule, that is, to put itself up and against anything which is given to experience. Marcuse writes that: "Reason establishes an authority and reality which is... antagonistic to the immediately given facts" (Marcuse, 1965, p. 281). The characteristic cleavage in modern temporal realism, between immediate experienced reality and the world as it "is" in objective scientific truth, is as old as philosophy itself.

The set of beliefs we find within modern temporal realism is identical to the set of beliefs that are part and parcel of the modern realistic interpretation of physics. This realistic interpretation was, broadly speaking, developed in the Seventeenth century. The goal of this interpretation was to pierce through all our common sense deceptions, to disclose the real mathematical structure of the universe; its deterministic and non-temporal nature, to open up a reality that is manifest in mathematics. Modern science starts by refusing to accept our common sense experiences at face value (Gurwitsch, 1965, p. 293).[5] What is of concern here is, as Aron Gurwitsch says, "the problem of the very existence and the sense of science ... [which is] the conception of nature as in reality possessing a mathematical structure." (Gurwitsch, 1965, p. 294).

We can conceive of a nature that is disclosed as mathematical, without the aid of our immediate experiences, because experiences are taken to be deceptions or illusions. The world is not believed to be what it looks like. Only mathematical construction can discover the true condition of the world. This overlooked the fact that there are several mental operations involved in the performed conceptualizations. It also omitted mental processes such as, for instance, idealization, or formalization, which are crucial for the generalization of the conceptualized content. What happened was exactly the same as when our modern day temporal realist

---

[5] Others with similar viewpoints are Feyerabend (1993) and of course Husserl (1970).

attempts to hide subjectivity behind the product of his formalizations. Focusing on the formalized product, one can discard the producing activity or the originating qualities from which the products spring. It is natural that the failure to refer such products and results to the mental operations from which they derive makes one-self the captive of one's own creations.(Gurwitsch, 1965, p. 300). It is not enough that we loose sight of our own creativity when we create, because, as Einstein has commented, we want to regard the products of our imagination as nature in itself, since they appear necessary and natural. We would also like others to regard them, accordingly, as given realities (Einstein, 1954, p. 270). Therefore "a cloak of mathematical ideas and symbols, metaphysical ideas, is cast upon the world of experience so as to conceal it to the point of being substituted for it." (Gurwitsch, 1965, p. 300). Method becomes reality.[6] This means, in Husserl's context, that the ideas and symbols that are involved in the constitution of mathematical theories – ideas that thus facilitate the application of mathematics to the science of nature, become the "whole thing". This new mathematical science encompasses every-thing that represents the "life-world", which is the world that is found in everyday experience. In fact "it (mathematical science) dresses it up", as Husserl says, "as 'objectively actual and true' nature." (Husserl, 1970, p. 51). Through this complex of ideas we start to believe that it is true being – as opposed to a method. Thus, we arrive at the conception of reality as being a mathematical manifold (Gurwitsch, 1965, p. 300). Of special interest in the present context, temporality must be looked upon as one of the most important customs or habits of nature; time is a typical fea-ture of natural behaviour. This is not that kind of behaviour that we experience, but which we otherwise, i.e. by way of experimental science, know are fundamental to processes at the microscopic and, by implication, "constitutive" level of nature. We have a "theoretical" world of physics that works as a framework or ideal world picture for the thinker that operates with the given theory or ideas. Husserl writes that they (physicists) are "constantly oriented in their work toward ideal poles, toward numerical magnitudes and general formulae." (Husserl, 1970, p. 48).

Historically speaking this specialization of a narrow and restricting scientific thought began with Galileo Galilei and his invention of the universal law. Husserl states: "The 'a priori form' of the 'true' (idealized and mathematized) world, the 'law of exact lawfulness' according to which every occurrence in 'nature' – ide-alized nature – must come under exact laws." (Husserl, 1970, p. 53). The "ideal poles" are at the center of interest in all physical inquiry. What is discovered is discovered in the "formula world", which thereafter is coordinated with nature (Husserl, 1970, p. 48).

---

[6] (Gurwitsch, 1965, p. 300). See also (Husserl, 1970, p. 51)

The coordination with nature is, of course, coordination with the whole set of metaphysical-epistemological ideas making up the notion of nature as mathematically structured, or, one could say, with a suitable ontology.

For Galilei the course of action was to abstract from individually lived life; be it spiritual or mental; from cultural aspects as well as from those aspects of existence that are attached to things in human praxis (Husserl, 1970, p. 60). Along with the mathematization of nature, we also find the idea which is so crucial to the idea of a deterministic non-temporal universe. This is the idea that reawakens the Parmenidean notion of uncreated being and Democritean or atomistic self-enclosed natural causality. This is a causality in which every occurrence is predetermined, both necessarily and unequivocally. Thus we see that Galilei has opened the path for dualism to enter the arena of natural philosophy. The notion was to have a separation of reality in two worlds: nature and the psychic world. The first division was Platonic, the second one was Cartesian. It is important to understand that the consequence of the separation of the objective world from that of the subjective is that the latter, psychic world, does not achieve the status of an "independent world". On the contrary, the "psychic world" is dependent upon the world of matter, as it was conceived in a scientific-theoretical construction. What is more, this separation led to a belief in an absolute distinction between the subjective and objective realms of being. The absolute line of demarcation thought to "exist" between them, which renders the two worlds apart. From the point of view of objectivism, this was necessary because the real mathematical world of science should not be linked to the mental and relative world of subjectivity. In any case, as Husserl points out, "natural science possessed the highest rationality because it was guided by pure mathematics and achieved through inductions, mathematical results." (Husserl, 1970, p. 61).

The rational scientific world consists of bodies, a "world" that exists in itself. As we have pointed out before, a world that exists in itself must be a strangely "split" world. It is strange in comparison with our common sense experiences of the world because this one is split into the realist notion of "nature-in-itself" and a mode of being that is absolutely distinct or different from this, namely of what exists psychically. In the years after Descartes, subjectivity became more and more separated from the rational scientific sphere.

The amputation of the psychic from "the scientific real" causes difficulties whenever we are trying to determine the true source of "time". The problem consists of the intuitive knowledge that the natural philosopher has of the source of his own knowledge, namely his own experiences and thoughts. These subjective manifestations clash against the nexus of assumptions and notions that constitute his rational scientific ontology. What legitimates, and thus removes the doubts that the natural philosopher might have about the independence of his "objective and rational knowledge", is an escape into the new psychology facilitated by the division of nature and spirit. This is the subjective-objective distinction, which is a presupposition for the specialization of the sciences, and thus also the foundation of naturalistic psychology, which holds subjectivity to be a nest of illusions. The Cartesian doctrine states that bodily and psychic "substances" are characterized by radically different attributes found to be fundamental to that kind of rationality which holds nature to be determined and non-temporal, since it is believed to be causally law-governed and mathematically representable. We see that Husserl, who claimed that "the naturalization of the psychic comes down through John Locke to the whole modern period up to the present day", has pointed out the further historical development (Husserl, 1970, p. 62). As Capek points out, "what was relatively new in Locke was his interest in the introspective basis of our awareness of time. From this time on, the distinction between subjective, psychological and objective, physical time gradually became common." (Capek, 1976, p. xxxv).

The project of scientific rationalism looked upon as a whole, which includes dualism, non-temporalism, determinism, naturalism, and scientism/psychologism, is surely an attempt to extrapolate an epistemological-ontological model. It attempted "to classify thought in particular cases or situations, to the whole of reality..." as the philosopher Owen St. John writes (St. John, 1974, p. 76). It was necessary for the concept of rationality that rationality was to be uniform and conventional, that there was no room left for subjective whims to enter the arena. However, as St. John says, in arguing that some particular thoughts are universal while others are not is to pass over from science to metaphysics. He writes:

> We can never extrapolate from a deliberately restricted sphere to all possible spheres, to all aspects and levels of existence... We can never arrive in science at an unconditional generalization that everything, under all possible conditions, everything that is, or will be, or has been, is of such and such nature and behaves in such and such a way. (St. John, 1974, p. 76)

If a deterministic, non-temporal universe is real, then we have a science that can transcend all possible experience. It can transcend experience because it can go beyond the temporal limits that are somehow put on experience. It can assert unconditional knowledge about a universe that does not conform to the conditional thinking that is based upon experience. The metaphysics behind this kind of science have no temporal limits to knowledge, and are, therefore, in deep disagreement with the empirical and temporal limits that we, the experiencing individuals of the world, have to obey in order to have coherent and corresponding knowledge. Or as Andrew Pickering writes:"Atemporal knowledge is marked by the processes of its emergent becoming, but it cannot itself explicitly register the existence of truly emergent phenomena, nor can it thematize the shocks and the struggles that their emergence precipitates. Becoming is actively obscured in the way we use atemporal knowledge in the world. The price to pay for a metaphysics of becoming is recognition of this fact." (Pickering, 2003, pp. 102–103).

3 Becoming, dissipation and the temporal mind

The fundamental characteristic of becoming is transience. In the process of actualization of potentialities to a particular thing there is not one moment that can be singled out as the defining moment in this process. This is a moment that would, thus, be more real than the process itself. All the "moments" that pass by are but "phases" or "fleeting" images of this something as it is changing continuously. From the observer's point of view, the present moment presents the real, since the process has evolved only so far as when it appears "now" for the observer. All the other phases of this something have been leading up to this present "moment". Yet development does not halt, and it will always take place in a moment that, in principle, is present to someone. "Phases" and "stages" have succeeded each other, or followed straight after another, yet the substance in question retains its identity over time. The obvious temporal direction here is primitive, yet it is assumed "it has some unknown causal source". What can this unknown source be? The answer that many scientists have given to this question is "entropy".

The problems of reducing our experience of, say, direction, to the entropy gradient does not establish a link between internal time and external time; between the time of mind and physical time. The kind of reduction that we should object to is that we do not access the direction of time outside our immediate experience of it. This is to say that it is not by "awareness of entropic or other causal processes that we know of events in our immediate experience what their time order is." (Sklar, 1995, p. 218). Since we have an immediate experience which is temporally structured we also have a direct access to temporal direction. The reduction that is wanted by those who wish to establish a link between physical time and temporal experience, is the reduction of temporal experience to a conceptual construct, such as entropic order. This entropic order is less fundamental than the temporal experience itself. This is, according to Sklar, a "scientific reductionism" (Sklar, 1995, p. 219).[7] The claim is, Sklar points out, that we do not determine temporal order and

---

[7] We are thus fighting against tradition and the habits of thought that have become second nature to *most scientifically trained* persons of our time. The mathematization of secondary qualities marks the turning point in our thinking about reality in the sense that it defines how to define nature as an object of science. This new way of thinking about reality and how to get correct scientific knowledge about it can be called "the program of the scientific objectification of the experienced and non-experienced domains". The "program" emphasizes the faculty of abstraction. Thus, it removes "the phenomenal precept"; it suspends "every *experiencing subject* and, simultaneously, of any transient

direction by knowing about how the entropic order is working, but that we instead discover that temporal order is identical to entropic order (Sklar, 1995, p. 219).

I believe, in accordance with L. Sklar (1995) and P.V.C. Davies (1997), that Sir Arthur Eddington's illumination of this problem is to the point (Eddington, 1946, pp. 87–110). As Sklar says, "there is something about time that makes a treatment of its relation to entropic asymmetry. . . implausible." (Sklar, 1995, p. 223). What is implausible is not that entropy has an ordering of events that must obey the order of time, but that time has been "reduced" to signify entropic order. Here we are again faced with a theoretical domain and its relation to human temporal experience. For Eddington it is evident that when we are talking about real time we have to differentiate between theory and experience. Meaning comes out differently for terms in the sense that in one aspect meaning is secured through identification through experience and in another aspect by location in theoretical structure. These are two separate things. Time seems to be a feature that we wish to attribute both to the realm of perception, or experience, and to the realm of the theoretically inferred. Sklar points out that "it is just a confusion to think that the spatial relations' visual percepts bear to one another are the same sort of relations that physical objects bear to one another." (Sklar, 1995, p. 224). First of all, we know from our experience what the former relation is like. Secondly, we can only talk about knowledge concerning the latter relations from what our theoretical structures say about them (Sklar, 1995, p. 224). But does this mean that we can dismiss entropic order as merely theoretical and not in any way as part of reality? This is to take things too far. Eddington was of the opinion that time is given to us twice, once in our immediate experience and secondly in our theoretical reflection about the irreversibility of external processes. It is the same time that is given in both of the modes. We should, not, however, confuse them.[8]

We are not seeking the replacement of entropic order by experienced order, or the reduction of experienced order to entropic order. Rather, we are trying to see how the two spheres are attached to the same time. We will have to face the fundamental role of temporal experience, in the sense that it is our starting point in any theoretical construction of the world. In this sense it is important to admit that we have direct epistemic access to the relation of temporal succession of the world because, alternatively, what happens if we "radically distinguish" between time in experience and the time of physics? The problem of not having any relation between the "time-spheres" is equally as bad as the reductionistic claims pointed out above. This would mean that we do not have any grasp of the nature of the

---

modality of time experience", as it is expressed by the physicist Massimo Pauri (1997, p. 280). Pauri continues by stating "this epochal transformation of the very conception of *subjectivity* soon became stabilized and shaped many general features of modern thinking." (Pauri, 1997, p. 280).

[8] See also (Sklar, 1995, p. 226).

physical world itself, since, in the claim which separates physical and psychical domains too radically, there is absolutely no correspondence between the ways we perceive things and the nature of the objects as physical entities. "We are left with merely the 'instrumental' understanding of theory in that posits about nature bring with them predicted structural constraints upon the known world of experience," as Sklar writes (Sklar, 1995, p. 224). Furthermore, if we omit consciousness and experienced temporality as a necessary point of reference and instead attempt to render an objective (external) time that is mathematical, we will perhaps end up with infinite regress.

In order to understand this we need to study Eddington's thinking about time a little closer. He presents us with an interesting idea, suggesting that there is a necessary linkage of physical time to the world of experience.

Why is it, Eddington asks, that we cannot immediately identify the "becoming" of temporal experience with the increasing "disorder" of the universe called entropy? Entropy is a concept about unidirectional physical processes and, as such, it could also be symbolizing a type of "unidirectionality" like the one we know of – meaning, the temporal, or transient one-way ordering of our experiences. There are, nevertheless, fundamental differences between the two approaches to the question of the nature of time, which have to be given some thought. The reason for this, Eddington states, is that a symbol is something (well, in this case at least) that only "exists" – where this "existence" is given its sole meaning through a theory.[9] It is "an elaborate mathematical construct" (Eddington, 1946, p. 88). When we want to locate becoming within nature, a symbol is simply not good enough. What we would like to have is something of significance, something that allows recognition of a deep dynamic quality in nature that the symbol of the metrical type cannot disclose. We do not create sense by stipulating that one end is more chaotic; we need, according to Eddington, "a genuine significance of 'becoming'... not an artificial symbolic substitute." (Eddington, 1946, p. 88).

Now, how do we proceed in order to come up with this genuine dynamical significance? Eddington provides support for the view, that our most fundamental and primitive concept of time is identical to that time which is the most descriptive of all empirically accessible natural processes. He describes it as an ontological acknowledgement of primitive experience in the sense that, "we must regard the feeling of becoming as a true mental insight into the physical condition which determines it." (Eddington, 1946, p. 89).

When Eddington writes that "insight into the physical condition [which] determines it", he is saying that the subjective mind can "recognize" an objective con-

---

[9] As in all cases where physics are trying to explore the "ultimate significance of time" solely within the framework of physical theory, and where this particular framework is taken to be more fundamental – since it is *physics*? – than experience and primitive concepts are developed from real-life situations.

dition which cannot only be an "external" condition. This "objective condition" must somehow also be an "internal" property, a condition of mind. That is to say, a condition equally integrated into the mind as it is integrated into the rest of nature. Why should the human mind conceive of temporality in a form that is totally apart from the time of nature? Our conception of time which is based upon experience is as close as we get to a conceptualization of the physical condition which supports experience – as a condition for it. We simply loose hold of the connection between the "physical" element and mental statement of the temporal element in the sense that we do not "see" the physical element at all, but only the mental expression of it. We never have a grip on the physical aspect at all; we only infer it from the fact that our temporal experience is so fundamental that we cannot ourselves be the source of this breathtaking perspective of time. As Eddington claims, we will always be able to recognize "becoming" because it is not "image-building", but insight. It is insight because our elaborate nerve-mechanisms do not intervene: "That which consciousness is reading off when it feels the passing moments lies just outside its door." (Eddington, 1946, p. 89).

So then, we must simply come to terms with the idea that the mental insight into the time of physical nature is fundamental for any conceptualization about the objective nature of time. The realism of the objective concept of time depends on the mental insight into the flux that appears to us in experience; we simply "see" it as it is, that is, in its "pre-conceptualized purity". This experience also brings with it the realization of the significance of the experience; that, for example, we cannot reverse what appears in transience. In this sense we also have an "insight" into time's nature as "a kind of one-way texture involved fundamentally in the structure of nature." (Eddington, 1946, p. 90). We can know about this "texture" as we can also know about other properties of the external or physical world. We conceive of this transience as the passing of time, says Eddington, and furthermore, this is a "fairly correct appreciation of its actual nature." (Eddington, 1946, p. 90). We have one way in which we experience time directly. In order to "bridge the domains of experience belonging to the spiritual and physical sides of our nature," we need access to the world through our sense organs. We gain access to the temporal properties of external processes. We are able, through our sense organs, to relate time to other entities in the physical world. Eddington calls this "time's dual entry into our consciousness" (Eddington, 1946, p. 91).

"Becoming" – with its transitory properties – will not easily fit into the overall scheme of nomological explanation which characterizes physics. Physical time, at least as it is posited by Einstein in his Special Theory of Relativity, cannot be transient. In STR local time is necessarily represented in "non-transient" modality as soon as it is objectified within the space-time description of physics. The transitory property of time soon gets lost when the nomological structure of physical explanation is applied to the matter. It is interesting to note, however, that entropy

– as the only physical symbol – gives us a specified direction to external processes that no other physical theory is able too. The Second Law of Thermodynamics is a "law" that in fact presupposes transitory properties of external nature. This means that transience is an actual part of objective (external) time. That this is a presupposition hidden in the structure of this law does not make it any less physical than the other and more causal (deterministic) laws that "presuppose" other non-empirical non-temporal "properties". Quite the opposite, the relatedness to experienced properties characteristic of time give the law an empirical basis that no other law can claim. We should, however, be careful not to claim too much.

We should state that, as Eddington writes, "Entropy had secured a firm place in physics before it was discovered that it was a measure of the random element in arrangement." (Eddington, 1946, p. 104). Without it we are faced with a physical world that is, in Eddington's words, "upside down". It simply does not make any sense in relation to our understanding of time, to have our complete inventory of concepts discarded just because they do not correspond to those in physics. As Arthur Eddington writes "For that reason I am interested in entropy not only because it shortens calculations which can be made by other methods, but because it determines an orientation which cannot be found by other methods" (Eddington, 1946, p. 109). This still does not establish any identity between "becoming" (experienced time order) and entropy, but it can be used as an indicator of orientation in external nature that corresponds symbolically with both the macroscopically perceived "order" of things and events, as well as with the direction in our temporal experience. In order to experience nature's processes as asymmetrical and irreversible in time, asymmetrical processes and human observers must presuppose both the objective anisotropy and direction of time. In order to retain some "realism" to the temporal framework that will always accompany questions about time's role in nature, in such a way that we truly are talking about a synthetic time, we always have to start with the foundation. This foundation is the experience of time, and it is the experience of real time.

4 The comparison of time and entropy deepened and some technology added

"I grasp the notion of becoming because I myself become." (Eddington, 1946, 96). What is fundamentally involved in this expression of becoming? Clearly one aspect is that I have a body. In one respect we are acting in a world that is constantly changing, and thus corresponding to the flux of sense-experiences. In another respect, I realize, sadly, that I myself am gradually becoming older. In my activity

I produce "something" that can just as well be called "entropy".[10] "Entropy" is a construction based upon perceived facts about a world which change irreversibly. If I had a twin who traveled through space with high velocity, he would not travel in time but only trough space. I would have aged considerably on his return 20 years later. On the other hand, my twin brother would not have had the time to produce so much entropy as I would have; he would not have aged as much as I would have. Hence, I have – through my activity in the world I live in – brought time out in the open through my activity. This time is local and irreversible. That I produce entropy through my activity and, consequently, spend my energy, is an objective measure for the transition of time. An objective time consists of local time in which the observers have dual access. As proposed by Eddington:

1. The experience of the irreversible direction after which other experiences are ordered;
2. The sense-experiences containing information of the external world, i.e., of things changing, coming and going, of births and deaths, of fires and floods, of conversations and studies, of our own aging, and of our expectations about life that are realized or not, through our actions. The time measured by the clock makes sense only because our experience of the world gives direction to the measure.

Time in this respect is local, and to achieve objectivity we need a "field of simultaneity", something stretching beyond the here and now of my actual experience. In an important sense we already possess such a field, as the Danish physicist Peder Voetmann Christiansen has pointed out, in the capabilities of modern media technology, i.e., the Internet, all kinds of phones, television, radio, etc. What we could wish is for time in the local point of view to be related to other frames of reference. Einstein created an opportunity for invariant transformation of data, but what, in fact, is the relationship between local time and invariant data? First of all, local time, as a property of experience of external processes, yields asymmetry and unidirectionality. Unidirectionality and asymmetry are facts about every local point of view, so it is not these universal aspects about time that need to be transformed. What need to be transformed are the measurements, or relations, that is, the data about the external events obtained in the local point of view. The reason for this is that the measurement apparatus, the clock and the measuring rod, undergo changes locally. The behaviour of clocks and measuring rods corresponds to the presupposed asymmetric and unidirectional nature of what is being measured, but is itself a local "behaviour", since it is not coordinated with the behaviour of what is being measured. These aspects are local particulars, since the behaviour

---

[10] The following points of argument have been put forward by the Danish physicist Peder Voetmann Christiansen in two papers: (Christiansen, 1987, 1988).

of the measuring devices points out in which direction in time one is conducting one's measuring. Nevertheless, the opinion is that accurate measurements depend on the flawlessness of the measuring devices. In relation to our sense of the nature of real time this is clearly wrong, because even the very best of clocks are inaccurate. This is a fact because even if a known process like friction, which is energy that becomes chaotic or random, is eliminated as far as humanly possible, there will inevitably be heterogenic interruptions of the wanted homogeneity. This also means, as Eddington has pointed out, that, "the more perfect the instrument is as a measurer of time, the more completely does it conceal time's arrow." (Eddington, 1946, p. 99)

Einstein's many definitions of the clock show a gradually increasing emphasis on the ideal "non-friction", which is a "mass-less" clock where nothing should be left that could indicate the asymmetry and unidirectionality of real objects and things in the real world.[11] It is difficult to see what was to be obtained by this, except for a formulation of a pure theoretical entity, that is, an ideal and perfectly accurate time measure. From the point of view of theoretical physics one might assume that real world clocks are imperfect, yet, from the point of view of the real world, they are actually perfect. Any global time is an expansion of local time through a communication of the results, data and methods of obtaining the data. In the case of time the transformations applied and the invariance achieved are not the objective aspects. Only the universal characteristics involved in every local point of view indicate or point to an objective foundation of time.

This "objective" foundation is what is necessarily excluded in global transformations, where measurement data are invariant with respect to any local point of view. Here we have obtained epistemological objectivity of the measurements – they are invariant with respect to any local point of reference. The data are not about time; they are about a relation of light signals between objects, in other words, a relation of distance in time. This is what is measured. We cannot measure time, in the sense that physics applies it, as "time" is the measure and not what is being measured. But since the measure has lost its direction – because as a measure it has become a "particular" of some specific theoretical framework – one could, in order to account for its objective grounding in reality, introduce into the theoretical framework, as an explanation and justification of the measure and its real empirical context, the experiential characteristics of direction pertaining to what is being measured. To obtain such directional data one can implement the experience of the human observer, and in addition supplement it with external devices, such as, say, a thermometer.[12] Consciousness has no problem establishing an arrow since it is itself directed in its awareness of changing perceptions that appear ordered and

---

[11] See (Kostro, 2000, pp. 88–89).

[12] As suggested by Eddington (1946, p. 100).

irreversible. One wishes an arrow in the world – for the sake of having something external to the mind that can indicate the same irreversibility. This arrow, which has to be found in the world, must be a local phenomenon and analogous to the unidirectional arrow of the mind. The arrow will be found because it is an inextricable part of all external processes on the perceivable macroscopic level of reality. It is hidden in the "messages from the outside", as Eddington says, but never in the messages from clocks. It is, however, found "in messages from thermometers and the like, instruments, which do not ordinarily pretend to measure time." (Eddington, 1946, p. 100).

Global time is an extension of our local time perspective. As we understand it today, physical time, as the sole time concept within physics that symbolizes the irreversibility of processes in time, is in fact an extension of locally experienced irreversibility to the global or objective perspective on external matter.

Normally one thinks that objectivity is achieved when we remove ourselves from what is going to be explained – like when we ask ourselves: "What would it look like if I was not present, if there were no people around to experience the phenomena in question?" The normal procedure is then to contemplate what we bring into the picture, and about those aspects, which, perhaps, do not belong to it in the sense of being subjective aspects, which is brought in with experience. It is, as argued, here – within the domain of "subjectivity" that the mistakes come about, that is, we overemphasize the "subjective character"[13] of what we bring into the picture of the world. In fact, in the context of traditional deterministic physics, epistemology and metaphysics have asserted the view that irreversibility is "subjective". The emergence of this new trend within philosophy, and science in particular, "gave rise to a subtle transference of ideas from 'randomness' to 'lack of knowledge' and from there to 'subjectivity'."[14] This line of thought is highly biased, since the determination of what is to count as subjective or not has to be based on highly "theory-laden" assumptions. These are assumptions that are always based on notions that operate tacitly as the "background" of our everyday thinking, or awareness (which we talked about in the beginning of this essay). In this case the background is the ontological framework of classical physics. The point made here is that we do not know what the world would look like without

---

[13] In the sense of being something "private" either individually or as part of the cognitive apparatus of human beings.

[14] Denbigh and Denbigh (1987) write that "there developed a marked tendency among the 19th century scientists to attribute any apparent randomness in natural phenomena to a lack of sufficient knowledge about those phenomena rather than to any real chance element in nature. And there remains at the present time a strongly entrenched view to the effect that entropy is a subjective concept precisely because it is taken as a measure of 'missing information' – information which we *might* use but don't, due to thermodynamic systems being incompletely specified." (Denbigh and Denbigh, 1987, p. 1)

the locally situated experiencing observer. We only know what it looks like if we add our local point of view.

There is nothing that points to locality as something fundamentally flawed or "wrong", i.e. that our experiences should not be in accordance with reality. Time's irreversibility and unidirectionality are objective, in the sense that we can all agree upon these properties as aspects of the world at large. Time cannot be a solipsistic phenomenon. Neither can time have properties that cannot be experienced; otherwise we could not talk about time at all. The objectivity of time is not that it is independent of us human beings, but that we experience, internally and externally, certain properties that we all agree upon as being properties of time.

The true "realist" would be one that could give an account of time based on the fact that we access temporal reality trough temporal experiences and sense-experiences of external phenomena of the world. Time's irreversibility must be explained both from the "inside" and from the "outside". We begin with experience and proceed to the concept and then to the formalized symbol. But even this course of experience is partly conditioned by our present awareness of the world and the continuous "production" of memory. Memory can only be produced by a subject that experiences things, objects, relations and phenomena, all which are "objects" that are themselves under transformation by the "fluctuation" and "dissipation" of the world. As experiencing subjects we are thus confronted with the visible traces of increasing entropy (increasing order and disorder) that are somehow "descriptive" for the flux presented to our senses. These traces are local changes. These are changes we experience. They are transformations of the external world, of the phenomena which can be experienced, as well as of the observer as an organism. This local "entropy", these experienced changes that take place externally, provide an opportunity to extend our viewpoint from the internal to the external, though it is still a local perspective.

Physical time is, in some respect, a product that has its reference to the experiencing subject. Physics represent a form of conceptualizing or symbolizing things that dismembers "head" from "body" with regard to the phenomenon of time. Still, the concepts of time within physics are, in reality, mere extensions of the unavoidable local viewpoint on time. From the subjective point of view, the process of extending the applicability of time begins with the experience of time and continues with the intellectual endeavour towards still greater external applicability for the experience. The imaginative mind strives towards "physical space" in an attempt to define an even more universal, global applicability of the conceptualization it has of time as an external aspect to itself.

5 The technological extension of local temporality without rejection of subjectivity

Now we have to consider something about clocks and measurement concerning the operational process of extending the local perspective on time to a global one, which does not deny its linkage to subjectivity. If we consider local time characteristics (for instance, ones own body, heart beat, droplets dripping from the roof, the sundial), one has, together with other irreversible processes, an instrument for considering the amount of time elapsed. For example, Galilei had to resort to his own pulse as a clock when he discovered the laws which explained falling objects. Other processes are possibly more up to the task, for instance the diffusion when we add some colour to a liquid starting t_0 and observe the diffusive spreading of the added colour (Christiansen, 1987, p. 38). Other technological phenomena that are based upon constant dissipation are easier to apply in this respect, for instance, a clock that applies water (clepsydrae) or sand (sandglass). Indeed, all our mechanical clocks are also of this type. All clocks use or spend energy and thus produce or increase entropy; albeit as a hidden process, for, of itself, there is no trace of "before" and "after", of what is past.[15] These types of timepieces are called diffusion clocks, and the time these timepieces measure is, according to Norbert Wiener, called Bergson-time as opposed to the Newtonian-time of classical mechanics (Christiansen, 1987, p. 38). The linkage between diffusion time and experienced time is obvious. The phenomena are conditioned by the same conditions which we are ourselves conditioned by. The phenomena are in specific aspects perceptual phenomena that are natural elements in everyday (macroscopic) experience of the world. All these types of phenomena represent the kind of time that is local.

The linkage between phenomenology and physics – as in the linkage between processes ordered by temporal experience and processes studied and explained by thermodynamics – offers, in fact, a lucid perspective on the question on time in physics; one which has been stated most thoroughly by the physicist Peder Voetmann Christiansen (1987, 1988).[16] It is never a question of reducing temporal awareness to something less fundamental originating, say; in the processes explained by the concept of entropy as it is stated in the second law of thermodynamics. Rather than reduction, it is an attempt to see a more profound identity between the ways we experience and how macroscopic nature behaves.

In 1905 Einstein discovered the Brownian movements (Einstein, 1956). In this theory a connection between diffusion and dissipation was established. Later Callen and Welton named the connection "the fluctuation-dissipation theorem"

---

[15] For readers interested in the *development* of timekeeping, clocks and the measurement of time, see (van Rossum, 1947) and (Landes, 2000).

[16] In the following I will try to give an account of Christiansen's thoughts.

(Christiansen, 1987, p. 38). Christiansen explains that the purpose of the theorem is to secure the "same time" whether we are using a diffusion clock or some other mechanism that dissipates. The connection expresses that random "forces" or "conditions", or some "influence" – that are responsible for the diffusion and the Brownian movements of small particles – are the same as those that are responsible for the dissipation or friction of macroscopic movements. There will always be some degree of "noise" involved in irreversible processes. The "noise" is not something that we – in practice – have to consider, since our surroundings, and we ourselves, are not in a state of thermodynamic equilibrium. For instance, if dissipative forces make a macroscopic body, (say, a pendulum) stop, i.e., reach equilibrium, it will not be at absolute rest but, rather, perform Brownian movements about the position of equilibrium. (Christiansen, 1987, p. 38) According to Christiansen, the energy of these movements at normal temperature is approximately $4 \cdot 10^{-21}$ Joule. This is so little that it almost disappears in comparison to the energies we apprehend in the, far-from-equilibrium, condition of the universe. What if we start the movement of the pendulum at, say, 1 joule? We can imagine that these Brownian movements, in themselves, could with time have enough energy produced to cause the pendulum to begin swinging "on its own accord", or rather spontaneously at 1 joule. This is improbable. For a pendulum to begin its movements at the point where the energy of 1 joule is reached, we must conclude with certainty, says Christiansen, that someone pushed it in the past and that it will again stop in the future, unless someone pushes it again. This is a retarded response in macroscopic physical systems: these systems are retarded since the activity always is caused by past stimuli – never by future stimuli. And in this sense we find irreversibility as the most characteristic aspect in our surroundings.

This "irreversibility" is to a very important degree a presupposition for our particular form of perceptions, cognition and thus for the experience of the world itself. It is so significant that without it we would loose our feeling of continuity; we would loose the "wholeness" we find in our own world picture. In short, we would loose our minds. Therefore, all talk of "advanced response" is idle talk (Christiansen, 1987)[17] – because what, in reality, forbids all talk of advanced response is precisely our temporal experiences and the "thermodynamics" of our perceptual surroundings.

Christiansen states that if we are to talk about stimuli and response we have to refer to memory, which is the only instance that secures that the system was undisturbed in the past, that is, before we introduced our "stimulus". We can state – with Christiansen – that memory is a presupposition for physical irreversibility, in the

---

[17] *Advanced response* is connected to the notion of *time-reversible processes*, processes that are – in the metaphysical cosmology of static time – "caused" by *future* events. What comes first in experience is illusion because "in reality" it is – so it goes – only the effect.

sense that it leads us to the selection of the retarded response-functions and the rejection of "advanced" ones. On the other hand, we have to state that irreversibility must be understood to be a presupposition for memory (Christiansen, 1987).[18] Memory depends on the fact that external processes leave behind some traces and evidence of what have been taking place in the past, that there is a "differentiation" or "transformation" taking place in nature, which result in visible aspects, traces of past processes. Eddington said that we are not ignorant of the nature of organization in the external world. And this goes for the concept of "becoming" as well. The quality of the external world is "so welded into our consciousness that a moving on of time is a condition of consciousness." (Eddington, 1946, p. 97).

The next step on the route to a global time is to accept that "thermodynamic", irreversible time, which is measured by local processes, is the fundamental perspective on time. And if this is so, then there should be no problem to proceed to mechanical time and from there on to astronomical time (Christiansen, 1987, 39). This means that we have to proceed from our use of sandglasses and other diffusion clocks, to the application of mechanical clocks that – although they dissipate – are more "precise".[19] This again means that we have already secured a fundamental identity for time in external nature, in nature's irreversible and unidirectional "processes". We have used these characteristics to establish a linkage[20] to the char-

---

[18] Our experience of time must itself presuppose *time*.

[19] Ferdinand Gonseth (1971, p. 277) has asked the following important question in his paper "From the Measurement of Time to the Method of Research": "Are we quite certain of what is exactly meant by the words *time* and *measurement* before a timepiece is constructed?" And his answer is as simple as it gets: "Generally speaking, *a clock is simply a very observable phenomenon, the temporal law of which is known*." However, this does not state that a "clock" that measures real (heterogeneous) time is less precise because it is not a mechanical clock. This is to say; it does not measure the time or processes *homogeneously* (in this sense precise) as the mechanical clock do, because it corresponds with the rhythm of the process itself. Thus, Gonseth states that any kind of observable "clock" or process will do as long as one agrees upon the use and application of it as a measure, that is, as a "temporal law" of that which is perceived. Thus, we have to differentiate between 1) ontological "precision", that is correspondence between the applied measurement and device and the heterogeneous rhythm of external process, and 2) scientific precision that aims at a result that is in all aspects homogeneous as a measure. The "homogeneous" measure does not and cannot "copy" the rhythm of the external process that is measured. Needless to say, it is this last measure – understood as a piece of information gained by applying, say, a mechanical clock – that is the type of measure-convention that we have become accustomed to for the sake of quantifiable precision.

[20] First by some "analogy" based on experience, but also – perhaps more so – by the necessity of experience to be (somehow) grounded in nature, at least adapted to the characteristics of external nature perhaps through evolution. Just consider John Cohen's opening

acteristic irreversibility and unidirectionality that is the essential characteristics of human temporal experience. It is from this point onwards that we have to worry about how we should proceed in order to create a measure that yields global invariant precision. As Richard Schlegel writes, "Our concept of time is based on two kinds of natural processes: those with progressive, non-cyclic change and those which undergo cyclic change. The former define a directed, increasing property of time, the latter the quantified measure of time." (Schlegel, 1971, p. 27). It should be noted, however, that cyclic change is still irreversible change, that is, as a process it does not repeat itself because each time a new process develops – each process has its own uniqueness. There is nothing homogeneous about these "cyclic" processes. Homogeneity is only obtained by abstracting from the differences in the processes and by isolating a generalized pattern of cycles.

All kinds of clocks need some kind of energy in order to tick. Clocks are either wound or driven by batteries or some other source of energy in order to function, and this specific function of clocks will again eventually create heat generated by friction or dissipation in the mechanical parts of the clock. The fundamental point to be made in this is that the time parameter t has to be defined by the fundamental irreversible time measure. It should not be a problem to apply mechanical clocks for precision.

The reason for this is that fundamental time is not precise seen, at least, from a strictly mathematical point of view. Our fundamental time is never "precise", certainly not in the same sense as in the function of the "time" that we derive from it. The sole purpose of the derived "time" concept is "precision". The precision of the symbolic time-measure is not something that can be found as a constituent property of the objective (external) world, but stems from the local point of view – as has been argued earlier. It stems from our local viewpoint because it is intimately connected to the interpretation and thus with subjectivity in its contemplative mode. It should be obvious that mechanical time is secondary; something that is merely derived from its primary, or more objective source. The objectivity of time measurements is evinced, according to Gonseth, only by "the practical exploitation of the temporal solidarity of the phenomena." (Gonseth, 1971, p. 284). Still, according to Gonseth, this approach only presents us with one particular point of view because, he writes, "from a certain level of technical capacity onward, this first aspect is apparently disguised and cloaked by another aspect, that of precision." (Gonseth, 1971, p. 284).

That precision and objectivity are "connected" to each other stems from the demand found in our need to communicate and to make every little thought public.

---

claim in his paper "Time in Psychology": "A scientific world picture with pretensions to comprehensiveness cannot refuse to reckon with human experience, which is itself part of nature, and, in particular, with the experience of time." (Cohen, 1971, p. 153). This is exactly what I have stated.

Where time is concerned we cannot do without the ideals to which we strive to adapt our practical-technical reality of time-measurement.[21]

In this sense one almost becomes suspicious of science and human intellectual striving in relation to the importance of time. The suspicion is that science and abstract thinking do not disclose anything of the mystery of time. Human interaction and communication emphasize "precision" in several respects, such as, for instance, precision in speech, so, in fact, it all comes down to inter-subjectivity, to that which all humans share as innate nature and which, thus, is as given, say, as temporal structure, through experience. In some aspects, for the sake of precision and inter-subjectivity, we sacrifice essential aspects of reality and no one doubts that technology and natural science have secured considerable "gain" for us.

The role of experiential time and derived time has changed places in the ontological scheme of things. The absurdity goes even further in that it is argued that, since temporal experience is not precise, in the manner time is presented on our clocks, it must be an illusion. But one should not, at least in ontology, exchange temporal reality, which must be presupposed in whatever epistemological context, with the demand for global invariance in universally applicable formalisms. This "invariance" only secures for us some kind of pragmatic utility. Or it gives something external to our related ideas of epistemological "objectivity" for the concept. However, we are thus made to believe that we can state sane things about the "real" nature of time simply by replacing the local with global "invariance".

The present is a characteristic of the observer (I am not stating that the present only belongs to the observer). The now of the observer can never be taken as something that can be isolated from its experiential context, that is, isolated from its necessary interconnectedness to the past. It is the wholeness or the totality of experience that yields something temporal, which we can recognize as corresponding to the temporality of the world at large; to the characteristic temporal properties related to real processes such as irreversibility, transience and unidirectionality. That these properties of time cannot be part of any existing physical theory does

---

[21] Gonseth (1971, p. 287) writes: "Of course, to certain and possibly essential extent, the progress of clock-making technology has been inspired and oriented by a theoretical ideal, by the abstract model of the isochronic oscillator. The word *abstract* should mean here that it is a question of a model of a mathematical character, conceived according to the principles of so-called rational mechanics. The efforts of technicians and practitioners have long tended, and still tend, to realize this model as perfectly as possible... all research was oriented... towards the realization of conditions, which, in the ideal model, ensured the correct functioning of the isochronous oscillator. The improvements and discoveries to be made on the technical level seemed to answer the need for a *guiding principle: that of seeking an ever-greater approximation of the theoretical model"*. To be more *precise*, the ideal in clock-making industry is that of *sustained isochronic oscillation* (Gonseth, 1971, p. 289).

not mean that they are not global (real, objective), ontic features of time. Even if these properties cannot be "measured", they are, nevertheless, global properties in the sense that they belong to every local frame of reference, with or without an observer present. The characteristics are global through the local. "Globality" is strictly speaking a theoretical term which refers to an epistemological context based upon the necessity of "inter-subjectivity" for the sake of communication. The local or real world, on the other hand, refers to actual and real experiences of a real relation between experience and the properties belonging to the world.

References

Capek, M. (1965). The myth of frozen passage: The status of becoming in the physical world, *in* R. S. Cohen and M. W. Wartofsky (eds), *Boston Studies in the Philosophy of Science*, Vol. 2, Reidel, New York, pp. 441–463.

Capek, M. (1976). Introduction, *in* M. Capek (ed.), *Boston Studies in the Philosophy of Science*, Vol. 12, Reidel, Dordrecht, pp. 441–46.

Christiansen, P. V. (1987). Har universet en tid?, *Paradigma* **2**(April): 33–41. Translated title: Does the Universe have a Time?

Christiansen, P. V. (1988). Absolut og relativ tid, *Profil* **3**: 36–44. Translated title: Absolute and relative time.

Cohen, J. (1971). From the measurement of time to the method of research, *in* J. Zeman (ed.), *Time in Science and Philosophy*, Elsevier Publishing Company, pp. 153–164.

Davies, P. V. C. (1997). *About Time, Einstein's Unfinished Revolution*, Touchstone: Simon & Schuster.

Denbigh, K. G. and Denbigh, J. (1987). *Entropy in Relation to Incomplete Knowledge*, Cambridge University Press, New York.

Eddington, A. (1946). *The Nature of the Physical World*, Cambridge, Cambridge University Press.

Einstein, A. (1954). On the method of theoretical physics, *Ideas and Opinions*, Wings Books, pp. 270–275.

Einstein, A. (1956). *Investigations on the Theory of the Brownian Movement*, Dover Publications, Inc.

Feyerabend, P. (1993). *Against Method*, Verso: London – New York.

Gonseth, F. (1971). From the measurement of time to the method of research, *in* J. Zeman (ed.), *Time in Science and Philosophy*, Elsevier Publishing Company, pp. 277–305.

Gurwitsch, A. (1965). Comment on the paper by H. Marcuse, *in* R. S. Cohen and M. W. Wartofsky (eds), *Boston Studies in the Philosophy of Science*, Vol. 2, Reidel, New York, pp. 291–306.

Husserl, E. (1970). *The Crisis of European Sciences*, Northwestern University Press, Evanston.

Kostro, L. (2000).   What is this: A clock in relativity theory?, *in* M. Duffy and M. Wegener (eds), *Recent Advances in Relativity Theory, Selected Papers From the Biennial Conferences on Physical Interpretations of Relativity Theory (1988-1996)*, Vol. 1: Formal Interpretations, Hadronic Press, pp. 84–90.

Landes, D. S. (2000). *Revolution in Time – Clocks and the making of the modern world*, Viking – Penguin Books Ltd.

Marcuse, H. (1965). On science and phenomenology, *in* R. S. Cohen and M. W. Wartofsky (eds), *Boston Studies in the Philosophy of Science*, Vol. 2, Reidel, New York, pp. 279–290.

Pauri, M. (1997). The physical worldview and the reality of becoming, *in* J. Faye, U. Scheffler and M. Urchs (eds), *Perspectives on Time. Boston Studies in the Philosophy of Science*, Kluwer Academic Publishers, pp. 267–297.

Pickering, A. (2003). On becoming: The mangle, imagination and metaphysics, *in* D. Ihde and E. Selinger (eds), *Chasing Technoscience*, Indiana University Press, Bloomington, Indiana, pp. 96–116.

Schlegel, R. (1971). Time and entropy, *in* J. Zeman (ed.), *Time in Science and Philosophy*, Elsevier Publishing Company.

Sklar, L. (1995). Time in experience and in theoretical description of the world, *in* S. F. Savitt (ed.), *Time's Arrows Today*, Cambridge University Press, Cambridge.

St. John, O. (1974). Nature, life and mind, *in* J. Lewis (ed.), *Beyond Chance and Necessity*, Carnstone Press.

van Rossum, G. D. (1947). *History of the Hour, Clocks and Modern Temporal Orders*, The University of Chicago Press.

Williams, B. (1988). Philosophy, *in* M. Finley (ed.), *The Legacy of Greece*, Oxford University Press, pp. 202–255.

Interdisciplinary philosophy – or the way Andur looks at it

Vincent F. Hendricks

Section for Philosophy and Science Studies, Roskilde University

> I worry about the worth of philosophy done by philosophers who have
> been trained in nothing else. (Papineau, 2000)

"Philosophy" means love of wisdom. The original meaning of the word "wisdom"
from Homer and Hesiodus relates specifically to the notion of a "skill", and thus
to the transition from thought to action. Love of wisdom is not just a philosophical province or the philosophers' monopoly. This love is likewise to be found in
the special sciences; all the way from the humanities and social science to natural science, medicine and technology. Here however, the aspect of skill is much
more pronounced. To become pertinent and relevant, philosophy has to gather its
material from the interdisciplinary interplay with the other sciences. This in turn
requires the philosopher to acquire skills – and plenty of them.[1]

In Antiquity and up through the Middle Ages the sciences were less divided
and specialized. The concept of *natural philosophy* in Aristotle thus covers mathematics, physics, astronomy and neighboring disciplines which in one sense or
the other relates to nature and its ways of working, its constitution and laws. Real
natural philosophers of those days had something qualified and important to say
about everything from algebra to poetry, from planetary motion to politics, ethics
and jurisprudence. They were not always right but they often enough had something novel and important to say which usually tied the philosophical reflections
together with the more specialized insights to a complete view of nature or man or
both.

As the general insight into nature and man rose, more and more refined, formal
and technological methods and tools were needed in order to predict, model and
explain the phenomena of nature. Different disciplines were distilled; chemistry

---

[1] I would like to thank V.J. Menshy for his pertinent comments on earlier versions of this
paper.

was no longer alchemy, other disciplines merged like for instance mathematical physics, mathematical astronomy etc. This meant that more and more technical skills were to be acquired in order to practice the science in question. As a result, less and less people could truly call themselves natural philosophers. A figure like Gottfried Wilhelm von Leibniz (1646–1716) who about the same time as Sir Isaac Newton (1646–1727) developed the differential and integral calculus but also wrote insightfully about metaphysics, logic, medicine and weaponry was truly a natural philosopher. Leibniz was also a very successful womanizer. Neither Karl Friedrich Gauss (1777–1855) nor Herman von Helmholtz (1821–1897) were the greatest womanizers as far as we know but they possessed the necessary skills to drive mathematics, physics, medicine, psychology and philosophy forward. More current natural philosophers include Norbert Wiener (1894–1964), who besides being the man behind the mathematical models for firing projectiles from a moving base, additionally developed what today is known as cybernetics and by this token has become quite important to the philosophy of biology. Herbert Simon, originally a psychologist, was a Nobel Prize laureate in economics and is known as one of the founding fathers of artificial intelligence and his theories have become of indispensable importance to the philosophy of mind. So natural philosophers are still around, but note that they seldom have a degree in philosophy.

Philosophy is thus *not* exempt from preconditions and the acquisition of skills but with the high technological level of most sciences these days it requires *quite* some skills for the philosophers' love of wisdom to materialize in practice. Given the etymology of the word philosophy it is virtually a contradiction to label oneself a philosopher without the proper skills.

If you are interested in the history of philosophy on a research level you have to be able to read the texts in the language in which they were originally written and you have to be knowledgeable of historiography and philology. To do philosophy of physics requires knowledge of the mathematical machinery of modern physics together with quantum mechanics and quantum field theory; epistemology requires skills in mathematical and philosophical logic, cognitive psychology and even theoretical computer science; ethics and moral philosophy demands knowledge of medicine, economics, game and decision theory, law and even biology; to do philosophy of mind you need a detailed understanding of neuroscience, psychiatry etc. To practice philosophy of science, knowledge of – and interaction with – science is required although it is unfortunately just very uncommon that these skills are around as the Nobel Prize laureate Richard P. Feynman (1918–1988) explains:

> Philosophers, incidentally, say a great deal about what is absolutely necessary for science, and it is always, so far as one can see, rather naïve and probably wrong. (Feynman, 1953)

Another Nobel Prize laureate in physics, Steven Weinberg has the following to say about the philosophers' interaction with science and their use of language:

> From time to time since then I have tried to read current work on the philosophy of science. Some of it I found to be written in a jargon so impenetrable that I can only think that it aimed at impressing those who confound obscurity with profundity. (Weinberg, 1992)

The book *Feisty Fragments: For Philosophy* (London: King's College Publications, 2004) is a collection of more than 550 quotations from people inside and outside philosophy. The main message from Beethoven to Woody Allen seems to be that:

1. Philosophy is admittedly important, but
2. loses itself, especially nowadays, in philosophical thought-experiments and toy-examples rather than focusing on scientific method and real-life problems, which implies that
3. philosophy looks like science fiction more than it looks like science,
4. written in an unnecessarily complicated and mostly incomprehensible language.

1 Intuition or method

When looking at current philosophical practice there is certainly support to be found for the critical tone voiced above. A common conception of modern, especially Anglo-American philosophy is that philosophy for the most part is concerned with intuition-driven conceptual analysis. This means that ordinary concepts like knowledge, rationality, mind, identity etc. are stretched to the maximum extend of their (hopefully plausible) usage by consulting our intuitions about "possible cases".

Here is an example from contemporary epistemology which also underlies the Wachowski-brothers three *Matrix* movies. Since Plato it has been the standard opinion of philosophers that knowledge is, or should be, true and infallible. When I say that I know that I'm reading the newspaper now, it means that right now there are no reasonably imaginable circumstances in which I'm not reading the newspaper. Had there been such a circumstance then it would possibly be false, thus knowledge would be fallible, and I could no longer be said to know that I'm reading the paper now.

How is knowledge possible if it is possible that we err? Imagine, as the philosopher Hilary Putnam did in 1983 in his book *Reason, Truth and History*, that a malicious scientist have taken the brain out of your skull, placed it in a vat of nutritious fluids, and rigged your brain with electrodes wired to a super-computer which stimulates your brain into believing that everything is normal, although it possibly couldn't get more abnormal. In this case you would not be able to know that you are not a brain in a vat because everything seems normal, and it is by the way false that you are reading the newspaper because this is something that the super-computer stimulates you into believing. Here is a possibility of error, and thus knowledge has been demonstrated impossible!

It is admittedly a relevant possibility of error that you have forgotten your glasses, and so you are not reading the newspaper but the *Yellow Pages*. But when the philosophers come to ask you whether it is a relevant possibility of error that you are brain in a vat and so not reading the paper just seems ridiculous. Asking a physicist whether it is a relevant possibility of error that his voltmeter is calibrated incorrectly while attempting to measure the voltage drop over some circuit seems reasonable enough, to ask him about the brain in vat story for a measurement error again seems stupid.

If we on the other hand accept the thought-experiment then we are left in the uncomfortable situation that none of us know whether we are brains in vats and thus just victims of the *Matrix* on a daily basis. The British zoologist Peter Brian Medawar (1915–1987) puts it very adequately:

> Medical scientists use the word 'iatrogenic' to refer to disabilities that are the consequence of medical treatment. We believe that some such word might be coined to refer to philosophical difficulties for which philosophers themselves are responsible. (Medawar, 1983)

Medawar as a zoologist is not the only one who has raised this complaint. Philosophers are and have been aware of this pitfall as George Berkeley (1685–1763) prophetically puts it:

> Upon the whole, I am inclined to think that the far greater part, if not all, of those difficulties which have hitherto amused philosophers, and blocked up the way to knowledge, are entirely owing to ourselves. We have first raised a dust, and then complain that we cannot see. (Berkeley, 1988)

Part of the story is, by the way, that Berkeley was a devoted idealist and thus thought that the only thing one could be really sure about was the existence of mind and that to exist is to be perceived by a mind. This is a troublesome stance; does a dust devil exist when it is not perceived? That's hard to see.

There are shelves and shelves worth of thought-experiments like the one cited above in contemporary philosophy and we are talking about some pretty bizarre philosophical beasts: Utility monsters and swamp men, devils and demons, ingenious mirror-installations and paper-mâché barns, archers and Zen-masters – the list goes on ad *nauseam*. In the book, *The Philosopher's Toolkit – A Compendium of Philosophical Concepts and Methods* (2001) which, as the title suggests is an introduction to the fundamental methods and distinctions in philosophy, it reads accordingly regarding thought-experiments in science and philosophy:

> The difference between the thought experiments in science and philosophy, however, is that those in science often lead to physical experimentation. For philosophers, however, in most cases physical experimentation is unnecessary because what one is exploring is not the terrain of the physical but the conceptual universe. Reasoning out the leads of our imagination is often sufficient for concepts. (Baggini and Fosl, 2001)

This sentiment does not seem entirely right. First of all, there are many thought-experiments in physics which are not amenable to physical experimentation. Second, there are still rules governing the use of thought-experiments in conceptual universes. The absurd world in which contradictions are true allows for all and nothing – *ex falso quidlibet*; from a contradiction everything follows. Some have called upon such absurd worlds to salvage or undermine various philosophical theses. These worlds are called "impossible possible worlds". Such worlds are worse than the worlds in which you are a brain in a vat because the latter are uncommon but at least logically possible, the former are just impossible! That something is physically impossible to carry out is acceptable, even in physics, when something is logically impossible, it is just impossible. If thought-experiments are not governed by at least what is logically possible, they are not ruled by anything other than your fantasy and imagination. And even if we restrict attention to what is logically possible it may just be that the justifying intuition is being outstripped by fantasia as Daniel Dennett and Douglas Hofstadter put it:

> When philosophical fantasies become too outlandish – involving time machines, say, or duplicate universes or infinitely powerful deceiving demons – we may wisely decline to conclude anything from them. Our conviction that we understand the issues involved may be unreliable, an illusion produced by the vividness of the fantasy. (Dennett and Hofstadter, 1982)

If one pays homage to the idea that philosophy is strictly about conceptual analyses then philosophy couldn't possibly get more detached from this world and irrelevant. The skills of the philosopher are exhausted by the ability to use intuition and visit fantasia ever so often without means and guidelines for thought. But the special sciences have whole arsenals of methods of inquiry publicly available. Without these arsenals the transition between thought and action is severed, the interaction with science is severed and it is then no wonder when Albert Einstein says:

> Whenever I study philosophical works I feel I am swallowing something which I don't have in my mouth. (Mackey, 1994)

Scientists often express a negative sentiment towards philosophy and there is something to it. Note however that both Einstein and Niels Bohr along with other scientists actively participated in philosophical debates of metaphysical, cosmological, logical, ethical, political and methodological nature. The ticket to participation is based on the interdisciplinary view of philosophy and from here on it is not difficult to get the scientists interested in philosophy as the physicist Max Planck exclaims:

> I am now convinced that theoretical physics is philosophy! (Born, 1951)

A philosopher should be an interdisciplinary scholar – one who shares the faith and fate of the scientist or expert. From this point of view it is better to be an expert than a philosopher as Niels Bohr summarizes:

> An expert is someone who starts out knowing something about some things, goes on to know more and more about less and less, and ends up knowing everything about nothing. Whereas a philosopher is someone who starts out knowing something about some things, goes on to know less and less about more and more, and ends up knowing nothing about everything. (Bohr, 1991)

2 Philosophy across disciplines

Plato and especially Aristotle referred to philosophy as *theoria* – the mother of all sciences, or the science of sciences to which all and every discipline should aspire to. Plato even thought that philosophers should rule as they are the most enlightened. The days when philosophers had the skills and wisdom to rule or call themselves natural philosophers are long gone. That is not only the fault of philosophers, but also due to the way things have gone in science and society.

In the age of information and technology the role of philosophy has not diminished but rather changed. To look ahead the philosopher has to look back and recall the etymological starting-point for philosophy focusing on skill while at the same time being interdisciplinary well-versed. With the development of science it has become harder to be a philosopher because there are so many things one has to acquire knowledge and skill of and sometimes it is pretty technical stuff. That's just the way it is. The scientists take it as a challenge, philosophers should do too. By way of example, today we know that there are fluctuations in vacuum, that is, movement in empty space. If that doesn't have pertinence for philosophical deliberations about being and thus for one of philosophy's cardinal disciplines, metaphysics, it is hard to see what does have pertinence.

Again, to participate in this exciting interdisciplinary metaphysical endeavour one needs to know something about physics and this again requires skills in mathematics and the like. Similarly, in the public eye today is the debate about the ethics and moral questions related the problems in the Middle East, but qualified inputs from the philosophers require knowledge of politics, economy, religion, history and social science. It does not suffice to consider the thought-experiment of whether Peter should smack Paul in the school-yard as the result of Paul's earlier threat with a baseball-bat, which Peter is not sure that Paul even has anymore.

Many a philosopher is about to leave, or has already left, the purist philosophy, and the reason is to be found in the words of Charles Taylor:

> In think that philosophy in most aspects is pretty well useless and hopeless unless it's done with other disciplines. And that's the way I like to do it. (Taylor, 2000)

All around one registers this tendency. Computer scientists have joined logicians and psychologists, epistemologists and even sociologists to study knowledge, its use in action, deliberation, decision-making, and the emergence of norms under the rubric "social software". Engineers have joined philosophers of science and sociologists to do the philosophy of engineering science and the role of technology in modern society. In bio-ethics today one similarly finds everything from vets and farmers to moral philosophers and philosophy of mind counts philosophers, psychologists, neuro-scientists. One could go on. The tendency is also reflected in the international journals like *Philosophy and Economics*, *Journal of Business Ethics*, *Journal of Biology and Philosophy*, *Synthese*, etc.

The close interaction with the special sciences is exactly what makes philosophy scientific. Without the skills of science philosophy becomes a speculative and irrelevant activity void of content and undermining its own original intention and ambition. The American engineer and educator Charles F. Kettering (1876–1958) once said:

> It is easy to build a philosophy... it doesn't have to run. (Boyd, 2002)

Sure it does, but to run it has to run across the disciplines. That's pretty much the way Andur looks at it.

References

Baggini, J. and Fosl, P. (2001). *The Philosopher's Toolkit – A Compendium of Philosophical Concepts and Methods*, Blackwell, Oxford.

Berkeley, G. (1988). *Principles of Human Knowledge / Three Dialogues*, Penguin Books, New York.

Bohr, N. (1991). *Niels Bohr's Times*, Oxford University Press, Oxford. Edited by Bais, A.

Born, M. (1951). Natural philosophy and cause and chance, *The Restless Universe*, Dover Publications.

Boyd, T. A. (ed.) (2002). *Charles F. Kettering*, Beard Books.

Dennett, D. and Hofstadter, D. (1982). *The Mind's Eye*, Penguin Books, New York.

Feynman, R. P. (1953). *Six Easy Pieces*, Helix Books.

Mackey, A. (ed.) (1994). *A Dictionary of Scientific Quotations*, Institute of Physics Publishing, London.

Medawar, P. B. (1983). *Aristotle to Zoos*, Harvard University Press, Cambridge, MA.

Papineau, D. (2000). *The Philosopher's Magazine*, Spring.

Taylor, C. (2000). *The Philosopher's Magazine*, Autumn.

Weinberg, S. (1992). *Dreams of a Final Theory*, Vintage Books, New York.

Part V

Risk, uncertainty and safety

On imprecise statistical reasoning

Igor Kozine

Systems Analysis Department, Risø National Laboratory

1 Introduction

> ...The predominant philosophy of statistics a century ahead will be a Bayes/non-Bayes synthesis or compromise, and that the Bayesian part will be mostly hierarchical.... Aristotelean logic is insufficient for reasoning in most circumstances, and probabilities must be incorporated. You are therefore forced to make probability judgements.... Thus subjective probabilities are required for reasoning. The probabilities cannot be sharp, in general. For it would be only a joke if you were to say that the probability of rain tomorrow (however sharply defined) is 0.3057876289. Therefore a theory of partially ordered subjective probabilities is a necessary ingredient of rationality. Such a theory is "a compromise between Bayesian and non-Bayesian ideas.... (Good, 1980)

Uncertainty is an ever-present accompaniment of knowledge. An army of researchers of every hue seek to cope with it, and myriad definitions of uncertainty have emerged. Philosophers, psychologists, environmentalists, engineers and laymen devise their own version, and although uncertainty has been characterised in many ways, new types are no doubt under way.

Uncertainty plays an important role in decision making. A decision-maker's possibility of making a well-founded decision improves if uncertainty that is relevant to the decision is carefully addressed. In effect, there are three scientific schools in decision theory. One is classical decision theory, as presented by Luce and Raiffa (1957). This approach was challenged by Bayesian decision theory which, in turn, has been challenged by philosophers and mathematicians who disagree with its foundations.

A central assumption of the Bayesian theory is that uncertainty should always be measured by a single (additive) probability measure, and values should always be measured by a precise utility function (Savage, 1954). This assumption has been referred to as the Bayesian dogma of precision (Walley, 1991).

Imprecise statistical reasoning is motivated by the ideas that the dogma of precision is mistaken, and that imprecise probabilities are needed in statistical reasoning and decision. Good (1965, p. 10) suggests that "one is more or less a Bayesian depending on the precision with which one is prepared to make intuitive probability estimates".

In 1991 two books were published on the same subject. One was 'Statistical Reasoning with Imprecise Probabilities' by Walley (1991), the other was 'Interval Statistical Models' by Kuznetsov (1991).[1] A few years later it was established that the two books describe the same reasoning and are complementary. In (Utkin and Kozine, 2001) it was proven that the main constructive tool of the theory, natural extension, is one-to-one identical in both Walley's and Kuznetsov's exposition. The authors had never collaborated but came to the same theory independently and at about the same time.

The fundamental ideas set out in the books have been picked up by many others, and have been further investigated and developed into applied techniques.[2]

Imprecise probability is a generic term for a range of mathematical models that measure chance or uncertainty without sharp numerical probabilities. These models include belief functions, Choquet capacities, comparative probability orderings, convex sets of probability measures, fuzzy measures, interval-valued probabilities, possibility measures, plausibility measures, and upper and lower expectations or previsions (Walley, 1991). My own research has been focused on issues that belong to the theory of upper and lower expectations or previsions as described Walley and Kuznetsov.

The quote at the very beginning of this paper is likely to suggest that the paper will be about a synthesis and compromise between the Bayesian and the non-Bayesian approach. This suggestion is not misleading.

2 Discrete case

Let us look first at what kind of a discrete problem can be solved in the framework of the theory of upper and lower expectations.

---

[1] Kutznetsov's work is in Russian and has so far not been translated, making it less accessible to international readers than Walley's work.

[2] A number of researchers, including the present author, sensed in the publications of Walley and Kuznetsov promising signs of a new approach to reliability and risk assessments. Since 1998, my areas of research has thus been partly concerned with generalising reliability models to imprecise probabilities and modelling uncertainty in risk assessments with imprecise statistical reasoning.

Assume there are three possible outcomes s_1, s_2 and s_3 in a subject matter of interest. This is an exhaustive set of events meaning that $P(s_1) + P(s_2) + P(s_3) = 1$, where $P(\cdot)$ stands for a probability. Information on the probabilities of the occurrences of these events is given as three pieces of evidence: (1) $P(s_1) \in [0.1, 0.3]$, (2) s_2 is at least two times as probable as s_3, and (3) s_2 and s_3 is at least as probable as s_1. What are the probabilities $P(s_2)$ and $P(s_3)$ one can derive based on the provided information?[3]

One can hardly expect that the source imprecise information can result in precise answers in the form of precise probabilities $P(s_2)$ and $P(s_3)$. What is the mechanism of arriving at an answer?

As we have three possible outcomes, the simplex representation can demonstrate well the basic ideas of the approach. In Fig. 1 the vertexes 1, 2 and 3 correspond to the three states s_1, s_2 and s_3. The probability simplex is an equilateral triangle with height one unit, in which the probabilities assigned to the three elements are identified with perpendicular distances from the three sides of the triangle. Adding up these three distances gives 1. Thus, each point inside of the simplex can be thought of as a precise probability distribution. The simplex representation is especially useful for depicting pieces of statistical evidence and studying their effects on the probabilities of outcomes.

The first piece of evidence, $P(s_1) \in [0.1, 0.3]$, is depicted in Fig. 1, while Fig. 2 depicts all the source information with the simplex representation.

The source evidence can be rewritten in the form of inequalities (1) $0.1 \leq P(s_1) \leq 0.3$, (2) $P(s_2) \geq 2P(s_3)$, and (3) $P(s_2) + P(s_3) \geq P(s_1)$. These inequalities and condition $P(s_1) + P(s_2) + P(s_3) = 1$ define a constrained area which is shown in black in Fig. 2. The calculation of upper and lower bounds for the probabilities of interest becomes a geometric task. The calculated values of the probabilities are $\underline{P}(s_2)=0.466$, $\overline{P}(s_2)=0.9$, $\underline{P}(s_3)=0$, $\overline{P}(s_3)=0.3$, while $\underline{P}(s_1)=0.1$ and $\overline{P}(s_1)=0.3$ remain unchanged.

It can be noticed from Fig. 2 that the evidence $P(s_2) + P(s_3) \geq P(s_1)$ does not contribute to the precision and can be discarded without influencing the result. That is, the black area, defining the lower and upper probabilities, does not change if this evidence is removed from the set of evidence. This simply supports the fact of common sense that not all information has a positive contribution to the precision of the result.[4]

The coherent imprecise probabilities are considered a particular case of the theory of imprecise coherent previsions and based on three fundamental principles: avoiding sure loss, coherence and natural extension. A probability model *avoids sure loss* if it cannot lead to behaviour that is certain to be harmful. This is a

---

[3] This example in a greater detail was demonstrated in (Kozine and Utkin, 2002a).

[4] Precision is considered the value of difference between the upper and lower bound of the probability of interest.

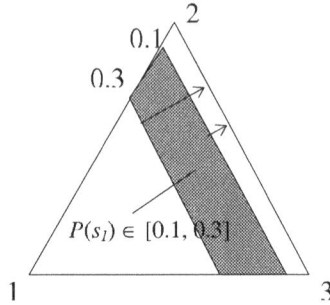

Fig. 1. Presentation of the statistical evidence $P(s_1) \in [0.1, 0.3]$ on the simplex.

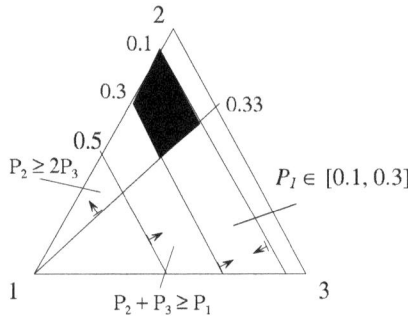

Fig. 2. Presentation of all available statistical evidence on the simplex.

basic principle of rationality. *Coherence* is a stronger principle which characterises a type of self-consistency. Coherent models can be constructed from any set of probability assessments that avoid sure loss through a mathematical procedure of *natural extension* which effectively calculates the behavioural implications of the assessments (Walley, 1991).

The principle of avoiding sure loss for the lower and upper probabilities is equivalent to holding the following inequalities:

$$0 \leq \underline{P}(A_i) \leq \overline{P}(A_i) \leq 1 \ \forall i = 1, \ldots, n, \tag{1}$$

$$\underline{P}(\Omega) = \sum_{i=1}^{n} \underline{P}(A_i) \leq 1$$

and

$$\overline{P}(\Omega) = \sum_{i=1}^{n} \overline{P}(A_i) \geq 1$$

where A_i are pairwise-disjoint subsets for any $i, j = 1, \ldots, n$ whose union is Ω, the possibility space.

The construction of coherent imprecise statistics and probabilities of events different from A_i is performed through the natural extension. The natural extension for this particular case is the solution of two linear programming problems

$$\underline{M}g = \sum_{i=1}^{n} g(A_i)P(A_i) \rightarrow \min \qquad (2)$$

$$\overline{M}g = \sum_{i=1}^{n} g(A_i)P(A_i) \rightarrow \max \qquad (3)$$

subject to

$$\left.\begin{array}{l} \sum_{i=1}^{n} P(A_i) = 1 \\ \underline{P}(A_i) \leq P(A_i) \leq \overline{P}(A_i), i = 1,\ldots,n \end{array}\right\} \qquad (4)$$

Function g can be, for example, $g = x$ and then $\underline{M}g$ and $\overline{M}g$ are lower and upper mean values of x.

If g is a characteristic function of an event B, i.e., $g = I_B(A_i) = 1$ if $A_i \in B$ and $I_B(A_i) = 0$ if $A_i \notin B$, then the natural extension is

$$\underline{P}(B) = \sum_{i=1}^{n} I_B(A_i)P(A_i) \rightarrow \min \qquad (5)$$

$$\overline{P}(B) = \sum_{i=1}^{n} I_B(A_i)P(A_i) \rightarrow \max \qquad (6)$$

subject to (4).

The lower and upper mean values $\underline{M}g$ and $\overline{M}g$ or $\underline{P}(B)$ and $\overline{P}(B)$ obtained as the solutions of linear programming problems (2) and (3) subject to (4) and (5) and (6) subject to (4) are referred to as coherent. In (Walley, 1991) and (Kuznetsov, 1991) other definitions of the natural extension can be found.

The sense of the natural extension in precise mathematical terms is to estimate the interval $[\underline{M}g,\overline{M}g]$ of possible values of Mg for all probability distributions for which $\underline{P}(A_i) \leq P(A_i) \leq \overline{P}(A_i)$, $i = 1,\ldots,n$. That is, we assume that any probability distribution consistent with the initial judgements $\underline{P}(A_i) \leq P(A_i) \leq \overline{P}(A_i)$ for $i = 1,\ldots,n$ is possible and base our inferences on this assumption without preferring a particular distribution.

The problem below is an example that could be made use of.

Problem: What are the probabilities of the outcomes of the following football match?[5]

---

[5] This problem was borrowed from (Walley, 1991) and slightly modified.

The three possible outcomes for a football match are win (W), draw (D) and loss (L) for the home team. Your beliefs about the match are expressed through the following simple probability judgements

> X1: *chance to win is less than 50%*
> X2: *win is at least as probable as draw*
> X3: *draw is at least as probable as loss*
> X4: *the odds against loss are no more than 4 to 1*

Check your results against those given on page 313.

3 Interpretation of upper and lower probabilities

> For many people, the first time they heard of the Pentagon's plan to ac-
> cept bets on terrorist activities was when the bizarre-sounding idea was
> abandoned. ... The Defence Advanced Research Projects Agency (Darpa)
> would have traded futures contracts that paid out if particular events, in-
> cluding terrorist attacks, took place. It was widely attacked as both ghoul-
> ish and nonsensical, *All bets are off at the Pentagon, by Tim Harford,
> Financial Times, September 2 2003*

Expressions (5) and (6) give us a formal definition (mathematical representation) of the upper and lower bounds for probabilities as maxima and minima of
the objective functions subject to a set of constraints. In turn, the set of constraints
includes also upper and lower probabilities. Where do they come from? How can
one acquire them?

To answer these questions we need to distinguish first the issue of interpretation from that of mathematical representation. There are many kinds of mathematical models for uncertainty, such as additive probability measures, upper and
lower probabilities, comparative probability ordering, etc. Any of these models can
be given various interpretations. Similarly, any single interpretation of probability
can be given various mathematical representations. De Finetti's work is a valuable
example of how interpretation can profoundly affect the mathematical theory. His
emphasis on finite (rather than countable) additivity and on exchangeability is a
consequence of his operational interpretation (Walley, 1991).

Let the possibility space Ω be the set of possible states of the world that are of
interest. The elements of Ω are assumed to be mutually exclusive and exhaustive.
A *gamble* is a bounded real-valued function defined on the domain Ω. A gamble
X should be interpreted as a random or uncertain reward; if the true state of the
world turns out to be ω, then the reward is $X(\omega)$ units of an appropriate asset. The
reward may be negative, in which case it represents a loss of $X(\omega)$ units. The value

of the reward X is uncertain, because it is uncertain which element of Ω is the true state.

Essentially, gambles are risky investments in which the utility values of the possible outcomes are known precisely (de Cooman, 1999). The subject's uncertainty about a domain can be measured through his attitudes to gambles X defined on that domain, and particularly by determining whether he will buy or sell a gamble X for a specified price x. In principle, we could measure the subject's uncertainty concerning Ω to any desired accuracy by offering him sufficiently many gambles and observing which of them are accepted. Equivalently, we could measure the subject's lower and upper previsions for a particular gamble X, which are defined to be the supremum acceptable buying price and infimum acceptable selling price for X. The transaction in which a gamble X is bought at price x has reward function $X - x$, a new gamble. A subject's supremum acceptable buying price for X is the largest real number c such that he is committed to accept the gamble $X - x$ for all $x < c$. Similarly, the transaction in which a gamble X is sold for price x has reward function $x - X$, and a subject's infimum acceptable selling price for X is the smallest real number d such that he is committed to accept the gamble $x - X$ for all $x > d$. This leads to the theory of upper and lower previsions in (Walley, 1991). The marginal buying and selling prices (lower and upper previsions) for a gamble may differ because the subject is indecisive or because he has little information about the gamble. As the amount of relevant information increases, the difference between the marginal buying and selling prices typically decreases. In the special case where every gamble X has a 'fair price', meaning that the supremum acceptable buying price agrees with the infimum acceptable selling price, one obtains the theory of linear previsions of (De Finetti, 1974).

Subsets of Ω, which are called events, can be identified with their indicator functions, which are gambles as well. When A is a subset of Ω, buying and selling prices (lower and upper previsions) for the indicator function A can be regarded as betting rates on and against A (lower and upper probabilities).

4 Judgements admitted in imprecise statistical reasoning: Continuous case

The thesis that "all available statistical evidence in risk and reliability analyses is to be utilised" is repeated in numerous guidelines in risk and reliability analysis. Everybody agrees but nobody knows how to make this true. As the remedy, Bayesian updating is usually brought up. Unfortunately, many people seem to believe that this is the only way of producing coherent statistical inferences. That is not so, for two reasons (Walley, 1997).

First, coherent statistical inferences need not be based on any assessment of prior probabilities. Second, even when inference proceeds by updating some kind of prior probabilities, imprecise prior probabilities can be presented by several mathematical models other than a set of prior probability distributions. In many problems it is difficult to identify a suitable prior distribution or set of prior distributions to perform Bayesian sensitivity analysis. Coherent imprecise previsions is an alternative method which in some problems is more convenient or more traceable.

In this section I will give some examples of the judgments that can be easily utilized by the method and that are relevant for a continuous set of possible outcomes (some more on admitted judgments can be found elsewhere in Kozine and Utkin, 2002b, 2003). The examples we may think of will usually involve the notion of time to failure (a continuous variable), this being a favorite target for reliability analysts. I will try to avoid giving too much mathematical formalism, but some of it cannot be avoided. To be able to utilize a judgment it has to be represented in a mathematical form that is then used as a constraint for a properly constructed objective function.

Direct judgements on the lower and upper probabilities of events or, in general, lower and upper previsions is a straightforward way to elicit the imprecise probability characteristics of interest. Constraint $\underline{a} \leq \int_{R_+} f(x)\rho(x)dx \leq \overline{a}$ is the model of a direct judgement. If, for instance, $f_i(X) = X$, then $\underline{a}_i, \overline{a}_i$ are the lower and upper expected values of X, correspondingly. If X is time to failure, then $\underline{a}_i, \overline{a}_i$ are the lower and upper bounds for the mean time to failure. If $f_i(X) = I_{[t,\infty]}(X)$, where $I_{[t,\infty]}(X)$ is an indicator function such that $I_{[t,\infty]}(X) = 1$ if $X \in [t,\infty]$ and $I_{[t,\infty]}(X) = 0$ otherwise, then $\underline{a}_i, \overline{a}_i$ are the lower and upper bounds for the probability of failure occurrence within $[t,\infty]$.

On a general note, direct judgements can be elicited and utilised on any probability characteristic that can be represented as an expectation to a properly chosen gamble.

Being able to utilise *comparative judgements* is a good feature of the theory of imprecise previsions. They could be, for example, "the failure of component A within the time interval [0,10] is at least as probable as the failure of component B within [0,20]", or "the mean time to failure of component B is less than the mean time to failure of component A". The first judgement is modelled as follows

$$\int_0^\infty \int_0^\infty (I_{[0,10]}(x_A - I_{[0,20]}(x_B))\rho(x_A,x_B)dx_A dx_B \geq 0, \tag{7}$$

and the second

$$\int_0^\infty \int_0^\infty (x_A - x_B)\rho(x_A,x_B)dx_A dx_B \geq 0. \tag{8}$$

Another kind of judgement is a *structural judgement*. Informally, a structural judgement is a hypothetical judgement that if you were willing to accept gamble X, then you would be willing also to accept gamble Y (Walley, 1991). Structural judgements may involve the notions [the properties] of independence and permutability, and both types can be modelled.

If the objective function for computing the lower bound of the expected value of a random function g appears in a form like this

$$\overline{M}(g) = \sup_P \int_{R_+^n} g(x)\rho(x)dx, \underline{M}(g) = \inf_P \int_{R_+^n} g(x)\rho(x)dx, \tag{9}$$

where $x = (x_1, \ldots, x_n)$, then this models the complete ignorance with regard to independence. The infimum is sought over the set P of all possible joint probability density functions $\rho(x)$. No structural judgement is introduced here. If there is a ground on which to judge independence among x_i, then $\rho(x) = \rho(x_1) \cdots \rho(x_n)$. It is clear that in this case set P is reduced and consists only of densities which can be represented as a product. As set P becomes smaller, then the imprecision, $\Delta = \overline{M}(g) - \underline{M}(g)$, is reduced.

In fact, the scope of judgements that can be utilised by the method is very wide (for more examples see (Walley, 1991, p. 169)). This, therefore, makes the thesis "all available statistical evidence in risk and reliability analyses is to be utilised" persuasive. This is because there really exists a tool that can utilise a wide spectrum of evidence.

One more class of judgements is the class of unreliable judgements which will now be addressed in the following section.

5 Unreliable judgements (hierarchical models)

> Good is prepared to define second order probability distributions..., and third order probability distributions over these, etc., until he gets tired (Levi, 1973)

The quality of information which a decision maker has concerning the possible states and outcomes of a decision situation is in many cases an important factor when making decisions. Experts providing judgements have different levels of expertise and their sources of information may not be equally reliable. So it is natural to assign different degrees of plausibility or probability to opinions by different experts. In order to be able to allow for this, a kind of hierarchical model can be used. In general, hierarchical models arise when there is a "correct" or "ideal" (first-order) uncertainty model about a phenomenon of interest, but the modeller is uncertain about what it is. The modeller's uncertainty is then called second-order uncertainty (de Cooman, 1999). The hierarchical model is, in many applications, a useful assessment strategy for constructing a first-order prior distribution (Walley, 1997).

The most common hierarchical model is the Bayesian one, where both the first and the second-order model are (precise) probability measures (Berger, 1985; Cooke, 1991; Good, 1980; von Winterfeldt and Edwards, 1986; Zellner, 1971). Other models allow imprecision in the second-order model, but still assume that the first-order model is precise. Examples are the robust Bayesian models (Berger, 1985), models involving second-order possibility distributions (de Cooman, 1998; Gilbert et al., 2000; Walley, 1997), and the Gärdenfors and Sahlin epistemic reliability model (Gärdenfors and Sahlin, 1982). G. de Cooman (1999) introduced and studied a particular type of imprecise behavioural second-order model in terms of so-called lower desirability functions.

We have studied hierarchical uncertainty models of a general form: imprecise first- and second- rder uncertainty models. Both models of uncertainty, first-order and second-order, are coherent interval statistical models.

Suppose that we have a set of unreliable interval-valued expert judgements on a parameter of interest b. To be more specific, we have n intervals $B_i = \left[b_1^i, b_2^i\right]$ provided by n experts, where b_1^i and b_2^i are the lower and upper bound of the interval B_i, respectively. The intervals provided are thought of as covering the true value of b, and are the models of uncertainty of the first order. The levels of confidence in the judgements depend on available information about experts' performance and their competences and may be subject to their own self-assessment. Suppose that each of n experts or each of their judgements is characterised by a subjective probability γ_i or, in general, by an interval-valued probability $\left[\underline{\gamma_i}, \overline{\gamma_i}\right]$, $i = 1, \ldots, n$. Now a hierarchical model can be written as follows:

$$\Pr\left\{b_1^i \leq b \leq b_2^i\right\} \in [\underline{\gamma_i}, \overline{\gamma_i}], i = 1, \ldots, n. \tag{10}$$

The hierarchical model is introduced to become a useful assessment strategy for constructing first-order uncertainty intervals. On how this is implemented is illustrated by the problem of combining expert opinions.

As given above, the information concerning a parameter b is given by a collection of n intervals B_i. Combined lower, $\underline{b}$, and upper, $\overline{b}$, bounds for b are the goals.

The result will definitely depend on the degree of credibility to each of the provided judgements. Say, the analyst is absolutely (100%) and equally confident about all the judgements. In terms of the formalism introduced above this means that $\Pr\left\{b_1^i \leq b \leq b_2^i\right\} = 1 \ \forall i = 1, \ldots, n$, that is, $\underline{\gamma_i} = \overline{\gamma_i} = 1 \ \forall i = 1, \ldots, n$. As proven in (Kozine and Utkin, 2002b), this case yields a simple rule of combination called the conjunction rule according to Walley (1991).

$$\underline{b} = \max_{i=1,\ldots,n} b_1^i \quad \text{and} \quad \overline{b} = \min_{i=1,\ldots,n} b_2^i \tag{11}$$

This rule is valid only for non-conflicting judgements ("consistent collection of intervals") and if the analyst is prepared to accept the modelling of the linguistic expression "equally credible" as $\underline{\gamma_i} = \overline{\gamma_i} = 1 \ \forall i = 1,\ldots,n$. Consistency as well as the absence of conflict mean that $\bigcap_i B_i \neq \emptyset$.

Another rule of combination is valid if all intervals in the collection are nested ("consonant"), that is, if

$$[b_1^1, b_2^1] \subseteq [b_1^2, b_2^2] \subseteq \ldots \subseteq [b_1^n, b_2^n] \tag{12}$$

and the credibility to the judgements is expressed in the different form $\underline{\gamma_i} = \gamma_i$, $\overline{\gamma_i} = 1$, $i = 1,\ldots,n$ and $\gamma_1 \leq \gamma_2 \leq \ldots \leq \gamma_n$. A closer look at this information gives a hint that the set-up of the source data of this kind is nothing other than a possibility distribution. This case of hierarchical models was described in (Walley, 1997) and (de Cooman, 1999).

The combination rule for this case looks as follows:

$$\underline{b} = \sum_{i=1}^{n} (\gamma_i - \gamma_{i-1}) b_1^i \tag{13}$$

$$\overline{b} = \sum_{i=1}^{n} (\gamma_i - \gamma_{i-1}) b_2^i$$

In this rule it is assumed that $\gamma_0 = 0$ and $\gamma_n = 1$.

A model for "equally credible" judgements could be differently constructed with the hierarchical model introduced. The modeller may choose to model equal credibility in the following way:

$$[\underline{\gamma_i}, \overline{\gamma_i}] = [\gamma_i, 1] \quad \text{and} \quad \gamma_1 = \gamma_2 = \ldots = \gamma_n = \gamma \tag{14}$$

then the last rule of combination degenerates to

$$\underline{b} = \gamma b_1^1 + (1-\gamma) b_1^n$$

$$\overline{b} = \gamma b_2^1 + (1-\gamma) b_2^n.$$

If γ tends to 1, then the results of this rule coincide with the results of the conjunction rule.

The conjunction rule can also be applied to consonant intervals as this rule is valid for a consistent collection of intervals, and it is clear that nested intervals are non-conflicting pieces of evidence. But it should be kept in mind that the use of the conjunction rule presupposes that the analyst is 100% confident about all the judgments, i.e., $\underline{\gamma_i} = \overline{\gamma_i} = 1$.

If the collection of intervals is conflicting (there is at least one pair of non-overlapping intervals), then one way of reconciling the conflict is to accept complete ignorance concerning the level of credibility in the judgments. That is, the analyst can assume $\underline{\gamma_i} = 0, \overline{\gamma_i} = 1, \forall i = 1, \ldots, n$. Using this assumption we arrive at the unanimity rule

$$\underline{b} = \min_{i=1,\ldots,n} b_1^i \quad \text{and} \quad \overline{b} = \max_{i=1,\ldots,n} b_2^i$$

These are simple combination rules that have been derived based on the hierarchical model, and the way they have been derived was fully predefined by the theoretical framework of coherent imprecise probabilities. This fact is worth stressing, since, in contrast, in the framework of purely Bayesian approach and point-valued probabilities only some ad hoc combination rules are possible. An example is the linear opinion pool which is one of many other devised to combine evidence.

6 Concluding notes: Is indecision good or bad?

Determinacy and decisiveness in decision making are favoured by the public and decision makers, while fuzziness and indecision in providing crisp answers are reckoned as signs of incompetence and meekness which are usually disliked. In this regard, both classical and Bayesian decision theory appear the right ones as providing a clear-cut answer to what action is to be preferred.

In contrast, the approach to decision making based on imprecise (interval-valued) probabilistic criteria will reach results that, generally, do not yield an 'optimal' action that is preferred to all others. In effect, this means that there is a third alternative answer under decision making. It is indecision in saying neither 'yes' nor 'no'. The failure to determine a uniquely optimal action simply reflects the absence of information about the set of possible actions which could be used to discriminate between actions.

What would be a strategy which could be used to make a decision in case there is more than one reasonable action? One of them is to search for more information concerning the set of possible actions to make the probabilities and utilities more precise. The other is to postpone a decision until a later time, when more information may be available. For more strategies see (Walley, 1991, pp. 239–240).

A small but growing number of authors have called for, and observed the development towards a paradigm shift in environmental decision making. As uncertainty becomes an accepted fact by scientists on the one side and the public and politicians on the other 'this requires a change of attitude on both sides: The politicians have to accept that fuzzy answers may be the best expression of expertise. The scientists have to learn that identification of the fuzzy borderline between knowledge and ignorance may be the sign of real competence' (Harremoës, 2003).

Imprecise statistical reasoning provides models to quantify scientific incerti-
tude which is a result of a lack of relevant information or/and a result of sizable
uncertainty. When there is little information on which to base our conclusions, we
cannot expect reasoning (no matter how clever or thorough) to reveal a most probable hypothesis or a uniquely reasonable course of action. There are limits to the
power of reasons (Walley, 1991). And an educated mind should provide answers
consistent with the relevant knowledge and uncertainty.

One of the important novelties of imprecise statistical reasoning approach is
that we now have a formal framework in which we can articulate uncertainty and
indecision.

Solution to the problem stated in Section 2

$$P(W) \in [0.33, 0.5]$$

$$P(D) \in [0.25, 0.4]$$

$$P(L) \in [0.2, 0.33]$$

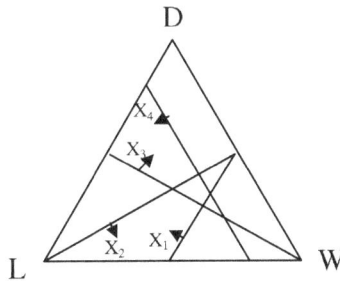

Fig. 3. Presentation of the four pieces of evidence in the football problem on the simplex.

References

Bedford, T. and Cooke, R. (2001). *Probabilistic Risk Analysis. Foundations and
methods*, Cambridge University Press.
Ben-Haim, Y. (2001). *Information Gap. Decision Theory. Decisions under Severe
Uncertainty*, Academic Press.

Berger, J. (1985). *Statistical Decision Theory and Bayesian Analysis*, Springer-Verlag, New York.

Cooke, R. (1991). *Experts in Uncertainty. Opinion and Subjective Probability in Science*, Oxford University Press, New York.

de Cooman, G. (1998). Possibilistic previsions, Proceedings of IPMU'98, *in* E.D.K. (ed.), *Seventh Conference on Information Processing and Management of Uncertainty in Knowledge-Based Systems*, Vol. 1, July 6–10, 1998, Paris, France, pp. 2–9.

de Cooman, G. (1999). *An Imprecise Hierarchical Model for Behaviour under Uncertainty*, Kluwer Academic Publishers.

De Finetti, B. (1974). *Theory of Probability. A critical Introductory Treatment*, John Willey & Sons.

Gärdenfors, P. and Sahlin, N.-E. (1982). Unreliable probabilities, risk taking, and decision making, *Synthese* **53**: 361–386.

Gilbert, L., de Cooman, G. and Kerre, E. (2000). Practical implementation of possibilistic probability mass functions, *Proceedings of Fifth Workshop on Uncertainty Processing*, WUPES, 21–24 June 2000, Jindvrichouv Hradec, Czech Republic, pp. 90–101.

Good, I. (1965). *The Estimation of Probabilities*, MIT, Cambridge (MA).

Good, I. (1980). Some History of the Hierarchical Bayesian Methodology, *in* Bernardo, J.M. et al (ed.), *Bayesian Statistics. Proceedings of the First International Meeting Held in Valencia (Spain), 1979*, University Press, pp. 481–519.

Harremoës, P. (2003). The need to account for uncertainty in public decision making related to technological change, *Integrated Assessment* **4**(1): 18–25.

Klir, G. (2003). Uncertainty-Based Information, *in* P. M. Pinto and H.-N. Teodorescu (eds), *Systematic Organization in Fuzzy Systems*, Vol. 184 of *NATO Science Series: Computer & Systems Sciences*.

Kozine, I. O. and Utkin, L. (2002b). Processing unreliable judgements with an imprecise hierarchical model, *Risk Decision Policy* **7**: 1–15.

Kozine, I. and Utkin, L. (2002a). Interval-Valued Finite Markov Chains, *Reliable Computing* **8**: 97–113.

Kozine, I. and Utkin, L. (2003). Variety of judgements admitted in imprecise statistical reasoning, *Risk Decision Policy* **8**: 111–120.

Kuznetsov, V. (1991). *Interval Statistical Models*, Radio and Sviaz, Moscow. [In Russian].

Levi, I. (1973). Inductive logic and the improvement of knowledge, *Technical report*, Columbia University.

Lewalle, P. (1999). Risk Assessment Terminology. Methodological considerations and provisional results. Report on a WHO experiment. Terminology Standardisation and Harmonisation, *Technical Report II(1–4)*, WHO/FN.

Luce, R. and Raiffa, H. (1957). *Games and Decisions. Introduction and Critical Survey*, Dover Publications, New York.

McColl, S., Hicks, J. et al. (2000). Environmental health risk management, *Report of Network for Environmental Risk Assessment and Management (NERAM)*, Institute for Risk Research, University of Waterloo.

Savage, L. J. (1954). *The Foundations of Statistics*, Wiley, New York.

Utkin, L. and Kozine, I. (2001). Different faces of the natural extension, *Proceedings of the Second International Symposium on Imprecise Probabilities and Their Applications, ISIPTA '01*, pp. 316–323.

von Winterfeldt, D. and Edwards, W. (1986). *Decision Analysis and Behavioral Research*, Cambridge University Press, Cambridge.

Walley, P. (1991). *Statistical Reasoning with Imprecise Probabilities*, Chapman and Hall, New York.

Walley, P. (1997). Statistical inferences based on a second-order possibility distribution, *International Journal of General Systems* **9**: 337–383.

Warren-Hicks, W. and DRJ, M. (1998). Uncertainty analysis in ecological risk assessment: Glossary, *Proceedings of the Pellston Workshop on Uncertainty Analysis in Ecological Risk Assessment*, Society of Environmental Toxicology and Chemistry (SETAC), Pellsten, Michigan, p. 276.

Zellner, A. (1971). *An Introduction to Bayesian Inference in Econometrics*, Willey, New York.

Zimmermann, H.-J. (1996). Uncertainty modelling and fuzzy sets, *Proceedings of the workshop "Uncertainty: models and measures"*, Lambrecht, Germany, July 22–24, pp. 84–98.

The case for in-depth uncertainty analysis in policy relevant science

Martin P. Krayer von Krauss

1 Introduction

Over the past 150 years, technology has had a profound impact on society.[1] The internet, computers and cellular phones are just a few recent examples of the wide variety of technological innovations produced by natural scientists and engineers that have dramatically changed the day-to-day lifestyle of people in industrialised countries. However, the development of technology is not the only function of scientists and engineers in modern society. Unfortunately, technological development has been accompanied by a number of unexpected, adverse impacts on human health and on the environment. Because of their knowledge and skills, natural scientists and engineers are intimately involved in the regulatory process put in place by society to manage technological development and minimize its adverse side-effects.

This essay argues that, given their role in the regulatory decision making process and the complexity of the problems they are often called upon to study, it is incumbent upon scientists and engineers involved in policy relevant science to communicate the uncertainty characterizing their assessments. Information about uncertainty contributes to the basis for deliberation on the precautionary measures warranted and the general desirability of technological innovations. It is also useful in informing adaptive decision making. Ultimately, uncertainty assessment should contribute to increasing reflexivity in the decision making process.

---

[1] This essay is based on Chapter 1 of the author's Ph.D. thesis (Krayer von Krauss, 2006).

2 Scientists and engineers as policy advisers

2.1 A brief historical perspective

Modern western thought has its origin in the age of rationalism and the Enlightenment of the 17th and 18th centuries. One of the major changes brought about by the Enlightenment was the displacement of the Church by science as the authoritative source of knowledge about man and nature.

The ideas of the early scientists Copernicus (1473-1543), Descartes (1596-1650) and Newton (1642-1727) were particularly instrumental in precipitating this change. Copernicus argued that the Earth revolved around the Sun, which at the time was a shocking theory because it contradicted the view that was presented in the Bible, according to which the Earth is at the center of creation and the Sun hangs from a celestial ceiling. If Copernicus was right, then the Bible could no longer be taken as a reliable source of knowledge. Scientific beliefs about the world would then need to be gathered in a radically new way. Descartes was instrumental in establishing how this should be done, calling for a methodological examination of knowledge before the "forum of Reason". Descartes believed that all material bodies, including the human body, are machines that operate by mechanical principles. Through Reason, man could understand these principles and use this understanding to improve his own condition. The new "scientific method" formulated by Descartes established procedures through which Reason could be applied to acquire knowledge that was free from arbitrary and unfounded or superstitious assumptions. Following this, the success of Newton in describing the laws that govern the motions of the planets in simple mathematical equations (his three laws of motion and his principle of universal gravitation) greatly bolstered man's confidence in his capacity to attain knowledge. Thus, by the end of the 18th century, science had replaced the Church as the chief source of man's understanding of the universe.

Between 1750 and the early 1900s, the technological innovations of the Industrial Revolution made it possible to greatly increase the transformation of natural resources and the mass production of manufactured goods. These innovations included the use of new raw materials (e.g. iron and steel), the use of new energy sources (e.g. coal, steam, petroleum and electricity), the advent of the factory system, and important developments in transportation and communication (e.g. steam locomotive, automobile, telegraph and radio).

The Industrial Revolution was accompanied by broad social changes such as the growth of cities and the explosion of urban populations. The problem of drinking water treatment received widespread public attention following the discovery that the London cholera epidemic of 1854 was due to a contaminated public well. Cholera epidemics forced most of Europe's cities to develop the means of providing clean drinking water and sanitation services to their populations. It was soon recognized that in order to deal with the many changes taking place in society, public officials possessing specialized technical knowledge were required. Thus, natural scientists and engineers became involved in the management of technological development. Before long, governments at all levels relied on scientists and engineers to occupy important positions within the civil service.

Two themes emerge from the brief historical account above. The first is that following the Enlightenment, scientists and engineers came to be regarded as the authoritative source of knowledge in society. When religion came to be perceived as being based on beliefs and superstitions, science came to be perceived as being unbiased and factual, the unique bearer of the True and therefore the Good. The second central theme is that scientists and engineers have become firmly anchored in the fabric of society. Through their research activities and their activities in the civil service, scientists and engineers function as both the source of technological innovation, and as key participants in the management of this technological innovation.

2.2 The involvement of science in the policy process

The broadening scope of technological applications, coupled with increasing public concern over the impacts of these technologies, led to the establishment of the modern regulatory agencies we know today (e.g. Environmental Protection Agencies). Because of their knowledge and skills, a myriad of different scientists and engineers are involved in the regulatory process. Some scientists and engineers are employed within the agencies themselves and perform the specialized tasks associated with the activities of the agency. Others are employed outside the agency (e.g. in industry, consulting or academia), but act as advisers to the agency in situations where important regulatory decisions must be made concerning issues where scientific expertise is required. The function of these scientists and engineers is to provide decision makers with scientific assessments upon which to base their decisions. An important goal of the regulatory process is to ensure that the health of the public and the environment are protected from the potential side-effects of technology. Thus, in many cases, the subject of scientific assessments is a matter of public interest, such as minimising the risks posed by chemicals to human health and the environment. Examples of frequently encountered assessment methods are risk assessment and cost-benefit analysis.

The function of scientific advisers is illustrated in Figure 1. The Figure presents an idealized model of the decision making process (OXERA, 2000). The model distinguishes four different roles in the decision making process: the decision maker, the policy maker, the scientific adviser and the stakeholder representative.

Their respective functions can be described as follows:

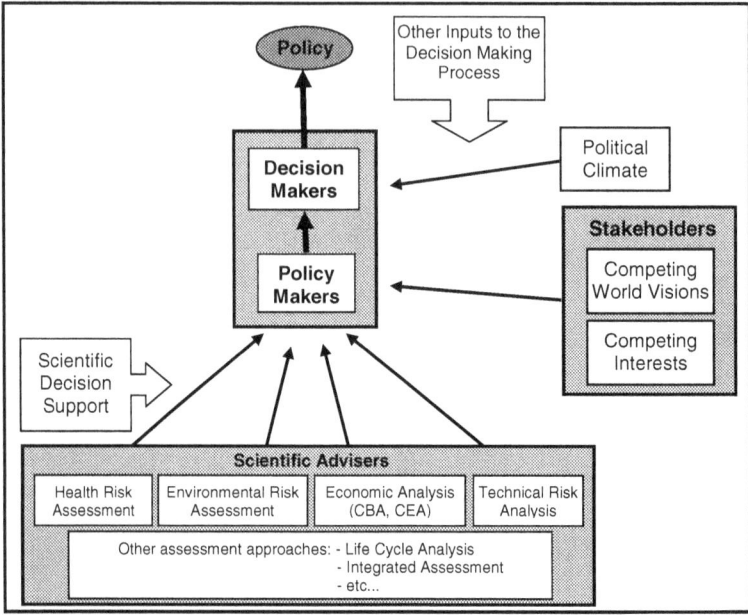

Fig. 1. Model of the regulatory process (adapted from OXERA, 2000)

- **Decision maker** – a person with the authority to take a policy decision. This may be a government minister, or a person or body with the delegated authority to take a decision in the name of a minister.
- **Policy maker** – a person or organization charged with assisting a decision taker in reaching a decision by providing policy analysis or generating policy options.
- **Scientific advisor** – a person or organization responsible for providing scientific input to policy-making or decision making, whether from within or outside the civil service. This includes both scientists expert in narrow disciplines relevant to the problem in question, and broader-based scientists able to integrate several disciplines.
- **Stakeholder representative** – a person or organization representing the interests and opinions of a group with an interest in the outcome of a particular policy decision.

The term *"decision support"* is used to describe the type of scientific activity conducted by Scientific Advisers. This type of activity is at times designated as policy relevant science, decision support sciences, or regulatory sciences. Decision support sciences can be thought of as *structured search processes that aim to provide knowledge that facilitates decision/policy making* (van Asselt, 2000). A number of different assessment approaches can be characterized as decision support, amongst which some commonly employed methods are environmental impact assessment, life-cycle analysis, ecological and health risk assessment, technical risk analysis, cost-benefit analysis (CBA) and cost-effectiveness analysis (CEA).

2.3 The liberal foundation of the regulatory system

Although the management of technological developments requires the expert knowledge and technical know-how possessed by scientists and engineers, the necessity of involving scientists and engineers in the regulatory process is more profound than this pragmatic requirement may suggest. To understand the role of science in the modern regulatory process, it is useful to consider some of the basic principles upon which it is based.

One of the founding principles of the modern regulatory process is the liberal principle of *state neutrality*. According to traditional liberalism, the state should be neutral with regards to particular attitudes and values, that is, conceptions of the good. Such conceptions are seen as private rather than public matters, and the law is supposed not to favour any particular conception. On the contrary, values are deemed to be illegitimate as justification for political action. Rather than being based on values, decisions should stem from a rational consideration of the facts. Thus, science is invested in the regulatory process in order to provide an impartial source of facts upon which policy decisions can be based.

A second founding principle of the regulatory system, the *harm principle*, was formulated more than a hundred years ago by John Stuart Mill. It basically states that persons should be free to do whatever they like, unless their activities are harmful to others. The principle was originally intended to protect individual freedom in matters of, for instance, religion and sexuality. Today, the principle is applied to many areas of regulation, including regulations on the application of new technologies.

The influence of the principle of state neutrality and the harm principle is to create a requirement for facts about harm. Harm is the trigger for regulatory intervention, and only facts can determine the existence of this harm. In practice, this means that in order to justify regulatory intervention, "threats" should be defined in as specific terms as possible and ideally in quantitative form. The basis for action should be a factual one, ideally developed through the use of a rigorous rational methodology (e.g. risk assessment) to ensure that the interpretation of the facts is the most objective possible (Fisher, 2005).

As a result of this, when a party disagrees with a particular technological enterprise, the most legitimate grounds for disagreement is to prove (or claim) the harmfulness of the enterprise in question. In other words, to be effective, opposition must be expressed in terms of risk, the existence of which is to be demonstrated using science (Jensen et al., 2003; Meyer et al., 2005). As will be explained further on, this can be quite problematic in situations where scientific knowledge is limited, facts are uncertain, stakes are high and values are conflicting.

3 Complexity

An important distinction between decision support sciences and conventional sciences lies in the complexity of the problems studied. Conventional sciences tend to focus on isolated systems under controlled circumstances. Decision support sciences must often deal with real world problems, the complexity of which far exceeds the complexity of the problems typically studied in conventional sciences. This realisation has lead to the designation of a new class of problems, *complex problems*, which includes issues such as global climate change, chemical pollution and stratospheric ozone depletion. Complex problems are characterized by one or more of the following properties (adapted from NRC, 1988; van Asselt, 2000; Funtowicz et al., 1999; Holling, 2001; UNESCO, 2005):

- There is not one problem, but a tangled web of related problems;
- The underlying processes interact with one another within some sort of hierarchy;
- The dynamics of the systems studied are not necessarily regular, but are characterized by synergistic and/or antagonistic relationships, indirect relationships, long delay periods between cause and effect, thresholds or nonlinear behaviours;
- The issue lies across or at the intersection of many disciplines, i.e., it has economic, environmental, socio-cultural and political dimensions;
- There are a number of different, equally legitimate and plausible, perspectives on how the problem should be conceived.

The hierarchical relationships encountered in complex systems may be hierarchies of inclusion and scale, as in a watershed that includes streams, ponds, rivers, lakes and the sea, at ascending levels. Alternatively, they may be hierarchies of function, as in an organism that is comprised of a number of separate organs, each performing a function subordinate to the overall function of the organism, which itself may be sub-ordinate to the overall function of an ecosystem. Environmental systems may also include human and institutional sub-systems, which are themselves systems.

In complex systems, causes and effects are not always obviously related. Pushing on a complex system "here" often has effects "over there" because the parts are indirectly dependent of one another. Similarly, the future conditions in a complex system may not always follow closely on the conditions in the past. When a particular threshold is exceeded, the system can abruptly shift away from a period of relative stability in one state, to another, fundamentally different state. An example is that of the impacts of climate change on thermohalyne circulation, the large-scale ocean circulation that currently transports heat from the mid-latitudes to the high latitudes. Geological analysis and model experiments suggest that these currents can be on or off, and that the two states are characterized by drastically different environmental conditions in Western Europe (Broecker, 1997; Cubasch et al., 2001). Similar dramatic regime shifts have now been documented for a wide range of environmental systems (Scheffer et al., 2001; Scheffer and Capenter, 2003).

Because of the hierarchical, indirect, synergistic and nonlinear relationships that can characterize complex systems, any attempts at reductionist analysis will be inherently incomplete. The concepts used to represent the functioning of the system will necessarily be rough approximations. The empirical data required may not be available, or may only be available in a form that requires interpreting or massaging to make it relevant to the problem at hand. Thus, in addition to the obvious uncertainties resulting from the operations of data collection and aggregation, the analysis will be characterized by deeper, structural uncertainties, not amenable to quantitative analysis. Similarly, every analyst of a complex system will operate at a certain level in the hierarchy of the system, with certain observational and analytical tools, and with certain prior experiences. The result of their separate observations and analyses are not at all arbitrary; but none of them can singly encompass the whole system. There is no unique, more plausible or more legitimate approach through which to analyse the system. The choice of which analysis approach to use is therefore value-laden. The values involved are those embodied in the societal or institutional system in which the science is being done.

An approach inherent to conventional sciences is that of reductionism, whereby an overall system is understood as an assembly of sub-systems. By studying and understanding each of the sub-systems, an understanding of the overall system is achieved. While the reductionist approach has led to many great achievements in Western science, the properties of complex problems, explained in more detail below, greatly reduce the effectiveness of the approach. Systems that are complex are not merely complicated; by their nature they involve deep uncertainties and a plurality of legitimate perspectives. Thus, rational, reductionist approaches to analyzing risks, such as risk assessment and cost-benefit analysis, will have only limited value.

Funtowicz et al. (1999) have illustrated the challenges related to the analysis of complex systems, using the Hindu parable of the six blind men touching an elephant. One blind man touched the side of the elephant and said it was a wall. Another blind man touched the ear and said it was a large leaf of a tree. Yet another blind man was holding a leg and thought it was a tree trunk. Still another blind man took hold of the elephant's trunk and said it was a snake. Someone else was touching the elephant's tusk and believed it was a spear. Another blind man had the elephant's tail in his hand and was calling it a rope. All of the blind men were touching the same reality but were understanding it differently. Each conceived the object on the basis of his own partial imagination process, but none of them could visualise the whole elephant. Only an outsider possessing the sense of sight was able to visualise the whole elephant. Similarly, in science, a number of different experts, each from a different discipline, may produce a number of different analysis of a complex system. While each of these analyses may be a correct partial description, they fall short of a holistic grasp of the system. Although a truly holistic grasp will always remain unachievable, policy relevant science must strive to integrate partial views into a richer view of the whole.

4 Policy relevant science

Since the 1960's, the social dimensions of science have increasingly become a subject of study by philosophers, sociologists and historians of science. One of the central themes that have emerged from these studies has been a strong critic of the view of science as the provider of the objective truth. Following the publication of the seminal work by Thomas Kuhn (1962) (*The Structure of Scientific Revolutions*), it became increasingly apparent that rather than being the result of value-neutral, dispassionate observations, scientific knowledge is constructed and negotiated according to a social process. That which is considered acceptable or unacceptable as knowledge is largely determined by negotiations within the scientific community (Kuhn, 1962; Knorr-Cetina and Mulkay, 1983; Latour and Woolgar, 1986). These negotiations are influenced by social factors such as rhetoric, politics, and personal reputations. Because of this, scientific conclusions, or the 'truth' perceived, will depend on the actors and circumstances involved, and are therefore not definitive accounts of the physical world. Rather, they are social constructions which may evolve as circumstances change.

Funtowicz and Ravetz (1990) were amongst the first scholars to interpret the constructivist paradigm described above in terms of policy relevant science. They introduced the term "*post-normal science*" to designate science conducted in contexts of high political pressure, large decision stakes, disputed values and pervasive uncertainty. The scientific nature of decision support activities derives from the fact that they are conducted in a structured, methodological way. However, while the results of conventional science most often serve to increase understanding of the natural universe or to contribute to technological advancement, the results of decision support are used to facilitate decision making in the policy process. In this context, the leading scientific problems are no longer derived from abstracted scientific curiosity or industrial imperatives. Rather, the research agenda for decision support sciences is set in the context of complex policy problems. Typically in such contexts, political and ethical considerations influence how the problem is defined, which research questions are to be given priority, and often which research groups should be mandated to investigate the problem. Thus, Funtowicz and Ravetz (1990), Wynne (1992) and others (e.g. see Krimsky and Golding, 1992; Jasanoff and Wynne, 1998) argue that decision support sciences are particularly prone to the "subjective pitfalls" highlighted by social constructivism. Rather than "hard" facts and "soft" values, decision support sciences should be considered "soft" sciences providing the basis for "hard" decisions.

Funtowicz (2004) proposed that there are (at least) two different visions of the relation between science and decision making in policy processes, which he labels as "modern" and "precautionary".

4.1 The modern vision of the science-policy interface

To a large extent, the vision of science that emerged following the Enlightenment, which is of science as a special and higher form of knowledge, the provider of objective facts about the natural world, still underlies the modern paradigm for science conducted at the science-policy interface.

In the modern paradigm, policy should follow from the consideration of objective scientific facts. This vision is "reductionist", in that it understands the world in terms of component parts that allow for abstracting the part from the whole, as well as increasingly specialized knowledge of each component part. The result of this "logic of component thinking" is a view of the world which is referred to as "technocratic", according to which the world is perceived as a system that can be technically redesigned in ways that make it more efficient and controllable. Typically, technocrats tend to see technical solutions as applicable to most social and cultural situations (Fischer, 2000). Problem solving, in short, is reduced to a technical matter of plugging solutions into different social contexts. In addition to being reductionist and technocratic, the modern paradigm can also be characterized as "positivistic", in that it accepts a separation of facts and values (Proctor, 1991), claiming that empirical research can and should be conducted without normative (value-dependant) references, and viewing uncertainty as a temporary and resolvable certainty deficit. In the modern paradigm, the role of science is to speak truth to power (Wildavsky, 1979).

4.2 The precautionary vision of the science-policy interface

The precautionary paradigm holds that through complex issues such as water management, climate change and endocrine disruptors, it has become apparent that in many cases, complete knowledge or 'truth' is not achievable. In such cases, there are relatively few simple 'facts' upon which decisions can be based. While numerical models are often used to investigate complex systems, the properties of such systems, the intricacies of modeling and the lack of transparency in complicated models imply that assessments involve value-laden problem definitions and assumptions. Assessments of policy problems involve uncertainties of many sorts, including irreducible lack of knowledge and ignorance. Within the precautionary paradigm, scientific knowledge is perceived as socially constructed to some extent, in that it necessarily offers an account of the physical world that is mediated through social processes; and that it therefore cannot be considered definitive. Scientific conclusions are claims that have been deemed to be adequate by a specific group of actors in a particular cultural and social context. Scientific assessments may facilitate decision making, but their conclusions cannot be defended with reference to objectivity and neutrality. Problem framing and the identification of relevant scientific disciplines and knowledge resources are political decisions. In the precautionary paradigm, uncertainty is a phenomenon inherent to science, which should be recognized and taken into consideration in the decision making process.

As will be seen below, the precautionary paradigm has been crystallised in more definite terms within the debate surrounding the Precautionary Principle.

5 The Precautionary Principle

To a significant degree, the environmental "surprises" witnessed in the past 50 years have been due to disregarded knowledge, uncertainty or ignorance (Harremoës et al., 2001). Too frequently, the growing innovative powers of science seem to outstrip its ability to predict the consequences of its applications. At the same time, the scale of human interventions in nature increases the chances that hazardous impacts may be both serious and global. Therefore, the Precautionary Principle has been proposed to guide decision making in situations of high uncertainty and potentially large-scale, irreversible impacts.

A central conclusion of the European Environmental Agency report on the Precautionary Principle (Harremoës et al., 2001) concerns the importance of recognizing and fully understanding the nature and limitations of scientific knowledge. No matter how sophisticated knowledge is, it will always be subject to some degree of uncertainty and ignorance. By their nature, complex problems have traditionally been inadequately addressed in the decision making process, and that which is commonly referred to under the umbrella term 'uncertainty' actually hides important technical distinctions. Engineers and scientists are taught at an early stage how common problems such as sampling errors and imprecise measurements generate uncertainty in experimental results, and how this uncertainty can be dealt with using statistical methods. However, this well-established approach to uncertainty analysis leaves out many important aspects of the uncertainty encountered when assessing complex policy problems (ESTO, 2001).

The regulatory process can at times be the stage of bitter disputes amongst stakeholders, all of whom are trying to steer the process in a direction that best serves their particular interests. Proponents of potentially harmful activities often tend to use uncertainty as an argument for postponing or waiving regulation (Michaels and Monforton, forthcoming; Michaels, 2005). They demand certain knowledge about the harm caused, as well as about the cause-effect relationship leading to the harm, to justify the need for regulation. Such a strategic behaviour towards uncertainty is not only observed among the defenders of business interests, but also among NGO's and other interest groups (Jasanoff, 1994; Fischer, 2000). While striving to ensure transparency and consistency in decision making, regulators themselves often become trapped in a quest for certainty (van Asselt and Vos, 2005). Experience demonstrates that often, conclusive evidence of harm only becomes available once harm has been done (Harremoës et al., 2001). In many cases, the politicisation of uncertainty (i.e., emphasizing or amplifying uncertainty to serve a specific interest) paralyses the environmental management process (Funtowicz and Ravetz, 1990; Michaels, 2005), a particularly undesirable outcome when the consequences of regulatory inaction could lead to irreversible harm.

There is no universally accepted definition of the Precautionary Principle. Gee and Krayer von Krauss (2005) present one formulation of the principle stating that it

> provides justification for public policy actions in situations of scientific complexity, uncertainty and ignorance, where there may be a need to act in order to avoid, or reduce, potentially serious or irreversible threats to health or the environment, using an appropriate level of scientific evidence, and taking into account the likely pros and cons of action and inaction (Gee and Krayer von Krauss, 2005)

This formulation does not clarify who has the burden of proving absence or presence of threats of harm, nor how or who is to determine the appropriateness of the level of scientific evidence. Another reading of the precautionary principle (Rogers, 2003) holds that

> the proponent of an activity posing uncertain risk bears the burden of proving that the activity poses "no" or an "acceptable" risk before the activity can go forward (Rogers, 2003)

6 The precautionary paradigm revisited

Two themes emerge from the above statements of the Precautionary Principle: one concerns the generation (and presentation) of policy relevant scientific knowledge including its uncertainty, while the other concerns the application of (uncertain and perhaps even partisan) scientific knowledge in political, regulatory and judicial decision making processes (Krayer von Krauss et al., 2005). In view of further articulating the precautionary paradigm, it is useful to consider the insights generated within three bodies of literature: i) the literature on deliberative decision making, ii) the literature on adaptive management, and iii) the literature on reflexivity.

6.1 Deliberative decision making

One of the profound implications of the precautionary paradigm is its bearing on the legitimacy of regulatory decisions. In a system where regulators are meant to be the value-neutral administrators who base all of their decisions on facts, what may justify regulatory interventions and what kinds of interventions are justifiable, in situations where the facts are uncertain? The problem is that this is the case with nearly all environmental and public health issues, which leads to a situation where regulators cannot act, and/or a façade of objectivity is constructed to justify action. The challenge is thus to conceive the regulatory process in a way that ensures the ability of regulators to act in an open and accountable manner in situations of uncertainty. In response to this challenge, many scholars argue for a regulatory decision making process where deliberation amongst actors plays a central role (Funtowicz and Ravetz, 1990; NRC, 1996; RCEP, 1998; Fischer, 2000; ESTO, 2001; Klinke and Renn, 2002; Harremoës et al., 2001; Fisher, 2005).

Deliberative decision making aims to achieve a synthesis of scientific expertise and public values on a specific issue. Here, the notion of the "threat" that justifies regulatory intervention is interpreted broadly, such that there is no pre-defined or precise definition of the acceptability nor the nature of the risk (Fisher, 2005). The legitimacy of regulatory decisions is restored through an increased democratisation of decision making, whereby a variety of actors, representing as wide a spectrum of perspectives as possible, are invited to participate in the decision making process. Here, conflicts are resolved in consultation rather than in confrontation (Webler, 1995; Jasanoff, 1994).

While deliberative decision making processes begin with the consideration of scientific inputs (i.e., risk assessments), this is only one activity in a more complex evaluation procedure. The scientific inputs are subsequently brought into a deliberative arena for debate in a wide forum which includes stakeholders and public interest groups in addition to scientists and decision makers. Funtowicz and Ravetz (1990) designate this process as *extended peer review*. The expectation is that the participants in the debate (i.e., the extended peers) will introduce additional factors to the decision making process. There are no methods or guidelines pre-determining which factors should serve as the basis for decision making, and these will vary from one context to another. The ambition is to create a synergistic process, whereby deliberation generates a body of considerations that is richer than that which would be generated if each of the stakeholders simply put forth their considerations individually, without reflecting on those of others. In other words, the relevant aspects of a decision will co-evolve as they are considered (Fisher, 2005).

It is not possible to provide general guidance as to what should form the basis of a deliberative decision. In nearly all circumstances, science will be an important consideration (Stirling, 2001), but other factors may also be relevant. Such factors may include *extended facts* (Funtowicz and Ravetz, 1992), i.e., information stakeholders and laypersons possess on the extent and nature of the risk. Examples of extended facts include anecdotes circulated verbally, edited collections of such materials prepared by citizens' groups and the media, the experiences of persons with a deep knowledge of a particular environment and its problems, or the materials discovered by investigative journalism.

While the training and employment of experts can socialise them to abstract, generalized conceptions, those whose livelihoods depend on the problem will have a keen awareness of how general principles are realised in their backyards. The traditional, intuitive and particularized knowledge the affected lay people possess gives them a firmer grounding in real world operational conditions. Often, too, their knowledge may be based on different perceptions about what is salient, or what degree of control is reasonable to expect or require, whereas the knowledge of technical specialists may simply be based on the common practice, without further reflection. Thus, in addition to contributing to the formulation of policy problems, local knowledge can also help determine which data, models and assumptions are relevant in particular cases.

One prominent example of a contribution from lay knowledge relevant to the regulatory process concerns workplace awareness of emerging patterns of ill health (Harremoës et al., 2001). The histories of usage of asbestos (Gee and Greenberg, 2001) and PCBs (Koppe and Keys, 2001) provide examples where workers were aware of what regulators subsequently recognised to be a serious problem. Similarly, local communities may become aware of unusual concentrations of ill health before the authorities, as occurred in the Love Canal case (Gilbertson, 2001).

In addition to extended facts, lay people and stakeholders can raise a number of ethical considerations that are relevant to the decision making process. These may include different perspectives on the acceptability of risk in the face of uncertainty, issues of fairness and environmental justice, visions on future technological developments and societal change, and preferences about desirable lifestyles and community life. While the extended facts provided by stakeholders and lay people may broaden the factual basis for decision making, it is the process of including their additional perspectives and values that legitimizes the decisions made in the face of complexity and uncertainty.

An important reflection on deliberative approaches is that public preferences do not necessarily match the real interests of the public since the preferences are clouded by misinformation, biases, and limited experience (Klinke and Renn, 2002). Of course, the input provided by stakeholders and lay persons should be subjected to the same intensity of critical scrutiny as specialist expertise (Harremoës et al., 2001). Just like expert knowledge, lay knowledge can be uncertain, partial, biased etc. Simply organizing a platform for mutual exchange of ideas, arguments, and concerns does not suffice to ensure fair and competent policies (Renn, 1999). Mixing all of these knowledge and value sources creates the danger that subjective perceptions supersede factual assessments, or that the rhetoric of powerful actors dominates the input of less powerful and organised actors, who may be those who must bear the risks. This has given rise to a growing number of formal methods to conduct deliberative decision making. Common to all of these methods is the aim to ensure that the contributions of different actors are embedded in a dialogue setting that guarantees mutual exchange of arguments and information, provides all participants with opportunities to insert and challenge claims, and creates an active understanding amongst all participants (Forester, 1999; Fischer, 2000). Examples of such methods and approaches include multi-criteria analysis (Jansson, 2001; Stirling and Mayer, 2001), constructive technology assessment (Rip et al., 1996), technical options analysis (Ashford, 1991; Tickner, 2000), consensus conferences and scenario workshops (Andersen and Jæger, 1999), cooperative discourse (Renn, 1999), and participatory policy analysis (Fischer, 2003).

6.2 Adaptive management

The basic idea of adaptive management – implementing policies as experiments – first emerged in the 1970s and 80s in the field of natural resource management (Holling, 1978; Walters, 1986). Adaptive management is grounded in the admission that humans do not know enough to manage ecosystems. Thus, adaptive management formulates management policies as experiments that probe the responses of (eco)systems as people's behaviour in them changes. Rather than thinking of ecosystem management as the task of managing nature, adaptive management aims to manage the *people* who interact with the ecosystem (Lee, 1999). In practice, it was proposed that this vision be enacted by developing computerised models, preferably using the best available inter-disciplinary knowledge, by using these models to test the impact of different policy measures, and by identifying key uncertainties in the models. These uncertainties could then be explored by conducting focused, large-scale management experiments in the field which, through monitoring efforts, would directly reveal the impacts of the policies tested, at the space and time scales where future resource management would actually occur, and the experience gained could then be applied on a large scale (Walters, 1997).

Adaptive management implies a change in the way uncertainty is perceived. In adaptive management, science and knowledge are considered intrinsically uncertain, and it is accepted that regulatory decisions must therefore be made in a context of uncertainty (Walker et al., 2001). Thus, information about uncertainty is used proactively, as a resource in the decision making process. The goal of the policy experiments is to learn something about the ecosystem's processes and structures, in view of designing better policies and experiments. Because ignorance of ecosystems is uneven, management policies should be chosen in light of the assumptions they aim to test, so that the most important uncertainties are tested rigorously and early (Lee, 1999).

Although adaptive management has been applied in many different contexts, to date, it has been much more influential as an idea than as a practice. Experience has shown that there are many human and institutional barriers that can hinder the proper implementation of adaptive management. For example, the complexity of the ecosystems and human behaviour implies that causal understanding is likely to emerge slowly (i.e., over several years). However, policy formulation is often driven by relatively short funding and election cycles, thereby making it difficult to treat long-term crises in the natural world effectively. In some cases, agency professionals may view admission of uncertainty as an admission of weakness (Gunderson et al., 1995). It can be very difficult to convince people who adopt such views that they will gain more credibility by openly admitting uncertainty, and then suggesting proactive ways of dealing with that uncertainty.

6.3 Reflexivity

Reflexivity is a concept that has been central to social scientific thought since the 1990s, after the writings of authors such as Beck, Giddens and Lash on modernity, risk and the cultural dimensions of contemporary environmental issues (Beck, 1992; Giddens, 1991; Beck et al., 1994; Lash et al., 1996). Beck introduced the term "reflexive modernization" to designate a new stage of development which, according to him and others, society has entered. This new era is characterized by a change in the way society perceives its relation to the risks to which it is subjected. In the previous era, known as "modernity", society was exposed to risks generated by external factors such as nature, which society responded to by developing technologies to overcome risks and increase welfare. In the current era of reflexive modernization, the principal risks to which society is exposed are no longer generated by nature, but by society itself: it is the unintended side-effects of technological development that currently pose the greatest threat to our welfare. Rather than producing ever-increasing security and welfare for people, industrial society has come to produce ever-greater risks for people and the environment.

The notion of reflexive modernization has led to calls for the development of institutions and approaches to decision making that foster reflexivity in the regulatory context (Hajer, 1995; Flyvbjerg, 2001). The extent to which a decision-making process can be considered reflexive hinges upon the ability of the policy community to recognize the limitations of the knowledge base underpinning a decision, draw upon the collective knowledge and experience of the past to design a policy, monitor and assess the effects of this policy, and adjust the policy accordingly. In this sense, reflexivity implies both "reflex" and "reflection" (Craye et al., 2005). Reflex in reference to the response of society to the unintended consequences of technological development. Reflection in reference to the careful consideration of the limitations of the information available, the diversity of viewpoints, and the multiplicity of possible policy options.

The rationale here is that the best way to cope with the reflexivity of the modernization process is to increase the reflexivity of the decision making process. In other words, expect surprises, seek them out, and deal with them as they arise. Reflexive decision making will require the actors of the policy community to change their attitude towards uncertainty, conflict and decision making. A pro-active attitude which recognises and accounts for uncertainty is required, and conflict must be perceived as a learning opportunity rather than as a battle to be fought and won.

In a reflexive policy community, conflict is treated as an important catalyst to reflection. Conflict may lie in the discrepancy between one's vision of the ideal and one's perception of reality, or in the discrepancy between one's vision and the vision of other actors in the policy process. They may have a different vision of the ideal, a different perception of reality, a different perception of the problem that is preventing the realization of the ideal, or, a different perception of how this problem should be solved. As has been mentioned previously, a frequent source of conflict lies in the interpretation of uncertainty. The actors in the policy process must collectively determine the quality of the information at hand, and the level of protection justified. In the process, the sources of uncertainty identified will be an important subject of reflection.

A reflexive policy community is one that is committed to learning about how decisions may affect society and the environment. Advocates of deliberation argue that reflection is provoked through deliberation, as the different actors of the reflexive policy community seek to develop a shared understanding of the issue under investigation. The insight gained is then applied and implemented through a policy decision, the effects of which are monitored. Inquiry is a transaction with the situation in which knowing and doing are inseparable. In such an approach, the role of the expert is deeply transformed (Schön, 1982; Fischer, 2000; Forester, 1999).

Rather than providing an optimal technical answer, the role of experts is to participate in the collective effort of producing, evaluating and applying knowledge, considering the interests at stake, and making a necessarily provisional decision.

7 Conclusion

Although many still perceive science as the provider of the objective truth, it is now increasingly being recognised that science is a social process of knowledge production, subject to its own social and cultural biases. As society becomes more aware of the complexity of the problems it faces and of the difficulties of studying these problems through the scientific approach, it can be expected that the precautionary paradigm will gain in influence, slowly displacing the modern paradigm. Under the precautionary paradigm, scientific conclusions are no longer considered definitive, and the legitimacy of regulatory decisions can no longer be defended solely by referring to scientific assessments. This legitimacy deficit is compensated through an increased democratisation of the decision making process. The precautionary paradigm also implies a shift towards a regulatory process that is more adaptive and reflexive. Adaptive in the sense that regulatory decisions are considered inherently uncertain, temporary and experimental, subject to revision as new information emerges. Reflexive in the sense that the actors in the policy process are perceived as members of a learning community that are capable of collective reflection, and collective responses to resolve conflicts between a vision of the ideal and the reality experienced.

The precautionary paradigm has profound implications for the role of experts in the decision-making process. Experts can no longer place messy factors such as the economic, social and political aspects of an issue beyond the boundaries of their narrowly defined technical field. Rather, they must accept these factors as part of their legitimate field of concern, opening up to complexity, instability, and uncertainty. Experts are now expected to reflect publicly on the quality of their knowledge, explicitly revealing their uncertainties and opening up to questioning and confrontation by other members of the policy community. They must emancipate themselves from the widespread statistical or probabilistic understanding of uncertainty, to recognize the full spectrum of the uncertainties characteristic of policy relevant science.

There is no reason to believe that it will be easy to shift from the modern to the precautionary paradigm. Professional cultures are not easily transformed, especially in the situations of high stakes and disputed values that characterize many regulatory decision-making processes. The competences required for scientists and engineers to function well in this new context will not be acquired simply as a result of deciding to do so. It is very likely that the shift, where it occurs, will proceed gradually and with difficulty as the different actors of the policy community increase their willingness to experiment with new modes of interaction and decision making. Institutional arrangements and new methodologies will help facilitate the transition required.

Formal methods for assessing a broad spectrum of uncertainties can help experts fulfill their role under the precautionary paradigm. Through the application of these methods, experts are brought to diagnose the uncertainty characterising their assessments, and explicitly communicate it to the other actors in the policy community. The results of uncertainty analysis contribute to a qualified discussion of the quality of the information underpinning a policy decision. The quality of the information available can then be considered in determining the extent of the precautionary measures that are warranted in a given situation, and how monitoring resources should be allocated.

References

Andersen, I. E. and Jæger, B. (1999). Scenario workshops and consensus conferences: Towards more democratic decision-making, *Science and Public Policy* **26**(5): 331–340.

Ashford, N. (1991). An innovation-based strategy for the environment and for the workplace, *in* A. Finkel and D. Golding (eds), *Worst Things first: The Debate over Risk-based National Environmental Priorities*, Resources for the future, Washington DC.

Beck, U., Lash, S. and Giddens, A. (1994). *Reflexive Modernization; Politics, Tradition and Aesthetics in the Modern Social Order*, Stanford University Press, Stanford, CA, USA.

Beck, U. (1992). *Risk society: Towards a New Modernity*, Sage, London.

Broecker, W. S. (1997). Thermohaline circulation, the achilles heel of our climate system: Will man-made CO_2 upset the current balance?, *Science* **278**: 1582.

Craye, M., Funtowicz, S. and van der Sluijs, J. P. (2005). A reflexive approach to dealing with uncertainties in environmental health risk science and policy, *Int. J. Risk Assessment and Management* **5**(2): 216–236.

Cubasch, U., Meehl, G., Boer, G. J., Stouffer, R. J., Dix, M., Noda, A., Senior, C., Raper, S. C. B. and Yap, K. (2001). Projections of future climate change, *in* J. Houghton, Y. Ding, D. Griggs, M. Noguer, P. Van der Linden, X. Dai, K. Maskell and C. Johnson (eds), *Climate Change 2001: The Scientific Basis*, Cambridge University Press, Cambridge, pp. 526–582.

ESTO (2001). *On Science and Precaution in the Management of Technological Risk*, European Commission Joint Research Centre and European Science and Technology Observatory. EUR 19056/EN/2. Edited by A. Stirling.

Fischer, F. (2000). *Citizens, Experts, and the Environment: The Politics of Local Knowledge*, Duke University Press, Durham and London.

Fischer, F. (2003). *Reframing Public Policy: Discursive Politics and Deliberative Practices*, Oxford University Press, New York.

Fisher, E. (2005). *Risk, Regulation and Administrative Constitutionalism*, Hart Publishing, Oxford.

Flyvbjerg, B. (2001). *Making Social Science matter: Why Social Inquiry fails and how it can succeed again*, Cambridge University Press, Cambridge, USA.

Forester, J. (1999). *The Deliberative Practitioner: Encouraging Participatory Planning Processes*, MIT Press, Cambridge, USA.

Funtowicz, S. O., Ravetz, J., Martinez-Alier, J. and Munda, G. (1999). *Information Tools for Environmental Policy under Conditions of Complexity*, Environmental Issues Series, no. 9, EEA.

Funtowicz, S. O. and Ravetz, J. R. (1990). *Uncertainty and Quality in Science for Policy*, Kluwer Academic Publishers, Dordrecht.

Funtowicz, S. O. and Ravetz, J. R. (1992). Three types of risk assessment and the emergence of post-normal science, *in* S. Krimsky and D. Golding (eds), *Social Theories of Risk*, Praeger, Westport, GB.

Funtowicz, S. O. (2004). Models of science and policy: From expert demonstration to post normal science. Oral presentation at the International Symposium Uncertainty and Precaution in Environmental Management, Copenhagen, Denmark, 7-9 June 2004.

Gee, D. and Greenberg, M. (2001). Asbestos: From 'magic' to malevolent mineral, *in* P. Harremoës, D. Gee, M. MacGarvin, A. Stirling, J. Keys, B. Wynne and S. G. Vaz (eds), *The Precautionary Principle in the 20th Century. Late Lessons from Early Warnings*, Earthscan Publications Ltd., London.

Gee, D. and Krayer von Krauss, M. (2005). Late lessons from early warnings: Towards precaution and realism in research and policy, *Water Science & Technology* **52**(6): 25–34.

Giddens, A. (1991). *Modernity and Self Identity: Self and Society in the Late Modern Age*, Polity Press, Oxford.

Gilbertson, M. (2001). The precautionary principle and early warnings of chemical contamination of the great lakes., *in* P. Harremoës, D. Gee, M. MacGarvin, A. Stirling, J. Keys, B. Wynne and S. G. Vaz (eds), *The Precautionary Principle in the 20th century. Late Lessons from Early Warnings*, Earthscan Publications Ltd., London.

Gunderson, L. H., Holling, C. S. and Light, S. S. (1995). *Barriers and Bridges to the Renewal of Ecosystems and Institutions*, Columbia University Press, New York.

Hajer, M. (1995). *The Politics of Environmental Discourse*, Oxford University Press, Oxford.

Harremoës, P., Gee, D., MacGarvin, M., Stirling, A., Keys, J., Wynne, B. and Vaz, S. G. (eds) (2001). *The Precautionary Principle in the 20th Century. Late Lessons from Early Warnings.*, Earthscan Publications Ltd., London.

Holling, C. S. (ed.) (1978). *Adaptive Environmental Assessment and Management*, John Wiley & Sons, New York.

Holling, C. S. (2001). Understanding the complexity of economic, ecological, and social systems, *Ecosystems* **4**: 390–405.

Jansson, R. (2001). On the use of multi-criteria analysis in environmental impact assessment in the Netherlands, *J. Multi-Crit. Decis. Anal.* **10**: 101–109.

Jasanoff, S. and Wynne, B. (1998). Science and decision making, *in* S. Rayner and E. L. Malone (eds), *Human Choice & Climate Change – The Societal Framework*, Colombus, USA.

Jasanoff, S. (1994). *The Fifth Branch: Science Advisors as Policy Makers*, Harvard University Press, Cambridge, USA.

Jensen, K. K., Gamborg, C., Madsen, K. H., Jørgensen, R. B., Krayer von Krauss, M., Folker, A. P. and Sandøe, P. (2003). Making the EU 'risk window' transparent: The normative foundations of the environmental risk assessment of GMOs, *Environmental Biosafety Research* **3**: 161–171.

Klinke, A. and Renn, O. (2002). A new approach to risk evaluation and management: Risk-based, precaution-based, and discourse-based, *Risk Analysis* **22**(6): 1071–1093.

Knorr-Cetina, K. and Mulkay, M. (1983). *Science Observed: Perspectives on the Social Studies of Science*, Sage Publications.

Koppe, J. and Keys, J. (2001). PCBs and the precautionary principle, *in* P. Harremoës, D. Gee, M. MacGarvin, A. Stirling, J. Keys, B. Wynne and S. G. Vaz (eds), *The Precautionary Principle in the 20th Century. Late Lessons from Early Warnings*, Earthscan Publications Ltd., London.

Krayer von Krauss, M. P. (2006). *Uncertainty in Policy Relevant Sciences*, PhD thesis, Institute of Environment & Resources, Technical University of Denmark.

Krayer von Krauss, M., van Asselt, M. B. A., Henze, M., Ravetz, J. and Beck, M. B. (2005). Uncertainty and precaution in environmental management, *Water Science and Technology* **52**(6): 1–9.

Krimsky, S. and Golding, D. (eds) (1992). *Social Theories of Risk*, Praeger, Westport, GB.

Kuhn, T. S. (1962). *The Structure of Scientific Revolutions*, The University of Chicago Press, Chicago.

Lash, S., Szerszynski, B. and Wynne, B. (eds) (1996). *Risk, Environment and Modernity: Towards a New Ecology*, Sage, London.

Latour, B. and Woolgar, S. (1986). *Laboratory Life: The Construction of Scientific Facts*, Princeton University Press, Oxford.

Lee, K. N. (1999).   Appraising adaptive management, *Conservation Ecology* **3**(2): 3.   URL: *http://www.consecol.org/vol3/iss2/art3*. Accessed March 19th, 2006.

Meyer, G., Folker, A. P., Jørgensen, R. B., Krayer von Krauss, M., Sandøe, P. and Tveit, G. (2005). The factualization of uncertainty: Risk, politics, and genetically modified crops – a case of rape, *Agriculture and Human Values* **22**(2): 235–242.

Michaels, D. and Monforton, C. (forthcoming). Scientific evidence in the regulatory system: Manufacturing uncertainty and the demise of the formal regulatory system, *Journal of Law and Policy* .

Michaels, D. (2005). Manufacturing uncertainty: Contested science and the protection of the public's health and environment, *Am J Public Health* **95**: 39–48.

NRC (1988). *Complex Mixtures*, National Research Council, National Academy Press, Washington DC.

NRC (1996). *Understanding Risk: Informing Decisions in a Democratic Society*, National Research Council, National Academy of Sciences, Washington DC.

OXERA (2000). *Policy, Risk and Science: Securing and using Scientific Advice*, Contract research report 295/2000, Oxford Economic Research Associates, Oxford.

Proctor, R. N. (1991). *Value-Free Science? Purity or Power in Modern Knowledge*, Harvard University Press, Cambridge, USA.

RCEP (1998). *Setting Environmental Standards*, Royal Commision on Environmental Pollution, CM4053, HMSO, London.

Renn, O. (1999). A model for an analytic-deliberative process in risk management, *Environmental Science & Technology* **33**(18): 3049–3055.

Rip, A., Misa, T. and Schot, J. (1996). *Managing Technology in Society*, Pinter, London.

Rogers, M. (2003). The european commission's white paper 'Strategy for a future chemicals policy': A review, *Risk Analysis* **23**(2): 381–388.

Scheffer, M. and Capenter, S. R. (2003). Catastrophic regime shifts in ecosystems: Linking theory to observation, *Trends in Ecology and Evolution* **18**(12): 648–667.

Scheffer, M., Carpenter, S., Foley, J. A., Folke, C. and Walker, B. (2001). Catastrophic shifts in ecosystems, *Nature* **413**: 591–596.

Schön, D. A. (1982). *The Reflective Practitioner: How Professionals Think in Action*, Basic Books, Cambridge, USA.

Stirling, A. and Mayer, S. (2001). A novel approach to the appraisal of technological risk, *Environment and Planning C* **19**: 529–555.

Stirling, A. (2001). The precautionary principle in science and technology, *in* T. Riordan, J. Cameron and A. Jordan (eds), *Reinterpreting the Precautionary Principle*, Cameron May Ltd., London.

Tickner, J. A. (2000). *Precaution in Practice: A Framework for implementing the Precautionary Principle*, Dissertation prepared for the Department of Work Environment, University of Massachusetts, Lowell, MA, USA.

UNESCO (2005). *Report of the Expert Group on the Precautionary Principle of the World Commission on the Ethics of Scientific Knowledge and Technology (COMEST)*, UNESCO COMEST, Paris.

van Asselt, M. B. A. and Vos, E. (2005). The precautionary principle in times of intermingled uncertainty and risk: Some regulatory complexities, *Water Science & Technology* **52**(6): 35–41.

van Asselt, M. B. A. (2000). *Perspectives on Uncertainty and Risk*, Kluwer Academic Publishers, Dordrecht.

Walker, W. E., Cave, J. and Rahman, S. A. (2001). Adaptive policies, policy analysis, and policymaking, *European Journal of Operational Research* **128**(2): 282–289.

Walters, C. (1986). *Adaptive Management of Renewable Resources*, Macmillan, New York.

Walters, C. (1997). Challenges in adaptive management of riparian and coastal ecosystems, *Conservation Ecology* **1**(2): 1. URL: *http://www.consecol.org/vol1/iss2/art1/*. Accessed March 19th, 2006.

Webler, T. (1995). 'Right' discourse in citizen participation. An evaluative yardstick., *in* O. Renn, T. Webler and P. Wiedemann (eds), *Fairness and Competence in Citizen Participation. Evaluating new Models for Environmental Discourse.*, Kluwer Academic Press, Dordecht.

Wildavsky, A. B. (1979). *Speaking Truth to Power: The Art and Craft of Policy Analysis*, Little, Brown, Boston, USA.

Wynne, B. (1992). Uncertainty and environmental learning – reconceiving science and policy in the preventive paradigm, *Global Environmental Change* **2**(2): 111–117.

Understanding the nature of apology in the context of healthcare

Marlene Dyrløv Madsen

Systems Analysis Department, Risø National Laboratory and Section for Philosophy and Science Studies, Roskilde University

In recent years, a growing number of authors have explored the notion of apology in various contexts, thereby bringing to light how the discourse, politics and processes surrounding apology can be extremely complex and critical (see for instance Govier and Verwoerd, 2002; Pettigrove, 2003; Taft, 2000). In healthcare too, 'apology' has been getting increased attention, primarily because it has been proven to play a significant role in the aftermath of adverse events, affecting patients and staff, but also because it may have financial consequences for the healthcare provider organization (see for instance Gallagher et al., 2003; Kraman and Hamm, 1999; Manser and Staender, 2005).

In this paper I will investigate the nature of apology and its internal logic in the context of healthcare. I will begin by defining apology and, in line with other authors, suggest that 'apology' in its primary meaning is a moral act (Govier and Verwoerd, 2002; Taft, 2004). I shall review the theoretical background of apology in order to illustrate its nature and function, and I will examine the different functions of apology in healthcare to (1) investigate when apology can be morally justified and (2) discuss the necessary conditions for an apology to work effectively and ethically in healthcare. Related to this, I will discuss when and how it is justified and necessary to apologize, and acknowledge or express regret after harm, in order to distinguish a spectrum of possible acknowledging actions. I will review arguments of different theorists, discussing some of these in terms of utilitarianism and deontology. I shall use cases to illustrate the possibility of apologizing in healthcare in the aftermath of harm, and the negative consequences when apologies are not given or are given in the "wrong" manner.

1 Background

As mentioned, apology is getting increased attention in healthcare. For instance, a Danish study, using focus group interviews and performing a questionnaire survey of how patients want healthcare staff to handle mistakes following adverse events, showed that 81% of patients find it extremely/very important that regrets and an explanation are given, and that 84 % find it extremely/very important that healthcare personal admit their error, if an error was made (Andersen et al., 2004; Freil et al., 2004; Østergaard et al., 2005). These findings are in line with what is reported in the literature: patients want physicians to acknowledge adverse events (Manser and Staender, 2005; Witman et al., 1996) and in some cases even minor mistakes (Gallagher et al., 2003), and when physicians decline to do so, patients will be more likely to file lawsuits (Witman et al., 1996). It is however not completely evident whether it is an actual "apology" or an "acknowledgement" that is most important to patients. A study by Kraman and Hamm (1999) illustrate that extreme honesty is the best policy regarding patients' interest while, at the same time, it is most likely to minimize cost of litigation (Hickson et al., 1992; Lamb et al., 2003; Liang, 1999; Vincent et al., 1994; Witman et al., 1996). In the long run, openness also improves the possibilities for learning about and preventing medical harm, hence saving the overall cost in healthcare (Berlinger, 2005). Should the moral arguments for truth telling and apologizing after harm therefore fail to bring action, there are evidently economic incentives that support openness and apology.

It is often said that healthcare has fostered a culture of secretiveness, and indeed, several studies and personal accounts point to the need for healthcare staff, and especially doctors, being able to disclose and apologize honestly after avoidable medical harm (Berlinger, 2003a; Finkelstein et al., 1997; Wu et al., 1993).[1] Many doctors may find themselves in a dilemma, wishing themselves to express sincere apology to the patient but finding several obstacles against doing so (Taft, 2004). A specific theme in this area concerns the severe emotional effects of medical harm on staff and the lack of support they experience – a theme which is discussed in a number of publications (Christensen et al., 1992; Newman, 1996; Smith and Forster, 2000) and often referred to by the tag "the second victim", a memorable phrase coined by Wu (2000). One of the first doctors to write on this subject, Hilfiker (1984), published an emotionally charged and moving paper "Facing our mistakes" in the *New England Journal of Medicine*, and later expanded this into a book (Hilfiker, 1985). Some specialties have created rituals for dealing with mistakes (Bosk, 1979) but they have never included the patient, as pointed out by Berlinger (2004). In her recent overview, Berlinger devotes a whole chapter to

---

[1] Also (Berlinger, 2005; Hilfiker, 1984; Manser and Staender, 2005; McNeill and Walton, 2002; Newman, 1996; Sharpe, 2004; Woods, 2004; Wu et al., 1991; Wu, 2000).

analyzing the literature and the spreading phenomenon of doctors telling about error (Berlinger, 2005). This trend is not confined, of course, to the English language literature. For instance, a Danish primary sector doctor made her "confession" after she stopped practicing, calling for more openness, understanding and support from colleagues in cases of medical harm, and she describes the pain, shame and feeling of loneliness following medical harm (Bærentsen, 1997). There is now an increasing number of websites where healthcare staff share experiences of harm with each other.

If apology and the act of acknowledging medical error are of such great importance – to patients as well as staff – why is it then that healthcare staff do not apologize when a mistake leading to harm has been made? There are of course several obvious reasons for healthcare staff to withhold apology (Finkelstein et al., 1997). First of all, healthcare staff may not find that they are personally at fault, but rather that the adverse event is caused by a systems failure, e.g., under-staffing, faulty equipment. Or the mistake may have been made by a colleague, perhaps an unknown one in another department. If a staff member has not personally made any mistake, is it at all appropriate to apologize?

Second, staff members may be afraid of becoming caught up in a lawsuit or be reported to the complaint board if they admit that they have made a mistake (Andersen et al., 2002; Finkelstein et al., 1997; Gallagher et al., 2003; Vincent et al., 1999). Some believe that apologizing or saying "I am sorry" might give the patient the impression that the apologizer is in fact personally at fault (Gallagher et al., 2003).[2] But does an apology necessarily entail acknowledgement of personal fault?

Third, there are cases where hospital risk managers and management admonish staff not to apologize in order to minimize the potentials of economic compensation (Gallagher et al., 2003; Wears and Wu, 2002).[3] In such cases it appears that healthcare staff feelings and their wish to assume moral responsibility by openly expressing their role and their regret, and perhaps even seeking forgiveness, are

---

[2] Robbennolt (2003) notes that in civil disputes legal actors have viewed apology as an admission of responsibility that will lead to increased legal liability, which is why apology should be avoided (Cohen, 1999). It has often been observed that in the American legal context the possibility that a sincere apology will be taken as admission is inhibiting people from apologizing (Wagatsuma and Rosett, 1986).

[3] These examples pertain mostly to a context as in the US where a risk manager's role is to avoid litigation. However, even in a context such as the Danish characterized by public hospitals and a no-fault compensation scheme, where there is no fear of litigation, the risk of being reported to the complaints board will apparently sometimes induce management to recommend that their staff do not apologize. Moreover, this recommendation seems to be followed in some cases.

overruled by management.[4] Can this be justified ethically? Is there not a moral obligation to support truth telling and apologies when appropriate? As Leape notes, "dishonesty is corrosive not only to the patient's trust, but to the physician's integrity. It is not surprising that many doctors have felt "unclean" after following advice not to admit responsibility for a serious error. Honesty is not just the best policy; it is also essential to our mental health" (Leape, 2005).

Fourth, the professional culture of doctors has sustained an idea of infallibility: a "good" doctor does not make mistakes, and it is a sign of incompetence to make mistakes; therefore, the mere thought that one has made a mistake is difficult to face, let alone admit to colleagues or, even worse, patients (Blumenthal, 1994; Finkelstein et al., 1997; Hingorani et al., 1999; Leape, 1994; Smith and Forster, 2000; Wu et al., 1991). Doctors may also fear losing the patients trust (Hingorani et al., 1999). Psychologists maintain that most people find it difficult to apologize – often because they simply do not know how (Lazare, 2004) – and for doctors it is believed to be even more difficult, not just because they are inexperienced in apologizing, but because of the strong traditions of the infallibility of the profession (Blumenthal, 1994). A recent review article on patients' expectations in the aftermath of adverse events and the practice of open disclosure describe several of the barriers (Manser and Staender, 2005).

Not until recently have the professions laid down an obligation to tell the patients about adverse events (Lamb et al., 2003), nor have integration of disclosure training been part of the curriculum of doctors (Davies, 2005; Leape, 2005; Manser and Staender, 2005; Sharpe, 2000). This tendency is rapidly changing as the traditional doctor-patient relationship has shifted and according to Taft (2004) the regulatory and ethical movements "reveal a philosophical shift in the very nature of communication between patient and care provider".[5] The medical curriculum is also changing internationally, where patient safety is prompting the establishment of courses in, e.g., communication and teamwork and modules that address medical mistakes, including in particular the difficult and sensitive subject of communicating, disclosing and apologizing to patients and families following medical harm (Berlinger and Wu, 2005; Crigger, 2004; Hobgood et al., 2004; Manser and Staender, 2005; Østergaard et al., 2005; Smith and Forster, 2000).

---

[4] The author recalls a few instances when, during seminars, a doctor has related how he or she has made a medical error leading to mortality or a patient becoming disabled, and where the immediate inclination of the doctor involved was to apologize and express regret to the relatives; however, following the advice of the risk manager, the doctors did not apologize to the patient or the relatives. All of them say that afterwards they have not themselves been able to come to terms with the event and its aftermath.

[5] There have been especially pronounced changes in the US with the new standards of Accreditation (JCAHO, 2006), the American Medical Association's ethics guideline (American Medical Association, 1994) and the American College of Physicians (2002).

To overcome some of the above-mentioned obstacles several organizations have provided guidelines for when and how to apologize (see for instance Doctors in Touch, 2004; Lamb et al., 2003; NHS, 2005), US states have enacted policies to enforce apology as standard of care (JCAHO, 2004; Sorry works, 2006), and some of them have so-called "I'm Sorry" laws according to which a person can express regret or convey sympathy without it being used as evidence of liability; these laws are thus intended to encourage apology (Robbennolt, 2003).[6] In effect it is becoming widely recognized that disclosing and apologizing to patients is a reasonable action following the event of medical harm (see for instance Davies, 2005; Finkelstein et al., 1997; Rosner et al., 2000; Sharpe, 2004), while the arguments for doing so, still vary. Some authors argue that apologizing is the right policy because it is prudent from an economic point of view, i.e., fewer patients will sue after receiving an apology (Cohen, 1999, 2002). Others argue that it is our moral responsibility to acknowledge and take responsibility for the actions that have caused harm to others, for instance through apology (Berlinger, 2004; Crigger, 2004; Finkelstein et al., 1997; Smith and Forster, 2000).

Most of the literature on apology is created on a disciplinary background of sociology, psychology and law and most of the philosophical literature deals with the political issues of apology, specifically on apologies between nations or ethnic groups for harms done in the past. There have been a couple of central philosophical articles on apology, one by Kort (2002), originally from 1975, and another by Gill (2000), each of the authors seeking to determine the necessary conditions for apologizing. In theology, discussions of apology are often made in relation to and from the perspective of forgiveness (Berlinger, 2003c). The literature on apology in the context of healthcare is mainly addressed at the duty of disclosure after medical harm and the ethical value of truth telling (Crigger, 2004; Finkelstein et al., 1997; Hébert et al., 2001; Rosner et al., 2000; Smith and Forster, 2000; Sweet and Bernat, 1997; Wu et al., 1997), and several of them also argue that apology should "be one of the ethical responsibilities of the profession of medicine" (Finkelstein et al., 1997). There seems to have appeared no philosophical analysis of the moral arguments about the justifications and role of apology in healthcare.[7]

---

[6] There is, however, wide differences between the enacted laws in the different states in terms of the types of expressions that are protected. For more detail and discussion about the contents, history and effects of these laws see (Berlinger, 2005; Cohen, 1999, 2002; Robbennolt, 2003; Taft, 2004). Interestingly for healthcare is the rule adopted by Colorado in 2003 that is limited to expressions made by health-care providers and that explicitly protects statements expressing fault (the first of its kind) (Robbennolt, 2003).

[7] Taft (2004) provides an exceptionally engaging and philosophically relevant analysis of apology and medical mistake, although his discussion is focused mainly on law related issues.

In healthcare most adverse events are caused by unintentional acts, errors of omission or commission, brought about in a very complex system, and medical harm is therefore often accidental and non-personal.[8] The fact that medical harm in most cases are caused by error and as such by acts that in themselves are not moral wrongdoings creates a slightly different case for apology. Although most apologies are in fact given after unintended behavior or negligence that cause some kind of harm to the victim most of the literature on apology discusses the conditions, appropriateness and effects of apology following especially group and racial atrocities that are intentional and person oriented, often made in the past and now condemned in the present. But what about harm caused neither by negligence nor by moral wrongs? How should we respond in such cases?

2 What is an apology?

What is an apology? A simple question, and perhaps one that invites an immediate and simple answer. However, the literature reveals that the language of apology is nuanced and its nature ambiguous. The word apology stems from Greek "apologia" and means a defense, a justification, an explanation or an excuse, all of which are not part of the primary meaning of the modern use of apology (AHD, 2004). In fact most authors agree that for anything to be a true apology, it must be offered without any excuse or justification (Gill, 2000; Govier and Verwoerd, 2002; Tavuchis, 1991). Dictionaries reveal three basic meanings of 'apology'. The primary definition of 'apology' according to several dictionaries concerns sincere regret. Thus, "a regretful acknowledgement of an offence or failure" in the Concise Oxford English Dictionary (COED, 2004) and "an acknowledgment expressing regret or asking pardon for a fault or offense" in the American Heritage Dictionary (AHD, 2004). The secondary meanings of 'apology' are: "A: a formal justification or defense; B: an explanation or excuse" (AHD, 2004), or just "a justification or defense" (COED, 2004).

Additionally the primary meaning can and is sometimes distinguished into two separate meanings in the literature, although the dictionaries do not discriminate these specifically: the "partial apology" or "sympathetic apology" refers to statements that convey sympathy but do not admit responsibility, and the "full apology" or "authentic apology" refers to statements that both express sympathy or regret *and* accept responsibility (Robbennolt, 2003; Taft, 2004). As we shall see, the partial apology is incomplete in terms of the most common and acknowledged

---

[8] Of course, it can be argued that errors may be due to carelessness or negligence and they might even be repeated by the same doctor, who of course must be dealt with appropriately: However the vast majority of errors are made by well-trained, conscientious, and well-meaning healthcare staff members.

definitions of apology. According to Lazar, "[a]pology refers to an encounter between two parties in which one party, the offender, acknowledges responsibility for an offense or grievance and expresses regret or remorse to a second party, the aggrieved" (Lazare, 2004).

There are numerous examples of how words such as 'true', 'genuine', 'sincere' or 'authentic' are added to apology to define certain necessary conditions that must be attained for an apology to be that which it claims to be "effective". However, these words in their intended sense and in this context are very nearly synonyms of each other.[9] So it seems that when authors use these qualifying terms they are merely trying to emphasize that they are dealing with "apology" in the so-called primary and full meaning of the term. An example is Harvey (1995), who makes a distinction between a 'genuine apology' and 'apology', claiming that in a person-to-person apology you may speak the right words, "but it does not constitute a genuine apology if sincerity is lacking". Taft (2004) defines apology as a moral act, and he distinguishes between the "authentic apology" and the "sympathetic apology" as expressed above. The only problem arises when different theorist confuse the meaning of apology by using the same 'qualifying word' with diverse meaning, for instance, Joyce calls an apology "authentic" as long as there are expressions of regret and the recipient is satisfied (Joyce, 1999), which does not cover the necessary conditions that Taft asserts. We shall return to this discussion and the necessary conditions for apology.

Most often, the qualifications put on apology are attitudinal states of regret, remorse, and sincerity (see for instance Gill, 2000; Taft, 2000; Tavuchis, 1991) although, rather controversially, others will claim that a genuine apology can be given without feelings of remorse or regret (Cunningham, 1999; Joyce, 1999; Pettigrove, 2003). Pettigrove (2003) criticizes Gill (2000) and Harvey (1995) – but they are not alone in holding the view criticized – for being inclined to make attitudinal states like regret, remorse and sincerity necessary conditions of apology, as he states, "apologies lacking such attitudinal states may be morally deficient, we are not generally inclined to say they fail to be apologies" (Pettigrove, 2003). Pettigrove argues – and I tend to agree – that an insincere apology is still an apology and hence does not semantically fail to be one; it may, however be "an infelicitous one" (Pettigrove, 2003). In this case we may call Pettigrove's apology "partial". If for instance someone claims that "the best way to keep a marriage is to apologize even if you don't mean it" then the person is obviously using the functions of apology strategically as a means to an end, but the person is not performing a full apology. This is a central discussion which we will return to.

---

[9] Genuine is defined as "true to what is claimed, authentic or free from deception or pretence or sincere"; sincere is defined as "genuine, true, and unaffected, honest or earnest", and finally, authentic is defined as "genuine, like the real or original" (AHD, 2004).

Govier and Verwoerd (2002) call what we have termed a "full apology" a "moral apology", which "implies a request for forgiveness and is an initiative toward reconciliation". However, within the philosophical literature apology is not referred to as a 'moral apology'. This is probably because philosophers interpret apology as an inherently moral act, in which case the label 'moral' is redundant in 'moral apology' – except when used for emphasis perhaps. Nonetheless, the qualifications asserted by Gill, Harvey, Taft, Govier and Vervoerd and others can be interpreted as conditions for a "moral" apology, which might explain why the necessary conditions these authors claim for apology are stricter than those asserted by others. In this paper the terms "partial" and "full" will be used when necessary to distinguish the primary meaning of apology, while the focus of the paper is on the full apology exemplified by Lazare's definition (see page 345).

2.1 Theoretical background of apology

The sociologist Goffman (1971), who published a very influential analysis of the social dynamics of apology, characterizes apology as a type of remedial interchange. Remedial activity is undertaken by a person as a response to having given "the appearance of encroaching on another's various territories and preserves; or he finds himself to give a bad impression of himself; or both" , and when someone seeks by the remedial strategy "to reinforce a definition of himself that is satisfactory to him" (Goffman, 1971). Common and basic categories of remedial interchange comprise accounts, denials, excuses, justifications and apologies. We noted earlier the difference between the primary meaning of apology (expression of sincere regret etc.) and its secondary meaning (an excuse or justification), and it is obviously not the secondary meaning that is the target of Goffman's analysis. Thus, he also observes that "[i]n contrast to excuses and justifications, an apology involves both an acceptance of responsibility for the act and an acknowledgement of its wrongfulness" (Gill, 2000). However, in all cases "[t]he function of remedial work is to change the meaning that otherwise might be given to an act, transforming what could be seen as offensive into what can be seen as acceptable" (Goffman, 1971). Goffman argues that the "remedial activity is a constant feature of ordinary interaction that, indeed, through ritually closed interchanges, it provides the organizational framework for encounters" (Goffman, 1971). Goffman defines apology as: "a gesture through which an individual splits himself into two parts, the part that is guilty of an offence and the part that dissociates itself from the delict and affirms a belief in the offended rule." (Goffman, 1971). Goffman also notes that there are varying degrees of apology relative to the size of the offence. He observes that there are two distinct and independent processes involved in the corrective behavior of apology: the *ritualistic*, where the offender states his relationship to the rules, which he has broken and which the offended party should have been protected by, illustrating that he has changed his attitude to the rule he violated, and the *restitutive*, where the offended party receives some sort of compensation for the offence, and implicitly for not being protected by the rules in place (Goffman, 1971). Depending on the offence the weight to the two processes will vary from case to case, some will be engaged mainly in gaining compensation for loss, others will be more concerned with the principal of the offence. In healthcare this distinction in two parallel processes becomes important as we shall discuss later.

Goffman's approach is sociological in so far that he interprets apology as a social act that is adhered to as a response to breaking a social rule. However, social rules may as well be interpreted as social norms, where the apologizer wants "to show that whatever happened before, he now has a right relationship – a pious attitude – to the rule in question, *and this is a matter of indicating a relationship, not compensating a loss*" (Goffman, 1971). It is important to note that with an apology one is not compensating a loss – often this is not even possible - but acknowledging the fact of having *caused* a loss.

In order for the apology to work it is necessary for the offender to receive a response indicating that the apology has been clearly received and whether it has been accepted. According to Austin's famous analysis, an apology is a paradigmatic performative utterance, which stands in contrast to constative utterances that can be judged true or false. When "I apologize" I do not merely state something, I *do* something (Austin, 1962). In this sense there is a dynamic in the discourse of apology, and as another sociologist, Tavuchis, underscores, "if sorrow and regret are at the heart of apology, they must be expressed. It is simply not enough to feel sorry but to say so in order to convert a private condition into public communion" (Tavuchis 1991). An apology is an interchange between persons and, according to Tavuchis, it is a "relational concept" and a "social discourse", where the "bedrock structure of apology is binary, a product of a relationship between the Offender and the Offended" (Tavuchis, 1991). This relational condition is significant in relation to cases of litigation, since lawyers and mediators often tend to forget this dyadic relation and therefore loose the opportunity for the offender and offended to make apology a healing process[10] as Taft notes (Taft, 2000, 2004). In healthcare and especially in litigious societies patients and staff do not take or are not always given the opportunity for restoring balance or healing.

---

[10] Interestingly, this is much like the essence of restorative justice that involves bringing back the "conflict" to the offender and the offended instead of having it fought out in the courtroom between third parties. Zehr defines restorative justice in these terms: "Crime is a violation of people and relationships. It creates obligations to make things right. Justice involves the victim, the offender, and the community in a search for solutions which promote repair, reconciliation, and reassurance" (Zehr, 2005). See also (Braithwaite, 2002).

3 The different functions of apology in healthcare

As mentioned above, an apology might serve several functions in the healthcare context where a medical error has led to patient injury or discomfort. First of all, it can satisfy patients' wish for acknowledgement; second, it may contribute to a healing process for both patients and staff by restoring the balance; and finally, it may be used as a strategy to persuade patients not to engage in litigation or seek compensation.

In healthcare an apology's main function is to acknowledge the harm done to the patients and to indicate that the offender will take responsibility for further action. Apologizing after harm can potentially restore or re-establish a trusting relationship between the doctor and patient (Finkelstein et al., 1997). Refusing or neglecting to apologize for harm, when appropriate, may thus have the opposite effect – namely creating patient distrust towards the doctor and the system in general (Gallagher et al., 2003; Vincent, 2003). Apologizing to patients after harm is a way of acknowledging the patient as a *person* by expressing regret for the suffering inadvertently inflicted upon them.[11] According to the philosopher Kort, an apology "is a gesture of respect, assuring and recognizing that the victim shouldn't be treated as they were" (Kort, 2002). Apologizing is an opportunity to treat patients with respect (Finkelstein et al., 1997) and make sure that they will not feel "devalued, humiliated and disrespected" (Leape, 2005) which is what patients feel when harm is not acknowledged (Vincent, 2003). Although it is possible to acknowledge harm to patients without actually apologizing, there is in the speech act of apology something distinct from mere acknowledgement or expressions of regret, and in cases where responsibility for harm is explicit, apology may be the only right action to restore balance and trust (Robbennolt, 2003).

Robbennolt (2003) found, when she compared partial apology (expression of sympathy) with full apology (admitting responsibility) with their effects on settlement, that the full apology had a positive impact on settlement, while the partial apology could have a detrimental effect, especially in cases where responsibility for harm was clear. Equally, in cases were strong evidence of culpability or severe injury was followed by a partial apology 'no apology' would be better than a 'partial apology' in terms of settlement, and significantly the offender was perceived as being unlikely to be careful in the future. In most cases offering a 'partial apology' was no different from 'no apology', except when responsibility was ambiguous or injury was minor, then there was slight evidence that a partial apology could positively impact perceptions. In general participants expressed more sympathy and less anger towards offenders who offered full apology compared to partial or no

---

[11] Govier and Verwoerd (2002) discuss at length the effects of apology in restoring the moral worth of offended parties.

apology, as they indicated more willingness to forgive and expected less damage done to the relationship following.

According to Govier and Verwoerd, writing within the framework of social philosophy, "[t]he purpose of apology is to make amends, and in this regard there is a difference in moral and (material) practical amends" (Govier and Verwoerd, 2002).[12] Basically Goffman makes the same point, namely that the function of apologies is to restore a balance – the fact that the patient is harmed is somehow made good by the fact that the offender or another responsible part is taking responsibility and perhaps even trying to make further amends. We shall return to this distinction between moral and practical amends and discuss how they differ and the effect hereof on the apology process.

As part of the apology the patient can be assured that the harm was neither intentional nor personal, and that he or she will be taken care of at least medically. The effect of such assurance is the beginning of the psychological healing. In the frame of public affairs Joyce states that: "Reconciliation is the function of the apology" (Joyce, 1999). Although this might be true within the realms of political affairs, reconciliation is not the primary purpose of apology in healthcare, since harm is rarely caused intentionally. Often there is merely an accidental relationship between the offender and the offended, and consequently, the doctor and patient might never meet or have any kind of relationship after the event. However one may argue that the individual approach to harm and apology may have a general effect on the future relationships between the healthcare provider and the patients.

Taft, an ethicist and lawyer and mediator of ethical opportunities in the wake of error, argues that: "Apology is an important ingredient in the healing of a moral injury" (Taft, 2000). The healing process of apology is not only reserved the patient; potentially the act of apologizing can have as much of a healing effect on the person who apologizes, the second victim. In fact it may be a necessary action for the offender – for someone seeking forgiveness – to find peace at mind and healing. Taft argues that apology is a moral obligation and, inspired by Kort, he claims that "[t]he authentic apology has moral meaning for both the offender and the offended as a vehicle for restoring moral balance" (Kort, 2002).

Finally, an apology might serve the purpose of minimizing litigation and economic compensation, since most studies show that patients' incentive to sue doctors and hospitals decreases with disclosure and apology (see for instance Gallagher et al., 2003; Hickson et al., 1992; Vincent et al., 1994; Witman et al., 1996).

---

[12] Govier and Verwoerd (2002), who are inspired by Golding (1984), use the term "practical amends" instead of "material amends" to indicate that amends can be other than material.

3.1 Utilitarian versus deontological approaches to apology

Within moral philosophy, we find in both of the two major positions, the utilitarian and the deontological, arguments that support or justify apology, but they will give quite different types of moral reasons. The utilitarian perspective holds that one has an obligation to perform the act that generates the best overall consequences, whereas the deontological theory holds that one ought to perform the act that realizes one's obligations according to given values, in principle regardless of the consequences. An obligation to apologize may therefore be argued on quite different grounds.[13]

From a classical utilitarian point of view we should apologize if the overall good that stems from doing so is greater than not doing so. The question is whether this general approach can be justified even if the conditions for making an apology are not met satisfactorily. Let us say, e.g., that I am a doctor who has harmed a patient inadvertently. I do not feel entirely at fault, but things definitely did not go as expected. I know that there is a possibility that the patient will file for litigation, that this may become an economic burden for my organization if they have to pay compensation, and that it will become so for myself as well when my insurance premium will go up, not to mention the uncomfortable likelihood that people may question whether I am up to par if this case goes to court. Let us further assume that I have heard that patients generally prefer to be apologized to after harm and that apologizing reduces the risk of patient litigation. I reflect upon this; of course I feel sorry for the patient, although I do not regret my own actions, since I did nothing wrong. I realize that others might not see it this way. After careful consideration I decide to apologize, hoping that this will lay the case to rest. Of course, I do not involve the patient in my reflections.

If I apologize to the patient under these conditions, will the patient accept my apology? If the patient perceives it as sincere and finds consolation in my statement and therefore refrains from litigation, then the best consequences in utilitarian terms have ensued. Cunningham (1999), who writes on the politics of apology, argues that "[t]he apology in itself may have value if sincerely offered and accepted as such by its recipient". Cunningham further argues that the relation between apology and responsibility need not be established, that accepting responsibility is not required, since the apology, if sincere, has a symbolic quality and a utility. If the suffering is recognized and acknowledged by others "this in it self may act as a form of restitution or reparation", and this, he claims, is the symbolic quality (Cunningham, 1999). He further argues that apology has a practical element, viz. the utility associated with promoting "better contemporary interstate or intercommunal relations" (Cunningham, 1999). In this perspective the main interest is

---

[13] Wu et al. (1997) make a relevant and useful analysis in relation to the deontological and utilitarian approaches to disclosure, but not directly in relation to apology.

the immediate consequences of the apology and, specifically, the stabilizing effect on the political situation. However, there are several examples where political interests go beyond the interest of the offended parties, and in such circumstances one may question who the appropriate stakeholders are who are entitled to define what should count as acceptable.

One of the examples given by Cunningham illustrates this: in 1998, British World War II prisoners demanded an apology from the Japanese Emperor. They did not receive this, but they received somewhat vague expressions of regret about past actions. The veterans' representatives thought that the British government, driven by a wish to move on in order to strengthen commercial links with Japan, too willingly accepted – as they saw it – a vague apology, which the veterans did not accept as convincing. In this case, I think that the main reason why the apology could not be accepted by the veterans stems from the fact that it was put forward only for instrumental reasons. And not only this: There was no attempt to hide that this was so. There was no overt sign of sincerity, since the offender (the Japanese government) obviously does not acknowledge their part in the harm. As pointed out by e.g., Lazare, this is often a reason why such apologies do not work. In healthcare the question is if an apology can be regarded as sincere if it is perceived as nothing more than an expedient act to keep litigation costs down. Basically Cunningham (1999) argues that if we can accept apology's symbolic meaning or utility then there is no need to establish a link to responsibility. Any deontologist would disagree on this.

Joyce continues Cunningham's line of argument when he concludes his article on "apologizing" with the following summary:

> The function of apology is to reconcile discordant parties – in other words, although the content of an apology is oriented toward the past, the whole purpose of the act lies in the future consequences. And there can be no overestimating the importance of the gains that may be secured: the contentedness of a family, the well-being of a community, the political stability of a nation. I see no reason to doubt that sometimes such welcome ends may be served by an utterance that might be taken to be an apology, but which, upon careful consideration, falls short of being one. (Joyce, 1999)

In this understanding of apology, Joyce eloquently and unambiguously argues for the view that an apology can be a means to an end.

Returning to the case we described above, let us consider the possibility that the patient finds my apology insincere and somehow develops a feeling that I am trying to cover up. If so, much may be lost. This possibility is serious to the utilitarian position, and evidently Joyce too ignores it. Among the potential negative consequences of an insincere apology is the loss of trust between groups. If we can apologize insincerely or falsely, have we not compromised trust, the central value for inter-personal relations? In healthcare the doctor-patient relation is extremely important and is founded on mutual trust. If this trust is lost in the process of apologizing falsely it may have long-term negative consequences for the patient, the doctor, and the hospital. If so, the utilitarian argument will defeat itself, since the consequences are no longer better. If however the patient's desire for apology and acknowledgement is greater than the desire for sincerity, then the utilitarian stance would require us to apologize. Hence, we will need to distinguish between classical utilitarianism, also known as act-utilitarianism and rule-utilitarianism, to clear the arguments. Rule-utilitarianism shifts from the justification of acts to the justification of rules, claiming that we need to look at "what general rules of conduct tend to promote the greatest happiness?" (Rachels, 1993). According to rule-utilitarianism we would not be able to justify lying or deceiving, since these are not general rules that will better society, and accordingly, they, like the deontologists, would not accept insincerity. Nevertheless Joyce's arguments are those of the act-utilitarian.

But perhaps public life and in particular the political scene – to which Joyce's arguments are addressed – do not require trust to have the same fundamental value as for instance in a doctor-patient relation; perhaps we accept and even expect insincere actions and less than honest strategies in politics. To the extent that there is a difference, I would argue that this is a case where it is necessary to distinguish between "full" apology and a "partial" apology. So in this type of cases, the latter – a mere expression of sympathy – but not the former might well be appropriate. By expressing sympathy, but not acceptance of responsibility, when appropriate, the apologizer is spared from having to "fake" an apology. Taft, who discusses the commodification of apology, is by principle against the utilitarian approach to apology, and he argues that "[i]f the defendant is not contrite and does not feel that he has committed a wrong, a staged apology would be a moral wrong" (Taft, 2000). According to Taft it is our moral obligation to apologize under certain conditions – which we shall discuss in greater detail below – and he notes that "[a]uthentic apologetic discourse occurs in an environment where the participants respect apologetic discourse as a moral activity and resist subverting the discourse for strategic and instrumental purposes" (Taft, 2000). Taft believes that the apology in its essence can not and should not be without risk for the apologizer, since one is apologizing for an offence for which one is responsible. Responsibility should not be risk free. Thus, Taft argues that the moral content of apology is diminished, if the legal consequences of apologizing are avoided, either through a partial apology or because the apology is legally protected from admissibility. He observes that "[i]nstead of being perceived as a moral ritual, apology becomes a material entity, an "object of exchange"" (Taft, 2000).

In principle, however, the utilitarian position on apology works hand in hand with the otherwise commendable and prudent proactive strategies of the systems approach, viz., strategies that are more concerned with the prevention of future similar adverse events than about who exactly was at fault in the current event (see for instance Jensen and Madsen, 2001; Rosner et al., 2000). From this perspective Liang (2002) proposes that the appropriate way of expressing empathy in a systems contexts is by having a representatives of the system saying, "we are sorry..." since this reflects the system accountability. Although this makes sense from a systems approach, other studies show that saying "we" rather than "I" makes an apology very impersonal (Lazare, 2004; Woods, 2004) and even in some cases makes the patients feel that no one is taking responsibility (Kilpatrick, 2003; Woods, 2004). There are cases where patients still seek 'personal accountability' and do not find the systems apology satisfying. Following the death of an 11-year old child, which could have been avoided, the hospital made a full disclosure, provided an excuse to the parents and took full responsibility and, finally, issued changes to prevent future similar occurrences. But the parents were not satisfied with the systems apology, saying that, "there's really no gratification in it". They therefore pursued a lawsuit calling for personal responsibility, as they found "personal accountability and responsibility is nowhere in the system" (Kilpatrick, 2003). This case is interesting because the "full apology" that was offered did not satisfy the parents, because it did not respect the "bedrock structure of apology" that Tavuchis (1991) underscores as being the relationship between the offender and the offended. This case touches upon one of the great challenges in healthcare, namely the fact that the systems approach deliberately seeks to avoid laying responsibility on individuals. An approach that as mentioned is normally perceived as enlightened and as being a move forward in terms of patient safety work (Brennan et al., 1991; Leape, 1994; Madsen et al., 2006). However, in the eyes of some patients it is perceived as nothing more than disregarding accountability.

The leader of the British based Action Against Medical Accidents (AvMA, 2006), Peter Walsh, argues that an apology might be a beginning but it is not the total solution, and he finds it necessary for healthcare professionals to stand up to accountability. He argues that is not enough to talk about system failures, since high standards are often linked to feelings of responsibility.[14] In sum, he and

---

[14] Walsh also points out that the fear of malpractice – the most often cited reason for not disclosing – is overestimated, and he claims that the barriers for reporting lies within the culture (Walsh, 2004). The numbers offered on AvMa's website support this claim: While the Department of Health itself estimates that there are approximately 850,000 medical accidents in English hospitals alone each year, half of which should have been avoided (Department of Health, 2000), the NHS Litigation Authority recorded only 5,609 claims in 2004-2005 (AvMA, 2006). In a Danish questionnaire survey the strongest reasons for doctors not to report was the fear that the press would start writing about it and the perception that their department had no tradition for reporting (Madsen et al., forthcoming). AvMa is an independent charity established in 1982.

AvMA argue that patient safety and justice can and should work together (Walsh, 2004). This view is contrary to the "I'm sorry" laws adopted by several US states that allow doctors to express regret without being penalized by legal liability in medical malpractice suits. Opponents of the "I'm sorry" laws claim that they are ways of saying "sorry without regret" (Taft, 2000). Tavuchis states that "apology cannot come about and do its work under conditions where the primary function is defensive or purely instrumental and where legalities take precedence over moral imperatives" (Tavuchis, 1991). The US movement "Sorry works" is a coalition of doctors, insurers, patients, lawyers, hospital administrators and researchers who have joined together to provide a "middle ground" solution to the medical malpractice case (Sorry works, 2006). They point out that saying sorry will work in everybody's favor – not only will it benefit patients (who receive compassion and acknowledgement) and doctors and nurses (who may unburden themselves); but a no less strong incentive is the awareness that patients that have been disclosed facts and "apologized" to are less likely to sue. Clearly these different approaches are well meant attempts to provide the patients with what they desire. However, the downside is the commodification of apology (Taft, 2000). An example of this is provided on the Institute for Healthcare Improvement's *SaferHealthCare* web site (IHI, 2006) where a woman offers this criticism: "My sister has recently experienced the medical director, head of midwifery and a consultant obstetrician all saying they were sorry for something that went very seriously wrong in the management of the delivery of her son. I was present on all these occasions and in every one, we felt as if they were saying it because they knew it was the 'right thing to do' but either did not mean it or did not understand what it meant. I say this because the staff who dealt with her clearly lacked training in communications in this difficult situation, and because in every way – before and after the apologies – they have demonstrated an extraordinary lack of imagination and sensitivity in the way they handle the communications with her and her husband." It seems that an apparent obstacle for offended persons in accepting an apology is when they feel that they become a means for the offender's self-serving goals rather than being the actual "beneficiaries" of the apology.

I suggest that the utilitarian arguments are worth considering and that these in themselves can provide a defense for apology in healthcare. However, I do not think that apology is always the right act; it is certainly not the right act just *because* the immediate consequences seem optimal. Therefore, using apology as a utility may not only have a negative effect in healthcare in terms of destroying the trusting relationship between patient and caregiver, it tends to distort the essential meaning of apology. In this regard I suggest that the rule-utilitarian has more to offer in the context of healthcare. Honesty and trust are fundamental values in medical ethics that should not be jeopardized, and we must choose the right and appropriate actions accordingly.

In the following, therefore, I shall attempt to elucidate the conditions for apology, describing when it is appropriate to apologize and when just to express regret. I am going to suggest that the necessary conditions for apology put forth by the deontologist can provide a basis for a discussion about the conditions for apologizing effectively after medical harm; conditions that at least the act-utilitarian is not able to use as a foundation.

4 The possible conditions for apologizing in healthcare

The philosopher Gill (2000), basing her analysis on Kort (2002) and Goffman (1971), proposes five necessary conditions for apologizing:

1. an acknowledgement that the incident in question did in fact occur
2. an acknowledgement that the incident was inappropriate in some way
3. an acknowledgement of responsibility for the act
4. the expression of an attitude of regret and a feeling of remorse
5. the expression of an intention to refrain from similar acts in the future

Besides these necessary conditions I am going to argue that we should add yet two pragmatic conditions for apologizing effectively in healthcare:

6. an explanation of what happened
7. practical amends

Another set of conditions is provided by Tavuchis (1991) who argues that the basic formula of apology involves 1) acknowledging the violated rule, 2) admitting fault by violating the rule, 3) expressing genuine remorse and regret for the harm caused by the violation. Tavuchis claims that the mere expression of "I'm sorry" includes implicit offers of reparations and promises to reform, and he therefore does not include these aspects explicitly, maintaining that it will only complicate the essential message. I propose that this non-inclusion may cause problems in the aftermath of apology, and therefore see no reason for not including explicitly what he and many others believe to be the implications of an apology. Only by making these matters explicit is there a possibility that they will in fact be met by offenders.

In the following I will discuss the possible conditions for apology in healthcare. Whether it becomes necessary to explicate all the conditions in apologizing depends upon the offense, the more serious the more elaborate the apology must be, where as a minor offence may only need an undemanding "sorry" (Gill, 2000).

4.1 Recognition

For any apology to begin, "at least one of the parties involved must believe that the incident actually occurred" (Gill, 2000). This first step of recognizing that there has been an incident is crucial in healthcare, since many patients' experience that the hospital and staff often do not even acknowledge that an incident has occurred (Gibson and Singh, 2003). Empirical research illustrate that the mere acknowledgement of pain, suffering and perhaps even wrongdoing plays a significant role for victims. Govier and Verwoerd advance as their central thesis "that it is through acknowledgement that the importance of apology to victims, and their power as a step toward reconciliation, can be explained" (Govier and Verwoerd, 2002).

During a presentation the author gave on "necessary conditions for a moral apology" at the Hastings Center[15] (Madsen, 2005) one of the attendees expressed amusement when this condition on acknowledgement was introduced. She was a woman in her thirties and had been paralyzed from her waist down during delivery of her second child due to medical error. She explained after the presentation that neither this condition nor any of the other conditions, outlined above, were ever recognized or acted upon by the hospital at which she was harmed, a story we shall return to (Anonymous patient, 2005). Obviously it is not always sufficient that only one party believes that the incident occurred, although it is necessary, since it would not make sense to apologize for an incident that no one believe has occurred.

"The most essential part of an effective apology is acknowledging the offence. Clearly without such a foundation the apology process cannot begin" (Lazare, 2004). Acknowledging harm is one of the key elements allowing patients to find closure (Berlinger, 2003b, 2005; Gallagher et al., 2003; Gibson and Singh, 2003; Witman et al., 1996), even if the acknowledgement is not followed by a full apology. In this context "to acknowledge" means to recognize something to be a fact, to admit the existence, reality or truth of the issue. The issue, of course, is mainly related to the consequences following medical harm, but acknowledgment could also be relevant in regards to perceived lack of care, treatment or respect. The importance of acknowledging the incident and the negative effects for the patients is critical. By admitting that an incident has happened one is acknowledging that the patient is not delusional: T,he harm is a fact and should not have occurred. In some instances patients have felt they were being ignored and the harm not taken seriously (Gibson and Singh, 2003).

The acknowledgement becomes a key act and staff should in principle acknowledge the harm done to the patients even if they are not or do not think they are directly at fault. An acknowledgement in it self is not an admission of fault, nor is it an apology; it is however a condition for an effective apology, and is really

---

[15] The Hastings Center for Bioethics, Garrison, New York.

just a simple act of showing respect and sympathy for the patient. Of course, the consequences of the harm to the patient can be just as critical whether the harm is induced by the fault of an individual or the system, or induced by the underlying disease of the patient (i.e., a complication). The literature deals mostly with actions following medical mistakes (Finkelstein et al., 1997; Wu et al., 1997), but I suggest it is just as important to consider "appropriate reactions" following any kind of harm afflicted on the patient. Why should harm caused by known and distressful complications not be followed by appropriate actions such as acknowledgement and sympathy? There is no ethical argument for not acknowledging harm per se, be it because of negligence, error or mere complications in relations to procedures. Apologizing in all of these cases however might not be appropriate or justifiable.

4.2 Acknowledging the inappropriate act

"At least one of the parties involved believes that the act was inappropriate. If the person offering the apology does not believe the act inappropriate, she must be willing to accept the legitimacy of the addressee having taken offence" (Gill, 2000).

What is an inappropriate act? Is patient harm an inappropriate act - in all instances? Is it a moral offence to harm patients? In principle the reason for treating patients is to heal them and not to bring harm to them, however in healthcare there will always be known complications attached to certain procedures, and therefore calculated risks of harm to patients. In this sense, harm can not in it self be regarded as an inappropriate act, although certain types of actions leading to harm may be regarded as inappropriate. First of all, to act negligently would be inappropriate, since one is not acting according to standards of care. Second, to harm patients when it could have been prevented would also be inappropriate, even if the person, who causes the harm, is not aware of the existence of the preventive measures. In such a case the hospital is responsible for putting measures in place that may detect flaws in the system and obliged to introduce standards and procedures that will prevent adverse events. An example of such a measure could be an incident reporting system for learning and prevention. Third, harm that is caused by excessive workload or other performance shaping factors known to affect safety negatively is also inappropriate, and again the responsibility may lay with management. We may make a distinction between the inappropriate acts and the inappropriate consequences. Technically it does not necessarily change the fact that an apology may be justified, but it might resolve who should apologize.

4.3 Responsibility

"Someone is responsible for the offensive act. Either the party offering the apology takes responsibility for the act or there is some relationship between the respon-

358 Marlene Dyrløv Madsen

sible actor and the apologizer such that her taking responsibility for offering the apology is justifiable' (Gill, 2000).

When patients are harmed no matter the reason someone is responsible for these consequences. Taking responsibility for an act does not necessarily mean that one has acted in a blameworthy way. One may be causally responsible or even several may be causally responsible for the harm without being culpable. The responsibility for patients and harm done to patients in the healthcare organization must not be reduced to individuals and individuals' acts. Even though the doctor is bound by the Hippocratic Oath, *primum non nocere*, ("above all, do no harm"), health professionals are not alone in being responsible for giving the patient full care. We need to define responsibility in wider terms.

The reason for medical harm in healthcare is typically multi-causal and a result of both failures in latent conditions and active failures in the sharp end in other words "organizational accidents" (Reason, 1997). A healthcare organization is a complex system, which has been defined as a high reliability organization although one may argue that formally it is not (Roberts et al., 2005). Originally high reliability organizations (low-risk-high-hazard-domains) were defined as organizations which have a high level of safety and few accidents because the system is tightly coupled and redundant and the organization devotes a lot of resources to safety measures (Maurino et al., 1995; Reason, 1997; Rijpma, 1997). When accidents occur, which seldom happens, they usually have catastrophic consequences on persons and environment. High-reliability organizations include aviation, nuclear power plants, and off-shore oil platforms. Generally and traditionally healthcare is not a tightly coupled and redundant system with a main focus on safety, nor do incidents and harm happen rarely and to an undefined number of people, rather people are harmed every day because of failures in the system.[16] High-reliability theory and system theory is therefore successfully adapted to the medical field in order to enhance safety although there is still a long way.

Within a complex "high-reliable" system all members of the organization needs to be able to rely on each of their colleagues for taking full responsibility for their tasks and function. Everybody working in the healthcare organization has individual responsibility and shared responsibility while management also has objective responsibility. The hospital management is "objectively responsible" for the consequences of the performance of their staff. French (1979, 1981), arguing

---

[16] An important indication of patient safety is the rate of adverse events among hospital patients. Adverse events are unintended injuries or complications caused by medical care. Some of these lead to disability or death, others to prolonged hospital stay. Adverse events include avoidable events (mistakes) and unavoidable events (e.g., unforeseeable allergic reaction to antibiotics). For instance, a Danish study and a Canadian showed that 9.0% and 7.5%, respectively, of admissions involved adverse events, of which 40% in both studies were deemed to be avoidable (Baker et al., 2004; Schioler et al., 2001).

that organisations can be treated as moral persons, makes a distinction between the primary principle of accountability (PPA), in which the person who is directly responsible for wrongdoing can be held accountable, and the extended principle of accountability (EPA), in which any other staff member or management may be held accountable for colleagues' or sub-ordinates' wrongful act if he/they know or could have known that the consequences would be negative. It is well known in systems thinking that any negatively impacting performance shaping factors has a direct effect on performance. To have well-functioning performance shaping factors it is necessary to have "a system of rewards for reporting and discovering error" (Fassert, 2000). By refraining from implementing such systems of organizational learning, management will become indirectly responsible for harms that could have been prevented had such systems been in place.

In a systems context there is still a responsibility for honest mistakes or "slips and lapses" whenever such unintentional acts have negative consequences. Medical harm is not in itself a moral offence and may not in principle be caused by moral wrongs; however, morality is at issue in taking or not taking responsibility for the consequences. If a doctor, say, has acted according to current standards of care, then there might not be anything wrongful in the acts committed although they turned out to harm a patient. However if the consequences of the doctor's "right" acts lead to "wrongs" or harm, then he or she is still obliged to take responsibility for the outcome. Just because we do not intentionally seek to harm somebody does not mean that we cannot or should not take responsibility for the consequences of our actions. If for example I as a driver were to injure someone in a car accident, I would still be compelled to apologize, perhaps not for my moral wrongdoing but for the consequences of my act. Not every instance of medical harm will call for an apology, not even every instance of avoidable medical harm; but every instance calls for an acknowledgement of the harm.

Being causally responsible for an act is, in short, *being* responsible; and, in the end, the hospital is responsible for the acts of medical care that harms its patients. In this sense it may therefore also be justified that another than the offender apologizes, in particular if the cause of the mishap cannot be singled out. There are many different views on who should be responsible for apologizing in healthcare. Some believe it should be the person directly involved in the incident, others are convinced it is better if it is the responsibly doctor, and again others think it best to take the "issue" out of the direct context and instead make either the risk manager or the hospital management take the overall responsibility and make the apology (Kraman and Hamm, 1999). There are good arguments for each position, although I propose that if the person directly involved has the possibility to apologize he or she should do so in order to maintain some level of personal relation; if not, then it should be his or her immediate superior. In terms of the healing process for the parties involved it is essential to keep the conflict and its resolution in the arena in which it has taken place. In any case, it is the responsibility of the hospital that the incident is analyzed and a discussion takes place about what actions to take, who should talk to the patient and what should be said. Giving an apology is about taking responsibility for one's acts, and in principle it has nothing to do with legal culpability. Unfortunately, this is not so easy to distinguish in practice, and it is particularly difficult to keep apart if the overall framework is litigious (as in the US), where liability plays a significant role.

4.4 Regret and remorse

According to Gill "the apologizer must have an attitude of regret with respect to the offensive behavior and a feeling of remorse in response to the suffering of the victim [...] Expressing apologies in the absence of any genuine remorse is deceptive, and so is morally suspect" (Gill, 2000). This condition is not without its problems; for instance, how can we know if someone feels remorse or feels distress? Must one "feel" regret in order to express regret?

Pettigrove and Joyce will agree that there is a difference between expressing regret and actually feeling regret, but they will argue that in practice expressing regret still constitutes an apology, excluding the need for "feeling" regret. Of course people can and do express remorse without feeling remorse, just as they sometimes express regret without feeling regret. But what exactly does it imply to express regret?[17] The most common way of expressing regret is to say: "I'm sorry", but saying so may mean different things, as Berlinger points out: "To say 'I'm sorry your father died' is not at all the same thing as saying, 'I'm sorry I killed your father'" (Berlinger, 2005), and it is quite clear why not. In the first case, "I'm sorry" is a simple expression of sympathy in a situation in which the speaker has no direct responsibility, whereas the second is an expression of regret in a situation for which the speaker has direct responsibility. In the second case, however, it is not clear whether or not the killing was done with or without intent, which would seem to make a difference; and if it was done with intent, was it perhaps justifiable self-defense or was it manslaughter or murder. Depending on the situation, we might expect more from the offender, perhaps an expression of contrition and indications of sincere remorse for wrongdoing. So, if the speaker did the killing and should not have done so, an apology is due and restitution may be appropriate (Cunningham, 1999), whereas in the first example – the sympathetic response – this would not make sense.

First of all, these examples illustrate that the expression of regret – although semantically the same – may in fact have different implications: I can say that

---

[17] Regret means "Feel or express sorrow, repentance, or disappointment over" (COED, 2004) and "1. To feel sorry, disappointed, or distressed about; 2. To remember with a feeling of loss or sorrow; mourn" (AHD, 2004).

I'm sorry without being at fault, and I can do so and be at fault. This ambiguity explains some of the reasons why staff may be hesitant to say sorry, and why patients tend to believe fault is implied, when the words "sorry" are expressed. Because, as Goffman notes "this expression itself [sorry] may be *relatively* little open to gradation" (Goffman, 1971). Secondly, an apology must entail – or is in itself – an expression of regret, whereas an expression of regret does not necessarily constitute an apology. This distinction is not always made clear and too often expressions of sympathy are defined and offered as apologies, when they are not meant in this sense (see for instance Berlinger, 2005; Taft, 2004).

Gill argues that the feeling of empathy, distress or guilt is essential to apologizing and that for my apology to be convincing I must somehow wish that I had not done the act – I must have "an attitude of regret" (Gill, 2000). Gill sets this as a principal requirement on apology, but does not discuss the critical problem in securing this condition on the same premises as Joyce and Pettigrove. One may claim that in order to know whether the fulfilment conditions of 'feeling regret' are satisfied we need to know the apologizer's state of mind and intentions; and we might therefore slowly be moving into the realms of philosophy of language, of mind, and speech act theory. However, Gill stays within the field of normative ethics, contending that feeling regret entails (merely) that one will strive to refrain from similar acts in the future, observing that, "believing that an certain act is wrong, experiencing regret and remorse for having done it, is inconsistent with a carefree repetition of a similar act" (Gill, 2000). In other words, if the apologizer repeats his offence after having apologized we have little grounds for believing that he regretted his behaviour. In this way we are able to test the truth-value of the apology by using "post-apology behaviour as a test of sincerity" (Gill, 2000). Obviously this still contains some problems in the sense that we may not always have the possibility to check offenders' post-apology behaviour. Furthermore I may in principle refrain from future acts not because I have felt regret, but because I have strategic reasons, say fear of punishment or humiliation, for doing so.

The problem with the condition of "feeling" regret shows its inconsistency in relation to institution or group apologies. There is an essential difference between person-to-person apologies and institution or group apologies, which Tavuchis (1991) discusses in great detail. In the cases where apologies performed by a representative on behalf of a group, e.g., a representative of the hospital apologizes for harm done, the representative cannot be expected to "feel" regret (Gill, 2000; Lazare, 2004). The expression of regret may be taken as sincere based on "what the group feels and what the group intends" (Joyce, 1999), for instance, if the intention to learn from the event is clearly communicated and shown. But we cannot and should not require an actual feeling of regret when an institutional representative apologizes on behalf of the staff.

Theorists on apology tend to divide up into two groups, those who maintain the attitudinal states and those who do not. Gill tends to connect the feeling of regret with sincerity, whereas Joyce, as mentioned earlier, argues that "sincerity is not a necessary component of apology, though it is certainly usually a desirable feature, for both individual and group apologies" (Joyce, 1999). He argues that the "inner life" is "neither necessary nor sufficient for us to consider the apology to be a sincere one", especially in the context of institutional apologies. If a doctor harms a patient due to inadequate procedures and if the hospital makes no effort to change procedures after the incident, then the doctor's apology and feelings of distress and regret might be sincere on a personal level, but on an institutional level the apology fails according to Gills conditions. In this case Joyce would maintain that the state of mind of the doctor is irrelevant, since the institution continues to practice unsafe behavior, and the apology therefore must be taken as insincere. In healthcare, however, I do believe that it will make a difference to the injured patient that the doctor is personally regretful and that he himself will try not to cause harm again, even if the apology fails to be a full-fledged apology on an institutional level.

4.5 Refrain from similar acts

"The person to whom the apology is offered is justified in believing that the offender will try to refrain from similar offences in the future" (Gill, 2000). We have already discussed this condition at length in relation to the other conditions, where we concluded that it would be inconsistent to experience regret and remorse for an action that one would choose to repeat, and similarly, "if a person carelessly continues to offend in the same way, apparently not even trying to stop, we have reason to believe the apology was not sincere" (Gill, 2000). Refraining from future offences is an integral part of the apology process and for patients one of the most important features following harm. In a Danish survey about patient wishes following medical harm 89% found it extremely/very important that learning takes place in order that future patients may be spared (Andersen et al., 2004).

If staff or representatives of the hospital management apologize to patients without having the intention of changing harmful procedures, it can hardly be taken to be an apology. In a systems context to refrain from similar offences would not only apply on an individual level but on a group, unit and managerial level, and it would require that necessary precautions are taken to prevent future occurrences. This means that hospitals must be willing to learn from experience to maintain and improve safety. Only through such activities can patients take an apology seriously, and it is, as empirical evidence shows, of outmost importance to patients and a sign of respect (Finkelstein et al., 1997). "The learning opportunity presented by a mistake [...] is an integral to the ethics of being a responsible professional who upholds the physician-patient relationship even when it is not at all comfortable to do so" (Berlinger, 2005). To ensure that the sincerity of an apology is conveyed, an expression of an intention to refrain from similar acts in the future is or may be indispensable.

4.6 An explanation

In all cases of medical harm an explanation of what and why it happened is the least one can grant the patients. The offended has a right to know what has happened. Patients often cannot find peace at mind before they know what and especially why things have gone wrong. The explanation is, I will claim, a very important inclusion in the apology process.

The systems approach opens up for the possibility that no one feels directly responsible for negative outcomes (although in fact everybody has more responsibility than before). There may be a tendency for doctors to justify or excuse the incident in the old Greek sense of apologia, rather than explaining and informing about the incident to the patients, because there *are* often several reasonable explanations why the mistake has occurred – e.g. lack of sleep, being the only one on call or having inadequate equipment. But to excuse or justify harm while making an apology is what Lazare (2004) calls a "botched apology". Others call it an insult (Berlinger and Wu, 2005; Berlinger, 2005; Schneider, 2000) or a partial apology (Robbennolt, 2003). When one seeks to justify the actions leading to the harm one is denying one's responsibility for the harm and the consequences of one's actions and, therefore, only performing a "partial apology". Such an apology will not work, it might explain the incident but it does not justify as a "full apology". Berlinger cites a doctor for saying that "an error that can be rationalized is still a mistake, we must learn from them" (Berlinger, 2005). Just because we can explain why something happened does not take away our responsibility for the act nor our obligation to make bad outcomes good learning opportunities.

The purpose of giving an "explanation" is to give a plain and simple description of what went wrong and how and why it did so. In a Danish survey of patients' wishes following medical harm 90% of patients found it extremely/very important to get information about the consequences in terms of health (Andersen et al., 2004). In America COPIC quoted a physician who testified in support of the Colorado law: "Injured patients expect and deserve an explanation. They want to know what went wrong, and they want to be assured that steps are being taken to prevent similar occurrences to others. Yet [physicians'] fear of exposure to the tort system can act as a powerful deterrent to this communication" (Berlinger, 2005).[18] Most stories told by harmed patients or relatives following adverse events stress

---

[18] Testimony of Mark A. Levin, M.D. Quoted in COPIC Topics 86 June 2003, 4. Cited in (Berlinger, 2005).

the fact that the missing explanation is a burden that keeps them from being able to find closure and get on with their lives. The importance of "the explanation" is not unique to healthcare. In South Africa's famous Truth and Reconciliation Commission (TRC) one of the main aims was for relatives to get explanations and gain knowledge about what had happened to loved ones.

'Disclosure' means telling the truth about treatment and possible risks, complications and changes in treatment, so an "explanation" should be a natural part of disclosure. Following medical harm disclosure may be a vital part in the treatment of the patient, since the patient may need to become involved in the necessary medical precautions that must be taken (Finkelstein et al., 1997; Hébert et al., 1997).

4.7 Practical amends – the parallel process

Berlinger observes that "apology and compensation are intimately linked as responses to harm, despite efforts, sometimes well intentioned, sometimes calculated, to separate them" (Berlinger, 2005). The question of compensation in healthcare is controversial and, notwithstanding Berlinger's claim, it is not clear if it is part of the apology process. As we already mentioned, Goffman (1971) identified two distinct and independent processes involved in the corrective behaviour of apology, the ritualistic and the restituitive, and depending on the offence the weight of the two processes will differ. Govier and Verwoerd (2002) argue along the same lines when they claim that the purpose of apology is to make amends: both moral and practical amends. In criminal justice too there is a division between retribution and restoration, although, as a theologian has stated it "you cannot make a true apology without trying to make things right" (Camp, 2005). In this regard Joyce agrees: "Sincerity should not be assigned to the apology itself, but to the undertakings and the self-portrayals that accompany the act of apologizing" (Joyce, 1999).

Philosophically, compensation can hardly be proven to be part of the necessary conditions for apology. However, a number of authors (Goffman, 1971; Govier and Verwoerd, 2002; Pettigrove, 2003; Taft, 2004) hold – and I agree – that compensation is part of the practical amends and is closely connected with the apology process and should be dealt with parallel. The term "practical amends" implies that these can be other than economic as stressed by Govier and Verwoerd (2002). Ignoring practical amends, for instance in the form of compensation in cases of severe harm, may be considered unethical. Furthermore, I suggest that the right to compensation in severe cases of harm – or other practical amends in cases where economic amends do not make sense – may be justified through the principles of justice and fairness.

Suppose I am a patient who has suffered medical harm during hospitalisation and that I need re-surgery; suppose further that the re-surgery is not paid by a public or an insurance health service scheme, but that I will have to pay for it myself. Now, is it fair that I should pay for the treatment of someone else's mistake? Suppose I also will loose my pay check because of prolonged admittance to the hospital. Is it fair that I have to pay myself for my losses when the responsibility for the harm lies with the hospital? From a moral perspective it would seem fair that economic compensation is given when justified.

Govier and Verwoerd argue by examples from South Africa's Truth and Reconciliation Commission (TRC) that "an apology in which there is no willingness to undertake any practical measures of reparation is likely to seem insincere or hollow", and they go on to argue that "a full-fledged moral apology should include a commitment to practical amends" (Govier and Verwoerd, 2002). As Pettigrove notes "...the failure to offer reparation can prevent an action from being an apology. 'I'm sorry I stole your paycheck, but I'm not giving it back, even though I am able' would fall outside the parameters of apology" (Pettigrove, 2003). Pettigrove then concludes that "a locution of this sort is not merely infelicitous, in the manner of an insincere apology: It misfires altogether" (Pettigrove, 2003).

In cases following medical harm much attention is given to the question of economic compensation, which is of course essential and justifiable. Still, I suggest it is just as significant to think about practical amends in a broader sense. In healthcare practical amends could take various forms from simply informing patients about the consequences of the adverse event, information about and help to treatment, guidance and support following the event and ideally information about possibility for compensation or complaint. For instance, one way of making practical amends would be to take the patient's "story" and use this as a demonstration for the need of changing the parts of the system responsible for the event, possibly involving patients as partners in the process of change. The strong need for patients to experience 'meaningfulness in the midst of meaningless harm' can in fact be acquired through listening, learning and active system change; preventing harm to future patients. In contrast, these needs may not be met through economic relief.

In fact it is often observed that many patients do not at the outset wish or think in economic compensation (Levinson et al., 1997; Liang, 1999; Vincent et al., 1994), most often they just want to know what happened, why and what can be done (Andersen et al., 2004; Freil et al., 2004; Gallagher et al., 2003; Witman et al., 1996). When answers to these questions are not provided patients feel neglected, morally disparaged and hence seek other possibilities in their search for answers and redress (see for instance Berlinger, 2005; Levinson et al., 1997; Vincent et al., 1994). In fact several studies in healthcare illustrate that many patients and relatives seek legal recourse in the hope of effecting a change in future behavior of the wrongdoer and the organization (Gallagher et al., 2003; Hickson et al., 1992; Vincent et al., 1994; Witman et al., 1996). A hospital that omits to take medical harm as an opportunity to learn is indirectly allowing harm to happen to someone else.

Practical amends can be many things and satisfy different needs. There are examples where patients have not received an actual 'apology' but, through the actions taken by staff and hospital, have interpreted the gesture made by staff and management as such. A case from Norway illustrates this (Anonymous relative, 2005). During a birth delivery at a county hospital several unforeseen complications occur. After the delivery it turns out that the newborn has brain damage caused by bleeding, but the consequences cannot be determined at this point. The doctors are not sure if the baby is harmed in other ways. The hospital and staff are very supportive and explain to the parents that they have a right to file for compensation, which the parents never choose to do. Expressions of regret, but no formal apology, were given although never assigned to specific actions or wrongs. However the parents never found this to be a central issue, and in fact, when asked the father responded that they would rather prefer the right actions than the right words. Nine months after the incident it was discovered that the child had acquired numbness in the left hand as a result of the brain damage, and five years later the child still has difficulties in using the hand. Although it became clear that several crucial mistakes were made, the parents never considered holding anyone accountable. The parents had experienced a staff that took immediate responsibility in the aftermath and had helped them as best they could and this made all the difference.[19] Joyce discusses this phenomenon as well, i.e., the, perhaps strange, fact that it is possible to apologize by expressing all that constitutes an apology without saying the "magic words" (Joyce, 1999). In healthcare this is worth considering since much weight is being put on "communicating" the apology, and perhaps less on the supporting actions accompanying such an expression. In this relation, it is interesting to look at the earlier mentioned Danish study, where the questionnaire survey of how patients want healthcare staff to handle mistakes following adverse events, surprisingly showed that only 38% of patients find it extremely/very important that healthcare staff show "sympathy"[20] (Andersen et al., 2004). This seems to indicate that supporting actions should be understood in terms of apology or expressions of regret, an explanation, and information about the consequences in terms of health rather than "sympathetic feelings", and ultimately that learning takes place in order that future patients may be spared.

---

[19] To the knowledge of the parents of the child several procedures were changed at the hospital, as a result of this incident, to avoid it from happening again. However not all crucial and necessary changes have been implemented, which has made the parents consider contacting the hospital again in order to make them enforce such changes.

[20] In Danish: "medfølelse". Strangely, only about half of the respondents answered this question compared to responses to the other questions.

The five strict conditions for apology we have discussed do, as we have seen, provide the moral ground for claiming compensation (when this is otherwise justified), and this may possibly persuade doctors not to apologize for fear of prompting patients to seek compensation. However, one may hope that the moral arguments will convince them otherwise. Also, it is worth emphasizing that the fear of compensation in principle is only a problem in systems where no-fault compensations schemes are unavailable. Several countries – Denmark, Norway, Sweden and New Zealand, for instance – have national no-fault compensation systems in place and a few institutions in the US (Berlinger, 2005). Such a system largely defuses the fear of disclosure for economic reasons, thereby giving easier way to making apology. Of course, such systems do not fully secure apology since resistance to apologizing can have other reasons (Andersen et al., 2002; Madsen et al., forthcoming; Vincent et al., 1999). In fact the downside of no-fault-compensation systems may be that staff avoid apologizing, relying too much on the system taking care of the patient. Numerous observers have found that the USA is a very litigious society which, like Japan e.g., has no publicly supported compensation systems in place (Itoh et al., forthcoming). No doubt it is much more demanding for staff working in such systems to engage in making apologies even when justified, and there are also examples of risk managers encouraging them to refrain (Wears and Wu, 2002).

A number of studies have shown that – contrary to more cynical views among health care staff – patients do understand that things can go wrong and accept that "it is human to err" (see for instance Andersen et al., 2004; Freil et al., 2004; Manser and Staender, 2005). They are not vindictive when they are treated with respect, are given the opportunity to understand why things went wrong and told of precautions taken to prevent future patients from harm. What they do not understand, however, is when no one takes responsibility and especially when everyone ignores and refuses to recognize that an adverse event did happen. Which was the case in the story mentioned earlier about the woman who was paralyzed during delivery at a fine local hospital in the USA (Anonymous patient, 2005).

During delivery the patient was given an epidural due to pain, at which point she instantly felt an extreme pain going down her back and legs. The pain continued in her legs, which were paralyzed, after she had delivered. The child was not hurt during the event. The hospital showed no understanding and claimed that nothing was wrong and that everything had gone as planned. They began treating her as a nuisance patient and asked her to leave the hospital since they thought they could do no more for her, at which point she was still paralyzed. The patient and husband were not told by any staff member of any actions pertaining to her worsening conditions after delivery.

The patient chose to file for compensation as they were unable to get in dialogue with the hospital. In this process the patient found out that the anaesthetist who gave her the epidural was a known drug addict who had been given a second chance. The anaesthetist claimed she had been drug-free the night of the birth, although she could not recall what she had given the patient. The hospital finally chose to settle the case the day before the trial was to begin and five years after the incident had occurred. The patient and her family received an undisclosed amount in compensation from the hospital, but were never told *what* and *why* things went wrong; they never received an apology, expression of regret, information or any kind of help in regards to the treatment of the adverse event.

This is clearly unacceptable treatment of patients in the aftermath of harm and unfortunately they are not few (Berlinger, 2005; Gibson and Singh, 2003). It would be easy if we could say that they are only tied to litigious societies, but this would be untrue. Regrettably such cases also happen in countries where no-fault compensation systems are in place, although probably not as often. The case from Norway and the one just described illustrate to some extent the potential and barriers for making apology within different structural systems and the effects hereof.

5 Conclusion

The aim of this article was to investigate the nature of apology and its internal logic in the context of healthcare as 'apology' has been shown to play a significant role on several levels in the aftermath of adverse events. I have tried to take all the relevant issues that impact on the apology process into account as I have analysed 'apology' and its moral role and justification in the context of medical harm.

I have sought to show that each of the two major philosophical positions, deontology and utilitarianism, offers good arguments for how to justify apology in healthcare. The utilitarian approach can justify the act of apology through its good consequences; the satisfaction of both patients and doctors for healing and the economic incentive in avoiding litigation. As long as the good consequences of an apology exceed those of not apologizing then it is the obligation of the utilitarian to apologize. As we have seen, the utilitarian can justify the instrumental value of the process of apology, and claims that sincerity is not a necessary condition for an utterance to be an apology. However I conclude that using apology as a utility may not only have a negative effect in healthcare in terms of destroying the trusting relationship between patient and caregiver, it tends to distort the essential meaning of apology. In this regard it becomes important, as discussed, to distinguish between the different meanings of apology, when the classical utilitarian argues that "the relation between apology and responsibility need not be established", then he is only making a "partial apology" and not a "full apology". In this regard I conclude that the rule-utilitarian has more to offer in the context of healthcare, since they also would "value" honesty and trust as fundamental values in medical ethics that should not be jeopardized.

Contrary to classical utilitarianism, the deontologist claims that the apology has an inherent moral value which cannot be instrumental and which cannot be given without sincerity and at least showing, and perhaps nurturing, feelings of regret. The deontological position argues that it is our moral obligation to apologize when appropriate and justified and that we should do so in compliance with the five proposed conditions, to which I agree. However, some of these conditions, as I have discussed are difficult to maintain and secure.

Through the discussion of the five necessary conditions for apology in the context of healthcare I have shown the complexity of apologizing after harm, and the negative consequences when apologies are avoided or are given in the "wrong" manner. I have clarified the moral conditions for apology, describing when it is appropriate and justified to apologize, and when to acknowledge or express regret. Besides the five necessary conditions for an apology, I conclude that there are two additional issues of significant importance to the apology process in terms of making it effective. On the one hand providing the patient with an explanation surrounding the circumstances of the harm, and on the other, making practical amends thereby fulfilling what is also known as the parallel process of apology.

References

AHD (2004). *The American Heritage Dictionary of the English Language*, Houghton Mifflin Company. URL: *http://www.bartleby.com/61/*. Accessed March 17th, 2006.

American College of Physicians (2002). Ethics manual: Disclosure. URL: *http://www.acponline.org/ethics/ethicman.htm#disclose*. Accessed March 30th, 2006.

American Medical Association (1994). Code of medical ethics. URL: *http://www.ama-assn.org/ama/pub/category/8497.html*. Accessed March 30th, 2006.

Andersen, H. B., Madsen, M. D., Hermann, N., Schiøler, T. and Østergaard, D. (2002). Reporting adverse events in hospitals: A survey of the views of doctors and nurses on reporting practices and models of reporting., *in* C. Johnson (ed.), *Proceedings of the Workshop on the Investigation and Reporting of Incidents and Accidents (IRIA 2002),17th - 20th July 2002, University of Glasgow*, pp. 127–136.

Andersen, H. B., Madsen, M. D., Østergaard, D., Ruhnau, B., Freil, M. and Herman, N. (2004). *Spørgeskemaundersøgelse af patientholdninger til reaktioner efter utilsigtede hændelser.*, Risø, pp. 1–32. Delrapport 2 fra projekt om patientsikkerhed. Risø-R-1498(DA).

Anonymous patient (2005). Personal communication. October, 2005.

Anonymous relative (2005). Personal communication. October, 2005.

Austin, J. L. (1962). *How to do Things with Words*, Oxford University Press, Oxford, England.

AvMA (2006). Action against medical accidents: For patient safety and justice. URL: *http://www.avma.org.uk/*. Accessed March 30th, 2006.

Baker, G. R., Norton, P. G., Flintoft, V., Blais, R., Brown, A., Cox, J., Etchells, E., Ghali, W. A., Hébert, P. C., Majumdar, S. R., O'Beirne, M., Palacios-Derflingher, L., Reid, R. J., Sheps, S. and Tamblyn, R. (2004). The Canadian adverse events study: The incidence of adverse events among hospital patients in Canada, *CMAJ* **170**: 1678–1686.

Berlinger, N. and Wu, A. W. (2005). Subtracting insult from injury: Addressing cultural expectations in the disclosure of medical error, *Journal of Medical Ethics* **31**: 106–108.

Berlinger, N. (2003a). Broken stories: Patients, families, and clinicians after medical error, *Literature and Medicine* **22**: 230–240.

Berlinger, N. (2003b). Avoiding cheap grace: Medical harm, patient safety, and the culture(s) of forgiveness, *Hastings Center Report* **33**: 28–36.

Berlinger, N. (2003c). What is meant by telling the truth: Bonhoeffer on the ethics of disclosure, *Studies in Christian Ethics* **16**: 80–92.

Berlinger, N. (2004). Missing the mark: Medical error, forgiveness, and justice, *in* V. A. Sharpe (ed.), *Accountability: Patient Safety and Policy Reform*, Georgetown University Press, Washington, D.C., pp. 119–34.

Berlinger, N. (2005). *After Harm: Medical error and the Ethics of Forgiveness*, The Johns Hopkins University Press, Baltimore.

Blumenthal, D. (1994). Making medical error into "medical treasures", *JAMA* **272**: 1867–1868.

Bosk, C. L. (1979). *Forgive and Remember: Managing Medical Failure*, The University of Chicago Press, Chicago.

Braithwaite, J. (2002). *Restorative Justice & Responsive Regulation*, Oxford University Press, Oxford.

Brennan, T. A., Leape, L. L., Laird, N. M., Hebert, L., Localio, A. R., Lawthers, A. G., Newhouse, J. P., Weiler, P. C. and Hiatt, H. H. (1991). Incidence of adverse events and negligence in hospitalized patients. Results of the Harvard medical practice study I, *N. Engl. J. Med.* **324**(6): 370–376.

Bærentsen, H. (1997). En fejlbarlig læge, *Månedskrift for Praktisk Lægegerning* **75**: 441–444.

Camp, J. J. (2005). Personal communication. October, 2005.

Christensen, J. F., Levinson, W. and Dunn, P. M. (1992). The heart of darkness: the impact of percieved mistakes on physicians, *Journal of General Internal Medicine* **7**: 424–431.

COED (2004). *The Concise Oxford English Dictionary*, Vol. 11 Edition on CD-Rom, Oxford University Press.

Cohen, J. R. (1999). Advising clients to apologize, *Southern California Law Review* **72**: 1009–1069.

Cohen, J. R. (2002). Legislating apology: The pros and cons, *University of Cincinatti Law Review* **70**: 819–872.

Crigger, N. J. (2004). Always having to say you're sorry: An ethical response to making mistakes in professional practice, *Nursing Ethics* **11**: 568–576.

Cunningham, M. (1999). Saying sorry: The politics of apology, *The Political Quarterly* **70**: 285–293.

Davies, J. M. (2005). Disclosure, *Acta Anaesthesiologica Scandinavica* **49**: 725–727.

Department of Health (2000). *An Organisation with a Memory - Report of an Expert Group on learning from Adverse Events in the NHS*, The Stationery Office, London.

Doctors in Touch (2004). Apology and disclosure process and guidelines. URL: *http://www.doctorsintouch.com*. Accessed March 30th, 2006.

Fassert, C. (2000). "Automatic safety monitoring", air traffic control: Achievements and perspectives, extended abstract for ATM, Sorbonne, France.

Finkelstein, D., Wu, A. W., Holtzman, N. A. and Smith, K. M. (1997). When a physician harms a patient by a medical error: Ethical, legal, and risk-management considerations, *The Journal of Clinical Ethics* **8**: 330–335.

Freil, M., Ruhnau, B., Hermann, N., Østergaard, D., Madsen, M. D. and Andersen, H. B. (2004). *Resultater fra interviewundersøgelse af patienters holdninger til håndtering af utilsigtede hændelser.*, Risø, pp. 1–65. Delrapport 1 fra projekt om patientsikkerhed. Risø-R-1497(DA).

French, P. A. (1979). The organization as a moral person, *American Philosophical Quarterly* **16**(3): 177–186.

French, P. A. (1981). The DC-10 case: A study in applied ethics, technology, and society, *Business and Professional Ethics* **2**(Spring). Reprinted in J. C. Callahan (ed.) (1988), *Ethical Issues in Professional Life*, Oxford University Press.

Gallagher, T. H., Waterman, A. D., Ebers, A. G., Fraser, V. J. and Levinson, W. (2003). Patients' and physicians' attitudes regarding the disclosure of medical errors, *JAMA* **26**: 1001–1007.

Gibson, R. and Singh, J. P. (2003). *Wall of Silence: The Untold Stories of the Medical Mistakes that kill and injure Millions of Americans*, Lifeline Press.

Gill, K. (2000). The moral functions of an apology, *Philosophical Forum* **31**: 11–27.

Goffman, E. (1971). *Relations in Public: Microstudies of the Public Order*, Basic books, New York.

Golding, M. (1984). Foregiveness and regret, *Philosophical Forum* **16**: 121–137.

Govier, T. and Verwoerd, W. (2002). The promise and pitfalls of apology, *Journal of Social Philosophy* **33**: 67–82.

Harvey, J. (1995). The emerging practice of institutional apologies, *International Journal of Applied Philosophy* **9**: 57–65.

Hébert, P. C., Hoffmaster, B., Glass, K. C. and Singer, P. A. (1997). Bioethics for clinicians: 7. Truth telling, *CMAJ* **156**: 225–228.

Hébert, P. C., Levin, A. V. and Robertson, G. (2001). Bioethics for clinicians: 23. Disclosure of medical error, *Can.Med.Assoc.J.* **164**: 509–513.

Hickson, G. B., Clayton, E. W., Githens, P. B. and Sloan, F. A. (1992). Factors that promted families to file medical malpractice claims following perinatal injuries, *JAMA* **267**: 1359–1363.

Hilfiker, D. (1984). Facing our mistakes, *New England Journal of Medicine* **310**: 118–122.

Hilfiker, D. (1985). *Healing the Wounds: A physician looks at his Work*, Pantheon Books, New York.

Hingorani, M., Wong, T. and Vafidis, G. (1999). Patients' and doctors' attitudes to amount of information given after unintended injury during treatment: Cross sectional, questionnaire survey, *BMJ* **318**: 640–641.

Hobgood, C., Xie, J., Weiner, B. and Hooker, J. (2004). Error identification, disclosure, and reporting: Practice patterns of three emergency medicine provider types, *Acad Emerg Med* **11**: 196–199.

IHI (2006). Saferhealthcare. URL: *http://www.saferhealthcare.org.uk/ihi*. Accessed March 17th, 2006.

Itoh, K., Andersen, H. B., Madsen, M. D., Østergaard, D. and Ikeeno, M. (forthcoming). *Patient Views of Adverse Events: Comparisons with Self-reported Healthcare Staff Attitudes to Disclosure of Accident Information*, Applied Ergonomics, special issue.

JCAHO (2004). Hospital accreditation standards, Standard RI.2.90, Joint commision on the accreditation of healthcare organizations. URL: *http://www.jcaho.org/*. Accessed March 30th, 2006.

JCAHO (2006). Joint commission on accreditation of healthcare organizations. URL: *http://www.jcaho.org/*. Accessed March 30th, 2006.

Jensen, T. R. and Madsen, M. D. (2001). *Filosofi for flyveledere: En undersøgelse af hvilke moralske aspekter man bør tage hensyn til ved behandlingen af menneskelige fejl i sikkerhedskritiske organisationer*, Master's thesis, Department of Philosophy and Science Studies and Department of Communication, Roskilde University, Denmark.

Joyce, R. (1999). Apologizing, *Public Affairs Quarterly* **13**: 159–173.

Kilpatrick, K. (2003). Apology marks new era in response to medical error, hospital says, *CMAJ* **168**: 757–757.

Kort, L. F. (2002). What is an apology?, *in* R. C. Roberts (ed.), *Injustice and Rectification*, Peter Lang Publishing, New York, pp. 105–110. First published in *Philosophy Research Archives* **1**: 80-87 (1975).

Kraman, S. S. and Hamm, G. (1999). Risk management: Extreme honesty may be the best policy, *Annals of Internal Medicine* **131**: 963–967.

Lamb, R. M., Studdert, D. M., Bohmer, R. M. J., Berwick, D. M. and Brennan, T. A. (2003). Hospital disclosure practices: results of a national survey, *Health Affairs* **22**: 73–83.

Lazare, A. (2004). *On Apology*, Oxford University Press, New York.

Leape, L. L. (1994). Error in medicine, *JAMA* **272**: 1851–1857.

Leape, L. L. (2005). Understanding the power of apology: How saying "I'm sorry" helps heal patients and caregivers, *Focus on Patient Safety: A Newsletter from the National Patient Foundation* **8**(4): 1–3.

Levinson, W., Rooter, D., Mullooly, J., Dull, V. and Frankel, R. (1997). Physician-patient communication. the relationship with malpratice claims among primary care physicians and surgeons, *JAMA* **277**: 553–559.

Liang, B. A. (1999). Error in medicine: legal impediments to U.S. reform, *Journal of Health Political Policy Law* **24**: 27–58.

Liang, B. A. (2002). A system of medical error disclosure, *Quality and Safety in Healtcare* **11**: 68.

Madsen, M. D., Andersen, H. B. and Itoh, K. (2006). Assessing safety culture in healthcare, *in* P. Carayon and G. Salvendy (eds), *Handbook of Human Factors and Ergonomics in Healthcare*, Lawrence Erlbaum Assoc. Inc.

Madsen, M. D., Østergaard, D., Andersen, H. B., Hermann, N., Schiøler, T. and Freil, M. (forthcoming). Lægers og sygeplejerskers holdninger til rapportering og håndtering af fejl og andre utilsigtede hændelser, *Ugeskrift for Læger* .

Madsen, M. D. (2005). Apology after medical harm: What is a genuine / moral apology in a systems context? Presentation at Hastings Center for Bioethics, Garrison, New York, October 25th, 2005.

Manser, T. and Staender, S. (2005). Aftermath of an adverse event: supporting health care professionals to meet patient expectations through open disclosure, *Acta Anaesthesiologica Scandinavica* **49**: 728–734.

Maurino, D. E., Reason, J. T., Johnston, A. N. and Lee, R. B. (1995). *Beyond Aviation Human Factors*, Avebury Aviation.

McNeill, P. M. and Walton, M. (2002). Medical harm and the consequences of error for doctors, *MJA* **176**: 222–225.

Newman, M. C. (1996). The emotional impact of mistakes on family physicians, *Arch Fam Med* **5**: 71–75.

NHS (2005).   Being open policy, National Patient Safety Agency.   URL: *http://www.msnpsa.nhs.uk/boa*. Accessed March 30th, 2006.

Pettigrove, G. (2003). Apology, reparations, and the question of inherited guilt, *Public Affairs Quarterly* **17**: 319–348.

Rachels, J. (1993). *The Elements of Moral Philosophy*, McGraw-Hill Inc.

Reason, J. T. (1997). *Managing the Risks of Organizational Accidents*, Ashgate, England.

Rijpma, J. A. (1997). Complexity, tight-coupling and reliability: Connecting normal accidents theory and high reliability theory, *Journal of Contingencies and Crisis Management* **5**(1): 15–23.

Robbennolt, J. K. (2003). Apologies and legal settelement: An empirical examination, *Michigan Law Review* **102**: 460–516.

Roberts, K. H., Madsen, P., Desai, V. and Van Stralen, D. (2005). A case of the birth and death of a high reliability healthcare organisation, *Quality and Safety in Healtcare* **14**: 216–220.

Rosner, F., Berger, J. T., Kark, P., Potash, J. and Bennett, A. J. (2000). Disclosure and prevention of medical errors. Committee on bioethical issues of the medical society of the state of New York, *Arch.Intern.Med* **160**: 2089–2092.

Schioler, T., Lipczak, H., Pedersen, B. L., Mogensen, T. S., Bech, K. B., Stockmarr, A., Svenning, A. R. and Frolich, A. (2001). Incidence of adverse events in hospitals. a retrospective study of medical records, *Ugeskrift for Læger* **163**: 5370–5378.

Schneider, C. D. (2000). What it means to be sorry: The power of apology in mediation, *Mediation Quarterly* **17**(3): 265–280.

Sharpe, V. A. (2000). Taking responsibility for medical mistakes, *in* S. Rubin and L. Zoloth (eds), *In Margin of error: The ethics of mistakes in the practice of medicine*, University Publishing Group, Hagerstown, Md.

Sharpe, V. A. (2004). *Accountability: Patient Safety and Policy Reform*, Georgetown University Press, Washington, D.C.

Smith, M. L. and Forster, H. P. (2000). Morally managing medical mistakes, *Cambridge Quarterly of Healthcare Ethics* **9**: 38–53.

Sorry works (2006). *Sorry works! Coalition*. URL: *http://www.sorryworks.net*. Accessed March 17th 2006.

Sweet, M. P. and Bernat, J. L. (1997). A study of the ethical duty of physicians to disclose errors, *Journal of Clinical Ethics* **8**: 341–348.

Taft, L. (2000). Apology subverted: The commodification of apology, *The Yale Law Journal* **109**: 1135–1160.

Taft, L. (2004). Apology and medical mistake: Opportunity or foil?, *Annals of Health Law* **14**: 55–94.

Tavuchis, N. (1991). *Mea Culpa: A Sociology of Apology and Reconciliation*, Stanford University Press, Stanford.

Vincent, C., Stanhope, N. and Crowley-Murphy, M. (1999). Reasons for not reporting adverse incidents: an empirical study, *J Eval.Clin.Pract.* **5**: 13–21.

Vincent, C., Young, M. and Phillips, A. (1994). Why do people sue doctors?: A study of patients and relatives taking legal action, *Lancet* **343**: 1609–1613.

Vincent, C. (2003). Understanding and responding to adverse events, *N.Engl.J.Med.* **348**: 1051–1056.

Wagatsuma, H. and Rosett, A. (1986). The implications of apology: Law and culture in japan and the united states, *Law & Society Review* **20**: 461–498.

Walsh, P. (2004). What the patient wants, *Making Health Care Safer 2004 21-22nd October 2004*.

Wears, R. L. and Wu, A. W. (2002). Dealing with failure: The aftermath of errors and adverse events, *Ann Emerg Med* **39**: 344–346.

Witman, A. B., Park, D. M. and Hardin, S. B. (1996). How do patients want physicians to handle mistakes?, *Arch Intern Med* **156**: 2565–2569.

Woods, M. S. (2004). *Healing words: the Power of Apology in Medicine*, Doctors in Touch, Il.

Wu, A. W., Cavanaugh, T. A., McPhee, S. J., Lo, B. and Micco, G. P. (1997). To tell the truth: Ethical and practical issues in disclosing medical mistakes to patients, *Journal of General Internal Medicine* **12**: 770–775.

Wu, A. W., Folkman, S., McPhee, S. J. and Lo, B. (1991). Do house officers learn from their mistakes?, *JAMA* **265**: 2089–2094.

Wu, A. W., Folkman, S., McPhee, S. J. and Lo, B. (1993). How house officers cope with their mistakes: Doing better but feeling worse?, *Western journal of Medicin* **159**: 565–569.

Wu, A. W. (2000). Medical error: The second victim. The doctor who makes the mistake needs help too, *BMJ* **320**: 726–727.

Zehr, H. (2005). *Changing Lenses. A New Focus for Crime and Justice*, Vol. 3, Herald Press, Scotdale.

Østergaard, D., Hermann, N., Andersen, H. B., Freil, M., Madsen, M. D. and Ruhnau, B. (2005). *Rekommandationer om reaktioner efter utilsigtede hændelser på sygehuse*, Risø. Delrapport 3 fra projekt om reaktioner efter utilsigtede hændelser. Risø-R-1499(DA).

The indeterminacy of human error classification

Thomas Bove[1] and Henning Boje Andersen[2]

[1] Nokia Mobile Phones
[2] Systems Analysis Department, Risø National Laboratory

1 Introduction

In safety critical domains such as aviation, maritime operations, healthcare, nuclear power production, process industry etc., human error classification systems are useful, and possibly even essential, instruments for improving safety. They are needed because they help safety analysts to understand the mechanisms behind human errors, to identify patterns and trends in errors in a given domain and application, and to plan and implement barriers and means of error capture.

One of the main challenges for human error classification systems is to support reliable analysis of error events. This involves, first, that a human classifier obtains the same results across repeated trials on the basis of the same data material (intra-rater reliability), and second, that different classifiers obtain the same results when they work on the same cases (inter-rater reliability). In other words, the codings (classifications) should be independent of the time or persons involved in using the system.

In this article chapter we will try to shed light on some of the difficulties associated with the reliability of human error classification systems and how these could be alleviated.

To give concrete illustrations of the difficulties associated with classification systems we will focus on a specific safety critical domain, namely Air Traffic Control (ATC), which is an area that can provide a proper context for studying and analysing human error events. The overall goal of air traffic control is often described as ensuring a safe, expeditious and orderly flow of traffic at all times. Basically, this means that the over-riding goal of air traffic control is to ensure that safe separation is maintained (i.e. both between aircraft and between aircraft and other obstacles such as mountains and ground vehicles) and, at the same time, to ensure this so that the efficiency of the air-traffic system is maintained. Inevitably, human errors will occur and they will sometimes lead to loss of separation and in rare cases to fatal situations.

In the following we shall refer to human error classification systems as '[human error] taxonomies'. We include under this label not only the names and descriptions of human errors but also the criteria for singling out and distinguishing different types of errors.

2 Use of and requirements to error taxonomies

Error taxonomies are required to serve a number of purposes, all of them relating to improvements in safety. A human error taxonomy should support:

(a) investigation of single critical occurrences (an incident or accident) and especially the structuring and characterisation of events involved in order to explain what went wrong and why;[3]
(b) aggregated analysis of a number of accidents or accident reports to identify patterns and trends, for instance changes in error types and incidence when new procedures, work conditions, automation etc. have been implemented;
(c) identification of human error types and their likelihood as part of risk and reliability assessment when new procedures, work conditions, automation etc. are planned or are being re-designed.

At the same time, an error taxonomy must fulfil theoretical and practical requirements in order to be accepted. In the absence of a "model" and classification of human errors in safety critical operations, investigators will typically lump errors together in large umbrella categories, e.g., "lack of vigilance" or, even worse, by characterising the error only in terms of the rule or procedure that was breached.

The following requirements to a taxonomy are not exhaustive, but they contain the major ones (for greater detail see, e.g., Isaac et al., 2002).

First, a taxonomy must be shown to be *usable* under conditions in which it will normally be used. In other words, it must be reasonably intuitive and easy to apply by its intended practitioners; but usability also entails that it must be sufficiently *robust* to accommodate difficult cases.

---

[3] It is customary to distinguish between incidents and accidents in terms of the severity of the consequences of the event in question. An incident will normally be a critical event that involves no or very minor human injuries and no or little material or environmental damage. Events that involve fatalities or serious injuries or extensive damage are normally called accidents. Obviously, there may be political, economic and even legal issues involved in calling an event an accident or an incident. For reasons of brevity, we will sometimes just use one of the terms to refer Indiscriminately to both incidents and accidents in the following

Second, it must be shown to be *reliable* in the sense that different users will classify the same event in the same way (inter-rater reliability). This is a seemingly self-evident requirement, but taxonomies are rarely tested on this criterion (see examples in Isaac et al., 2003; Pedrali et al., 2002). Moreover, reliability may depend on the training that users receive when they begin to use the taxonomy. An important aspect of reliability is *consistency in use* across different cases and contexts. This means that the same event, in different contexts and at different times, should be classified in the same way.

Third, it should be sufficiently *comprehensive* so that it is able to classify all relevant types of human errors in its target domain. This requirement is, nevertheless, somewhat vague in so far as analysts' views about which types of human errors are relevant will in part be influenced by current theories of human performance.

Fourth, it should yield *insight* into and reflect the mechanisms and the conditions that make errors more likely to occur, i.e., the so-called performance shaping conditions (PSFs) such as fatigue, excessive workload, conflicting procedures, design of human-machine interfaces, organisational climate etc.

Fifth, it should be *consistent* with current knowledge about human performance as represented in cognitive science (cognitive psychology, group psychology, perhaps pragmatics).

In addition to the overall requirements, a number of practical requirements are relevant, in particular, that the system should be accompanied by adequate guidelines and be capable of being learned after a reasonably short training period for domain experts.

3 Human error and accidents

The preoccupation with risks and accidents has grown considerably in Western societies in the last century and especially after the Second World War. Until fairly recently (a few decades ago), accidents were seen in general as caused by human error or technical failures. Influenced by "error luminaries" (Weingart, 2005) such as Jens Rasmussen and James Reason (Reason, 1997; Rasmussen, 1986), the "systems" view gradually developed. According to this perspective, accident causation typically involves a contribution of or so-called "latent factors" (performance shaping factors); moreover, accidents often reflect a degree of complexity in the interdependence of conditions that may provoke an error or fail to capture the effects of possibly "trivial" errors. This view of accident causation is sometimes referred to as the systems view (systems model) of accident causation (Woods, 1988; Hollnagel, 1993).

More recently, starting in the late 80's, the systems view of accidents has been expanded and a much enhanced emphasis on learning from accidents and incidents has been added. It is a trivial point that the lessons one may derive from errors and their associated circumstances do not depend on the consequences of the error. Whether a given error has resulted in a near-miss situation, a minor incident or in a serious accident will typically depend on circumstances over which the erring operator has no control. Indeed, the presence of *moral luck* in safety critical operations is, from a philosophical and psychological point of view a source of fascination, but from an engineering point of view it is noise and useless for the purpose of controlling risks.

A simple definition of human error says that it is any action or omission that causes results that users neither foresee nor intend. But the most widely used definition of human error is the disjunction suggested by many authors: An error is a wrong action or an action carried out in a wrong way. Thus, as Reason has put it, an error is the failure of a planned action to be completed as intended (a so-called slip or error of execution) or the use of a wrong plan to achieve an aim. An action plan that is wrong could be a wrong judgement, decision or diagnosis. (Reason, 1997).

Still, the way the notion of error is used will sometimes involve a normative aspect – an error is an act that deviates from a professional norm or practice, however minutely. Therefore, a more adequate definition may include in the notion of error any action which (a) has negative consequences (actual or potential), (b) was unintended by the person, and (c) involved a deviation from an action norm. A deviation from an action norm may be a breach of written and therefore explicit procedures or it may be a breach of professional practice. Including the normative element in the definition of error does not entail that errors are careless behaviour, however. The academic literature on human error and safety critical work is unanimous that human error cannot be avoided. But errors may be captured by design of the conditions under which people accomplish their tasks, i.e., by proper design of performance shaping factors.

4 Indeterminacy and unreliability

There are several areas of active research in error taxonomies (Isaac et al., 2002), but in this paper we focus on one group of problems that relate to reliability and determinacy. Reliability is defined in terms of use and outcome: agreement or disagreement that event e belongs to category k. Determinacy (indeterminacy) is an aspect involved in reliability and has to do with (a) the well-definedness of membership in any given category and (b) the nature of the material to be classified – i.e., descriptions or narratives of acts of human error. We shall suggest that some acts are indeterminate with respect to even the most carefully defined and precisely described categories

Classifying human error will frequently be associated with difficulties. In contrast to, for example, coding technical failures it is often more difficult to identify the cognitive mechanisms behind a human error.

There are several potential sources of lack of reliability and determinacy. Here we review what we suggest are the basic types of indeterminacy, which in turn may affect the reliability of the analysis:

• *Data Source.* In some cases the information available in the data material (typically, incident or accident reports) is insufficient to allow unambiguous selection of relevant categories. This may, in turn, be due to a number of different factors. The accident investigators
  – may not have based their analysis on a theoretical or conceptual framework that allows them to express their findings in precise terms;
  – may not have had access to the operators involved (the operators may have died or are not able to provide interviews);
  – may not have been able to elicit an adequate account or a consistent one from the operators;
• *Cognitive properties of the target act.* Even though the classifier has all relevant information available it might still be difficult to make confident and reliable judgements about the psychological mechanisms involved in the act that is analysed. The taxonomy may not by itself be ambiguities or imprecise. Rather, the target act – say, a failure to transmit the contents of an oral message to a written output – may not be uniquely identified as a specific type of cognitive failure, e.g., memory, perception, manual control.

Below we shall provide a few concrete examples to illustrate the problems of producing reliable classifications. More specifically, an example selected from the relatively small family of well-established taxonomies will be applied to real-life incidents taken from air traffic control to demonstrate the inherent fuzziness and uncertainty associated with the classification task.

5 Error taxonomies

As was mentioned, there are many frameworks available within the area of human error. In this section we will briefly review two relevant frameworks. Each of the frameworks is recognised within the domain of Air Traffic Control (ATC). The HERA taxonomy (HERA-JANUS, as it was called in later versions) was developed as a collaborative effort between National Air Traffic Services (NATS) and Risø National Laboratory for the European organisation for air traffic control

(EUROCONTROL) and was based on prior work by Shorrock and Kirwan (2002). ("HERA" is an acronym of Human Error Reduction in Air Traffic Management).

Incidents and errors in ATC are typically events in which two aircraft cross the limits of separation dictated by procedures and regulations. A so-called loss of separation may not be dangerous in the actual circumstances, but some are highly dangerous and some even lead to collision and disaster.[4]

5.1 HERA – An information processing taxonomy

The HERA taxonomy is a comprehensive technique that has been developed to analyse the mechanisms and performance shaping factors behind human errors in the area of air traffic control. The HERA technique contains one of the most elaborated and detailed taxonomies for error analysis available in any domain. It has been developed on the basis of a review of both academic and industrial research of the past five decades (Shorrock and Kirwan, 2002; Isaac et al., 2002).

The taxonomy developed for analysing individual controller errors is based on a traditional cognitive information-processing model. Below are shown the cognitive failure types that were chosen within the HERA framework for the analysis of mechanisms behind individual errors.

- **Perception:** This cognitive domain concerns a subject's picking up and understanding information. A typical kind of error in ATC associated within perception is a so-called "hearback" error. That is, a controller fails to pay attention to the content of a pilot's "readback" and hears instead the expected message.[5]
- **Short-term memory:** This domain concerns short-term storage or retrieval of information. For practical reasons it has been decided that short-term memory errors are associated with information received during an operational shift.[6]

---

[4] A recent mid-air collision happened on July 1, 2002, when two aircraft hit each other near lake Constance in southern Germany. It was Bashkirian Airlines Flight 2937, a Tupolev 154, travelling from Moscow to Barcelona carrying 57 passengers and 12 crew. Most of the passengers were Russian children whose school had won them a trip to Spain. The other aircraft was a DHL-owned cargo plane going from Bergamo, Italy to Brussels, with two pilots aboard. All passengers and crew died in the accident.

[5] When a controller issues a clearance, the pilot reads it back (gives his "readback"), and the controller acknowledges (makes a "hearback") the readback. But this is in an ideal situation. When things become busy and work load increases, a pilot may expect a taxi clearance to runway 22R but receives a clearance for 22L instead. Because of a mind-set determined by expectation, the pilot "hears" and reads back 22R. The busy controller expects to hear 22L and does not catch the mistake.

[6] Please notice that the definition of short-term memory in the current context varies slightly from the one found in the research literature where it is normally said that information can only be maintained in the short-term memory store for about 10-15 sec (see

For example, a controller may forget to follow up on a potential conflict between two aircraft in spite of having intended to do so.

- **Long-term memory:** This domain concerns long-term storage or retrieval of more permanent information based on the person's training and experience. For example, a controller may forget to carry out a specific procedure because he or she has not been using it for a long time.

- **Judgement & decision-making:** Controllers are constantly required to make projection of trajectories, plan future actions and to make decisions. These activities may all be associated with errors. For example, the controller may mis-project the future position of two aircraft and consequently not consider any need to monitor them further.

- **Response execution:** While their intended action is appropriate, people will fail to carry out the action as planned. A typical example is when a controller gives a clearance to one flight level but had intended to give clearance to another. This is often referred to as a *slip-of-the-tongue*.

6 Case Examples of Uncertainty

In this section we go through some examples of incident reports that can used to illustrate some of the problems associated with making confident classifications, even with well-established error taxonomies. We shall review two incident reports, a Swedish and a British. Both countries are well-known for having a long tradition for high-quality reporting system.

Below we give an abbreviated version of the incidents. Each of the incident summaries is followed by an analysis of some of the classification problems associated with conducting error analysis based on the previously described error taxonomies.

6.1 Case 1 – Swedish incident

The following is a translation by the authors of the text of the Swedish report issued by the Swedish CAA after investigation of the incident (Luftfartsverket, 1998). Glosses are given in brackets.[7] The numbered error labels inserted in the text have been added by the authors to facilitate the discussion below and they are not part of the original report.

---

e.g. Wickens, 1992). However, in the current context this time span is limited and not useful insofar as controllers are normally expected to maintain task relevant information in the memory for a much longer period – e.g. 10-15 minutes (Hopkin, 1995).

[7] Based on the authors' study of about 80 Swedish ATC incidents stemming from several years, we should point out that this report is unusual in several respects. It is unusual because it describes an extended and entirely untypical lack of focus and concentration

Two aircraft, each at FL 350 [flight-level of 35,000 feet], were involved in the incident near Gothenburg, where separation minima were breached. ABC273 had departed from Arlanda for London/Gatwick and XYZ845 had departed from Trondheim for Copenhagen. When the tracks of the aircraft crossed near Gothenburg, separation minima were violated. The controller did not detect the conflict until it was too late to avert it. XYZ845 had visual contact with the other aircraft. The minimal distance between the two aircraft was estimated to be around 1.1 NM.

When ABC273 first called the Malmo controller, R1 ("Good morning, Speedbird 273K, 350 to AAL"), he did not underline the FL [flight level] on the strip. The call from ABC occurred at a time when R1 was occupied with coordination with Oslo ACC [area control centre]. ACC controllers usually put a line underneath the cleared flight level when they receive the report from the pilot and when the reported flight level corresponds with the one indicated on the strip. On the ABC273 strip no underlining of FL was made **[Error #1]**. The cause of this omission has not been discovered. Possibly, the coordination with Oslo distracted the controller so that he did not make the underlining or even failed to hear the report of 350. Just after the call of ABC273 the D position was opened and all R1's flight strips were moved to the shared work tray between R1 and D1

Oslo ACC initially cleared XYZ845 for FL 350. According to LOA (letter of agreement) between Oslo ACC and Malmo ACC the aircraft would then be transferred from Oslo to Malmo at "odd" flight level, in this case FL330. But XYZ845 had requested FL350 from Oslo and had received clearance for maintaining FL350. This information was coordinated with Malmo in connection with provision of estimates and was made *automatically* via [the Online Data Interchange system]. The agreement, however, stipulates that transfer of flights at "wrong" level [and this aircraft was at "wrong" level] shall be made only after verbal coordination **[Error #2]**.

When XYZ845 called R1 the pilot reported "FL 350" and R1 acknowledged "radar contact". R1 underlined FL350 on the strip, this corresponding to the reported flight level. The controller did not detect any conflict **[Error #3]**. Yet, the strips were not an ideal basis for identifying the conflict: first, while both strips had FL 350 the two aircraft were navigating via different waypoints and the controller would need to make an estimation based on times and the two neighbouring waypoints; and second, time variations would need to exceed 5 minutes before they would be reported.

---

in the control room. The reason for selecting the report is that it serves quite well to illustrate our point about the indeterminacy about the underlying cognitive mechanism behind the overt errors.

From the time when XYZ had called until the conflict was discovered, 15 minutes later, no critical examination of the situation involving the two aircraft was made - neither by R1 or D1 **[Error #4]**. The "dead cone" of the radar may have delayed the detection of the conflict.

Finally, D1 became aware of the conflict but not until separation minima had been violated. D1 immediately told R1 that two aircraft were at the same flight level. R1 was at first astonished, but then made efforts to re-cover the situation.

Error analysis: Information processing

In the above report the level of description does not allow us unambiguously to de-cide on the cause of the omitted underlining of the flight strip (Error #1). Nonethe-less, it is implied that the error occurred as a result of a distraction and, conse-quently, we can speculate that the error was a short-term memory failure. Later on, R1 has the opportunity to discover that the arriving aircraft from the Oslo sector will conflict with the other aircraft at FL350, but he fails to realise the problem (Error #3 and #4). Here, again, it is difficult to determine whether this is a case of forgetting that another aircraft is at the same level (i.e. a short-term memory failure). It could also be a failure to notice that the new aircraft is at FL350 (i.e. a perceptual error). Finally, it could be a failure to integrate the available infor-mation – that is, the controller fails to realise that the flights could conflict since the two aircraft were navigating through different waypoints and the strips were therefore organised in such a way on the stripboard that it was not apparent that the two aircraft would conflict (i.e. a judgement and decision-making error). Even if more information were available in the report it is very likely that some level of uncertainty would persist in the coding.

6.2 Case 2: A British incident

The following report is quoted from the British Civil Aviation Authorities' col-lection of so-called airproximity reports (Caa-Uk, 1997). Remarks and glosses in brackets are the authors'.

A B747 was en route from Zurich to New York, cruising at FL310 .... via Boulogne VOR [radio beacon] to Brookmans Park. A B767 from Paris (Orly) to New York was routeing UB376, also via Boulogne VOR [radio beacon] at FL180. Both aircraft were under the control of LUS [London upper sector]. The traffic situation was described as busy ...... The CSC [chief sector controller] decided, in consultation with the off-going sec-tor controller, that instead of splitting the sector into E[ast] and W[est],

the off-going controller would be used as a support controller to the relief controller. Most of the traffic was on the East side, so it was thought
that splitting the sector would be unproductive. The relief controller, who
had little experience of this mode of operation, agreed to the plan. The
B747 pilot reported level at FL310 on first contact, and was instructed
to maintain FL310 and given a routeing of Boulogne, Brookmans Park
and Trent. Shortly afterwards the B767 pilot established RTF [radio tele-
phony] contact with the LUS reporting approaching Boulogne at FL280 -
the expected level as indicated on the fps [flight progress strip]. However,
the Sector controller erroneously instructed the B767 pilot to "Maintain
FL310" **[Error #1]**. The controller then turned her attention to other traffic and did not note the B767 pilot's reply "up to 310" **[Error #2]**. The
support controller did not hear the sector controller's call because he was
concentrating his attention elsewhere (although there is no responsibility
for a support controller to hear all the calls). However, the support controller noticed that the B767 was at FL283 . . . , above FL280 as displayed
on the fps [flight progress strip]. When he drew this to the attention of
the Sector controller, she replied initially that the a/c [aircraft] was not on
frequency **[Error #3]**. Still concerned, the support controller continued to
prompt the Sector controller into taking action to resolve the problem. He
was convinced that the B767 pilot was on frequency because the Sector
controller had ticked the callsign on the fps. Both he and the [chief sector controller] tried to get the Sector controller's attention to contact the
B767 pilot but, because she was busy making calls, they found it difficult
to make her aware of the circumstances. The Sector controller still did not
believe that the B767 was on frequency, because she did not remember its
first call. Hence, she did not take any action, but they continued to prompt
her to call the a/c **[Error #4]**. About thirty seconds after the STCA [short
term conflict alert system] activated, she initiated avoiding action and the
conflict was resolved.

Error analysis: Information processing

It can be speculated that Error #1 was most likely a slip-of-the-tongue (i.e. response
execution error) by the controller and that this error was possibly made because
she had just instructed the B747 pilot to maintain FL310 and that this figure was
still in her mind. However, alternative explanations are possible that will affect the
error classification: It could be that the sector controller misheard the B767 pilot's
callsign and believed it was the B747 calling a second time, or that the controller
forgot that the B747's first call, and expected the call to be from the B747 pilot.
Hence, several error candidates are possible in this context.

An interesting feature of this incident is how the controller seems to have lost the "mental picture" and is unable to correct it in spite of receiving several pieces of contradictory pieces of information from colleagues as well as displayed on the radar (Error #3 and #4). Such errors are particularly troublesome for error analyses insofar as one underlying condition – namely loss of the mental picture – is the cause of several observable failures. The question is then how many errors should be counted on the basis of this condition. Since the error taxonomy based on situation awareness is aimed at discrete events such analytical problems are not easily solved.

7 Classification reliability – statistical data

The consequence of indeterminacy of human error classifications is that the reliability of a given taxonomy will be less than perfect. One way of assessing the extent of indeterminacy is by determining the reliability levels for different taxonomies. To do so involves having different raters (classifiers) assess independently the same event descriptions, e.g., descriptions contained in incident reports as above.

A commonly used statistics to measure the degree of reliability of classifications is the Cohen's kappa, which goes beyond "raw" agreement and gives a result which is corrected for chance agreement. (Kappa=(O-E)/(1-E), where O is the observed agreement among raters, and E is probability of chance agreement. (Cohen, 1960; Fleiss, 1971, 1981)).[8]

The interpretation of the level of agreement (above chance) obtained by independent raters is, by convention, nearly always stated along the lines suggested by Fleiss (1981) or Landis and Koch (1977), who differ only slightly. Landis and Koch suggest that levels below 0.40 show poor or merely fair agreement, and this figure remains a conventional cut-off point (rather as the interpretation of a p-value at or below 0.05). Fleiss proposes that levels above 0.75 show strong agreement, and Landis & Koch suggest that levels between 0.41 and 0.60 indicate moderate agreement, between 0.61 and 0.80 substantial., and above 0.80 almost perfect agreement.

It should be noted that even though Kappa statistic is widely used to determine the degree of agreement between classifiers some researchers have criticised this approach (Davies et al., 2003). One of the main points of criticism against using the Kappa statistic is its requirement that categories must be mutually exclusive and exhaustive (Cohen, 1960). As the analysis of the errors in the incident reports demonstrated such a requirement is hard to achieve even for well-established error

---

[8] Cohen's Kappa is the statistics most frequently used to measure of intra- and inter-rater agreement when the ratings are nominal. Kappa may, by weighting, be adapted to assessing rank-based classification. scale.

taxonomies where the categories are intended to be mutually exclusive. Another problem is that the events used in the Kappa analysis should be discrete and independent events, but as illustrated in Case 2 such conditions cannot always be fulfilled. Davies et al. (2003) suggest to simple use raw agreement based on the ratio between the number of agreed observations and the total number of observations.[9] However, on this approach one would abandon a crucial benefit of the Kappa statistic, namely that it is a measure of agreement which controls for chance. Another approach has been pursued by Pedrali et al. (2002) who have described and used a measure of pairwise agreement, i.e., the likelihood that any pair of raters will agree.

Still, few studies have been made to investigate classification reliability. Here we shall describe briefly experience from a study of the reliability of the HERA taxonomy (Bove and Andersen, 2003; Isaac et al., 2003). In the studies carried out by Bove and Andersen (2003) human errors in incidents and video-recorded simulator scenarios were analysed. Here Kappa coefficients between 0.52 and 0.81 were obtained. Using the qualitative interpretation recommended by Landis & Koch, this corresponds to an agreement ranging between 'moderate' and 'excellent'. In relation to the situation awareness taxonomy no data are currently available to evaluate its reliability.

8 Optimising reliability

Below we list and describe some factors that are known to influence the reliability of event classification. Some of these relate to the taxonomy itself whereas others concern its practical use.

System properties that are important to take into consideration in the development or selection of an error taxonomy include:

- *Domain fitting:* Although there are many general taxonomies that in principle are suitable to possibly a broad range of domains, it is not necessarily the case that one-taxonomy-fits-all. For example, the information-processing framework in the HERA taxonomy is highly suitable for analysing ATC errors, because the concepts within the information processing paradigm fits well with operator tasks in this domain; and, moreover, the so-called performance shaping conditions of HERA are tailored to the ATC domain. (Isaac et al., 2002). Other domains where possibly more subtle and expressive forms of team interaction take place might require other taxonomies.
- *Mutually exclusive categories.* The internal logical structure of the taxonomy is crucial to ensuring high reliability. In particular, it may be crucial to avoid

---

[9] For more information about "raw" agreement please refer to (Martin and Bateson, 1993).

overlap in the categories. The chances of improving mutual exclusivity are improved if the taxonomy is based on a performance model or if it is based on a strictly hierarchical structure. On the other hand, if the taxonomy is based on an ad-hoc developed system there is a risk of introducing unreliability (Wallace et al., 2002). Furthermore, it should be noted that a large number of categories will tend to increase the risk of overlapping categories and is therefore also a source of reduced reliability.

- *Variation in the use of categories.* A factor that might jeopardise the reliability of a taxonomy is when some categories are used infrequently and others very frequently. Not only does this reduce the diagnosticity of the taxonomy but there is also a risk of complacency or automation errors in the classification. Thus, classifiers become habituated to making certain classifications and might not consider all options. Such problems might not be exposed in reliability analyses.

- *No use of "other"-categories.* The use of an "other" category can jeopardise the reliability of the taxonomy (Wallace et al., 2002). The problem is that this catch-all category may refer to several things. In some situations it will be used to code errors that do not fit into the remainder of the categories, and in other situations the "other" category might be applied to cases which might not easily fit with a single category within the taxonomy. Hence, it is important to strive for an all-inclusive taxonomy. While an "other"-categories is useful in the development of a taxonomy (Wallace et al., 2002), it should be eliminated in a mature taxonomy.

In relation to the practical use of the taxonomy several things can be done to optimise the coding reliability. This includes improving the quality of the data source and minimising the influence of classifier's background:

- *Consensus in error identification.* An important first step in an error analysis is the identification of human errors in an event description. Basically, the identification process is a critical source of variance insofar as different people might disagree on some errors or simply not discover the errors. Empirical data have shown that this interpretation process could be a vital factor in reducing reliability that is independent of the logical structure of the codes themselves (Wallace et al., 2002). To minimize the problems of identifying errors, several things may be done: Ensure a common definition of human error and supplement it with practical examples from the domain in question; Generate practical principles for the error identification that can guide the process; Use several people – preferably both human factors experts and domain experts – in the identification of errors to produce a consensus list of errors.

- *Ensure adequate training.* Any diagnostic error taxonomy will require a certain level of familiarisation for the coder to get a thorough understanding of the system. Therefore, training time must be expected before a classifier becomes skilled in using the taxonomy. Typically, a few days of training will help classifiers to achieve a common understanding of classification criteria. Such training should not only be focused on the people responsible for the coding but also the people involved in incident investigations. By having a conceptual framework to guide the investigation the chances of uncovering relevant information to the error analysis is greatly increased.

- *Flowcharts.* In the case of complex taxonomies – and in particular during training of classifiers – it can be useful to convert the framework to flowcharts where the coder can decide on a category by answering yes or no at a series of branches. Hereby the complexity of the taxonomy is significantly reduced (for a good example of how the flowcharts can be used in the classification process please refer to Isaac et al., 2003). Needless to say, such a flowchart can only be built if the taxonomy is based on a hierarchical structure.

Ultimately, the final test of success will be the level of achieved statistical reliability across raters and time.

9 Conclusion

In this chapter we have explored some theoretical and practical aspects related to classifying human errors, focusing on reliability. Our analysis of errors described in non-theoretical terms in authentic ATC incident reports has revealed that not all errors are capable of being classified with certainty; in some cases, there is an inherent uncertainty and in others the uncertainty stems from practical but very real limitations. So, irrespective of how well-defined and developed a taxonomy is the classification of human performance will involve events that lie in the fuzzy areas of the taxonomic categories. There is, therefore, no such a thing as a perfect taxonomy that unambiguously can account for all situations.

Still, the fact that perfect agreement can not be achieved does not alter the fact that the use of error taxonomies can play a constructive role in uncovering patterns in human error databases and, ultimately, create a sound basis for preventive measures for safety-critical organisations. Still, it is vital to invest the necessary resources in establishing the reliability of the system, including training of personnel responsible for data collection and error coding.

While the focus of this article has been on reliability and sources of indeterminacy of error classification, it is important to stress that reliability should serve the overall requirement of validity. A framework is valid to the extent it reflects and captures the specific concepts that analysts attempt to measure, and it is useless to have a reliable but largely invalid system. We have argued that human error taxonomies need to be tested to ascertain to which degree they are reliable; however, their validity should also be examined. See for instance Bove and Andersen Bove and Andersen (2003) which illustrates how such a study may be undertaken. Issues of reliability is an important part of validity and is a critical feature for any taxonomy to be of practical use.

References

Bove, T. and Andersen, H. B. (2003). Validation of an error management taxonomy in ATC, *Int. J Appl. Aviation Stud* **3**: 39–60.

Caa-Uk (1997). *Aircraft Proximity Reports Airprox 24/96 (C) – Controller Reported*, Vol. 12 of *The Joint AIRPROX Asseesment Panel (JAAP). Nov. 1996-July 1997*, Civil Aviation Authority, London.

Cohen, J. (1960). A coefficient of agreement for nominal scales, *Educational and Psychological Measurement* **20**: 37–46.

Davies, J. B., Ross, A., Wallace, B. and Wright, L. (2003). *Safety Management: A Qualitative Systems Approach*, Taylor and Francis, London.

Fleiss, J. L. (1971). Measuring nominal scale agreement among many raters, *Psychological Bulleting* **76**: 378–382.

Fleiss, J. L. (1981). *Statistical Methods for Rates and Proportions*, second edn, John Wiley & Sons, Inc., New York.

Hollnagel, E. (1993). *Human reliability analysis: Context and control*, Academic Press, London.

Hopkin, V. D. (1995). *Human Factors in Air Traffic Control*, Taylor & Francis, London and Bristol.

Isaac, A., Shorrock, S. T., Kennedy, R., Kirwan, B., Andersen, H. B. and Bove, T. (2002). *Technical review of human performance models and taxonomies of human error in ATM (HERA)*, Eurocontrol HRS/HSP-002-REP-01. URL: *http://www.eurocontrol.int/humanfactors/public/site_preferences/display_library_list_public.html*. (Accessed March 28th, 2006).

Isaac, A., Shorrock, S. T., Kennedy, R., Kirwan, B., Andersen, H. B. and Bove, T. (2003). *The Human Error in ATM Technique (HERA-JANUS)*, Eurocontrol, HRS/HSP-002-REP-03, Brussels.   URL: *http://www.eurocontrol.int/humanfactors/public/site_preferences/display_library_list_public.html*. Accessed March 28th, 2006.

Landis, J. and Koch, G. G. (1977). The measurement of observer agreement for categorical data, *Biometrics* **33**: 159–174.

Luftfartsverket (1998). *Incident Report 980412*, Luftfartsverket, Norrköping. [In Swedish].

Martin, P. and Bateson, P. (1993). *Measuring Behaviour: An Introductory Guide*, second edn, Cambridge University Press, Cambridge.

Pedrali, M., Andersen, H. B. and Trucco, P. (2002). Are maritime accident causation taxonomies reliable? An experimental study of the human factors classification of the IMO, MAIB and CASMET systems, *Proceedings. RINA conference on Human factors in ship design and operation, London, 2–3 Oct 2002*, London.

Rasmussen, J. (1986). *Information Processing and Human-Machine Interaction. An Approach to Cognitive Engineering*, North Holland, New York.

Reason, J. (1997). *Managing the Risks of Organisational Accidents.*, Ashgate Publishing, Aldershot.

Shorrock, S. T. and Kirwan, B. (2002). Development and application of a human error identification tool for air traffic control, *Applied Ergonomics* **33**: 319–336.

Wallace, B., Ross, A., Davies, J. and Wright, L. (2002). The creation of a new minor event coding system, *Cognition, Technology & Work* **4**(1): 1–8.

Weingart, S. N. (2005). Beyond babel: Prospects for a universal patient safety taxonomy, *Int. J. Qual. Health Care* **17**(2): pp. 93–94.

Wickens, C. D. (1992). *Engineering Psychology and Human Performance*, second edn, Harper Collins Publishers, New York.

Woods, D. (1988). Coping with complexity: The psychology of human behavior in complex systems, *in* L. P. Goodstein, H. B. Anderson and S. E. Olsen (eds), *Tasks, Errors and Mental Models*, Taylor & Francis, New York.

Contributing authors

Biographies and relations to Stig Andur Pedersen

Andersen, Hanne

Steno Institute, University of Aarhus.

Most of my research addresses the cognitive aspects of the development of science, and includes historical case studies from 20th century physics and biology. Among my publications are a number of papers on conceptual structures, incommensurability and realism, and a textbook on the philosophy of Thomas Kuhn. A monograph on the cognitive structure of scientific revolutions coauthored with Peter Barker and Xiang Chen will appear on Cambridge University Press in spring 2006.

Andur daringly accepted to become my doctoral adviser only six months before my doctoral thesis was due for submission. I appreciated both his openness for a project that had been conceived and developed in another research group, and his insistence on clarifying and expanding a number of issues. The paper published in this volume originates in those very productive and challenging months under Andur's supervision.

Andersen, Henning Boje

Systems Analysis Department, Risø National Laboratory, Roskilde.

For the last 20 years my work has focused on human factors in safety critical domains: aviation, healthcare, maritime operations, process industry etc. My research interests include human error analysis, causal analysis of accidents and incidents, negligence, responsibility and error in the context of accidents, and modelling safety culture.

Andur has had a greater influence than any one else on the course of my professional life. He decided the topic of my thesis ("magisterkonferens") by suggesting that I should answer a prize essay question he himself had devised ("Epistemic logic") at Copenhagen University. After graduation, I was honoured to assist Andur in courses in medical ethics and philosophy of science, while still following some of his seminars in mathematical logic. And not least, his suggestions and support were crucial for my entering the lively human factors research environment around Jens Rasmussen at Risø.

Andreasen, Viggo

Department of Mathematics and Physics, Roskilde University.

His primary research interests are in mathematical models of population biology and genetics, in particular the transmission dynamics and evolution of influenza. He was one of the founders of the SIMA-project.

Bove, Thomas

Nokia Mobile Phones.

Since I graduated from University of Copenhagen in 1998 with a degree in psychology, my main interests have been with the interdisciplinary issues of man-machine interaction. After graduation I was able to pursue this interest at Risø National Laboratory where I have worked in the area of human factors in safety critical domains such as aviation, air traffic control and the process industry. My research included human error analysis and experimental evaluations man-machine interfaces. During the period at Risø I completed my Ph.D. dissertation about human errors and their recovery. More recently, I have been applying my knowledge of man-machine interaction to the development of electronic consumer products. In 2003 I started working as a usability expert at Bang & Olufsen where I was involved in products such as remote controls, audio equipment and telephones. Since 2005 I have been working for Nokia Mobile Phones within the areas of ergonomics, usability and user experience.

When writing my PhD dissertation in Human Factors at Risø National Laboratory and Roskilde University I was fortunate in having Stig Andur Pedersen as my supervision. I benefited from his broadmindedness, vast knowledge, and in general, his enthusiasm in discussing approaches and methods in the interface between engineering and psychology.

Christiansen, Frederik Voetmann

Department of Medicinal Chemistry, Danish University for the Pharmaceutical Sciences.

Following my graduation from Roskilde University, my interest in science education and educational change led to my Ph.D. study at the Centre for Educational Development at the Technical University of Denmark. My current interests include theories of learning, organizational learning and change, diagrammatic reasoning, philosophy of science and technology, and didactics of science and the scientific professions (particularly pharmacy and engineering).

During my graduate study in physics and philosophy I benefited from Andur's supervision both in a project on meta-ethics and in my master's thesis on the foundations of quantum mechanics. His "neo-Kantianism" was highly inspiring to myself and a number of students doing projects ranging from thermodynamics and statistical mechanics, astronomy to philosophy of mathematics. Having helped me getting a Ph.D. scholarship at the Technical University of Denmark, Andur became my supervisor and contributed greatly to our group at the Centre for Engineering Educational Development.

Faye, Jan

Department of Media, Cognition and Communication, University of Copenhagen.

I have been working mainly in the areas of philosophy of science, philosophy of physics, and metaphysics. I have published multiple papers and books including *The Reality of the Future* (1989), *Niels Bohr: His Heritage and Legacy* (1991), and *Rethinking Science* (2002). I also have co-edited several books such as *Niels Bohr and Contemporary Philosophy* (1994), *Logic and Causal Reasoning* (1994), *Perspectives on Time* (1997), *Things, Facts, and Events* (2000), and *Nature's Principle* (2005).

While Stig Andur Pedersen was still a teacher at the former Department of Philosophy, University of Copenhagen, I attended his seminar on modal logic. Since then I have had the pleasure of working with Stig Andur Pedersen in various academic contexts over the years.

Hendricks, Vincent F.

Section for Philosophy and Science Studies, Roskilde University.

He is the author of many books including *Mainstream and Formal Epistemology* (2006), *Masses of Formal Philosophy* (2006), *Formal Philosophy* (2005), *The Convergence of Scientific Knowledge* (2001), and *Feisty Fragments, Logical Lyrics* and *500 CC: Computer Citations* (2004-05). Editor-in-Chief of Synthese, Synthese Library and New Waves in Philosophy, he is also the founder of Φ-LOG – The Network for Philosophical Logic and Its Applications and founder and Editor-in-Chief of ΦNEWS – The Newsletter for Philosophical Logic and Its Applications. His work concentrates primarily on bringing mainstream and formal approaches to epistemology together.

Hohwy, Jakob

Department of Philosophy, University of Aarhus.

He works on philosophy of science as well as philosophy of mind and language, with a focus on the notions of explanation and reduction. He participates in inter-disciplinary work on the philosophy of neuroscience and cognitive neuropsychiatry. A representative publication in neuropsychiatry is Hohwy, J., Rosenberg, R. Unusual experiences, reality testing and delusions of alien control, *Mind & Language* **20**(2): 141–162 (2005).

Holm, Søren

Cardiff Law School, Cardiff and Section for Methical Ethics, University of Oslo.

I have been interested in the philosophy of medicine and in medical ethics since the mid-1980s, and have been employed to spend time on this interest since 1992 first at the University of Copenhagen, and later in Manchester and Cardiff.

I first met Andur when I was a student on the then optional course in "Medicinsk videnskabsteori" ("Theory of medical science") for medical students where he and others wetted my appetite for philosophical problems. He was one of the assessors of my submission for a prize essay question on "The ethical issues in prenatal diagnosis" and when I started to work at the Department of Medical Philosophy in Copenhagen we began to collaborate on the development of a simulation tool for analysing the ethical decision making of hospital doctors. The book "Philosophy in Medicine" that he co-authored with Henrik R. Wulff and Raben Rosenberg has had a very significant impact on my own understanding of and approach to the field.

Høyrup, Jens

Section for Philosophy and Science Studies, Roskilde University.

Studied physics and mathematics in 1962–69 and taught the same discipline at Danmarks Ingeniør Akademi 1971–73. Since 1973 at Roskilde University, first in social science, from 1978 in the humanities (since 1995 Philosophy and Science Studies, emeritus 2005). Much of his research has dealt with the cultural and conceptual history of pre-Modern mathematics, in particular with the conceptual structure and history of Babylonian mathematics and its interaction with practitioners' traditions. After finishing *Lengths, Widths, Surfaces: A Portrait of Old Babylonian Algebra and its Kin* (2002) he is now primarily working on late medieval Italian algebra.

Andur and I learned undergraduate mathematics together albeit without interacting much, Andur having already chosen his course whereas I ran down what for me became a pleasant but blind alley (physics). Our first professional contact arrived in *c.* 1975 when Andur and others created a group engaging in "integrated science studies" and I myself, being made responsible for the teaching of philosophy of science for social science students, was happy to find congenial but more enlightened discussion partners in this group. Intense interaction came with the creation of the course in Philosophy and Science Studies at Roskilde University in 1995.

Jakobsen, Arne

Formerly at the Technical University of Denmark (now retired).

My work and research has been in the field of engineering education, educational theory, engineering work and competence and philosophy of engineering science.

In the beginning of the eighties I worked together with Andur on educational theory and educational change in relation to engineering education. Later Andur had a decisive influence on our research into engineering work and on activities about philosophy of engineering science at the Technical University, including a series of Ph.D. courses.

Jørgensen, Klaus Frovin

Section for Philosophy and Science Studies, Roskilde University.

In 2001 I graduated with a cand.scient degree in Philosophy and Mathematics from Roskilde University. My primary interests have for a long time been centered around mathematical logic and the philosophy of mathematics. My Ph.D. study ended up with a study of Kant's notion of schematism and its relevance for the foundations of mathematics; it was consequently entitled *Kant's Schematism and the Foundations of Mathematics* (2005). More generally my research interests include general philosophy of science, diagrammatic reasoning and constructive epistemology, and I have co-edited books on logic and philosophy of mathematics and philosophy such as *Proof Theory: History and Philosophical Significance* (2000) and *Visualization, Explanation and Reasoning Styles in Mathematics* (2005), *Interactions: Mathematics, Physics and Philosophy, 1860-1930* (2006).

Both during my under-graduate and graduate studies Andur had a huge influence on me. Many years ago he introduced me to logic and proof theory and to a very fruitful view on science and mathematics. His view on these subjects and on the relation between philosophy and science has inspired me immensely. Moreover, he has been extremely generous with both his time and ideas, and I would never had thought the thoughts that I am thinking right now, had it not been for Andur. In 1995 he initiated The Danish Network for the History and Philosophy of Mathematics (supported by the National Science , which I became the secretary of and we made many interesting conferences and funny arrangements—such as sailing on Roskilde Fjord in an old Viking Ship with the president of the American Mathematical Society (who was actually rowing the ship).

Kauffmann, Oliver

Department of Media, Cognition and Communication, University of Copenhagen.

Since I graduated from the University of Odense in 1997 with a degree in philosophy, my main interests have been with issues of philosophy of mind. My occupation with empirically informed ways of doing philosophy led to a job as research assistant at The Royal Veterinary and Agricultural University in Copenhagen and a Ph.D. study at University of Copenhagen. The topic of my thesis is functional aspects of consciousness in the light of neuro-psychological case studies.

My relation to Stig Andur is primarily that of admiration at a distance. However, I have also benefited from his kind assistance as external examiner for my classes several times. And on one such occasion – examinations held after my first graduate course on 'neuro-philosophy' – he warmly encouraged me to pursue the interdisciplinary approach to the problems of philosophy of mind that interested me. Indeed, Stig Andur was the first philosopher I met who seriously strengthened my conviction that philosophers of mind should get out of the armchair and into the field.

Kelly, Kevin T.

Department of Philosophy, Carnegie Mellon University.

He is author of the book *Logic of Reliable Inquiry* and of many articles concerning computational learning theory and its relationship to traditional issues in the philosophy of science and induction.

The conventional, philosophical compliment to an esteemed colleague and mentor is to write a substantial scholarly paper for his festschrift, but something warmer seemed more appropriate in this case. Alternatively, mere personal reminiscences would be too insubstantial to honor a genuine man of science. Then a more venerable genre came to mind that allows for some of both.

Kjeldsen, Tinne Hoff Kjeldsen

Department of Mathematics and Physics, Roskilde University.

I first met Andur at an interview for a Ph.D. position in history of mathematics at IMFUFA in the fall of 1994. I had graduated from the University of Copenhagen four years earlier with a degree in mathematics and in between I spent some time at the Department for History and Philosophy of Science and Technology at University of Toronto and the Operational Research Centre at MIT. At the interview Andur asked whether I was interested in taking up some of the new trends in the history and philosophy of mathematics and during my Ph.D-study we, on Andur's initiative, formed the Danish Network for the History and Philosophy of Mathematics supported by the National Science Foundation. Even though he did not become my Ph.D. supervisor I benefited immensely from Andur's enthusiasm, huge knowledge, and his ability to encompass students in research activities. When Andur left IMFUFA to establish the philosophy department we continued to arrange international conferences and meetings through the network. My current research is focused on the history of the theory of convexity and, partly through Andur's inspiration, I am moving into the history and philosophy of mathematical models.

Kozine, Igor

Systems Analysis Department, Risø National Laboratory, Roskilde.

I studied at the Moscow Institute of Physics and Engineering (Technical University) at which I received my M.Sc. (1983) and my Ph.D. (1989) in Systems Analysis. Until 1999, I worked at the Institute of Nuclear Power Engineering, Obninsk, Russia as associate professor and senior scientist. My research concerns reliability and risk analysis, and current interests are uncertainty modelling and computer simulation of human performance. Having been a visiting scientist at Risø and State University of New York at Binghampton, I became a senior scientist at Risø's Systems Analysis Department in 1999.

My obsession about comprehending what uncertainty is led me to Andur some years ago. We have been working on the project "Coping with Uncertainty in Complex Technical Systems: An Epistemological Analysis of Uncertain and Imprecise Information". It was a great time of expectations and discussions, inspiring and full of challenges.

Kragh, Helge Stjernholm

Steno Institute, University of Aarhus.

After graduation from the University of Copenhagen (physics, chemistry) in 1970, I spent many years as a teacher at Gymnasium schools. In 1981 I defended my doctoral dissertation at Roskilde University. After positions as Associate Professor at Cornell University, Curator at the Steno Museum and full Professor at the University of Oslo (1987–89, 1995–97) I was appointed professor at the University of Aarhus. For the last several years I have focused on the history of cosmology and, more recently, the interaction between science and religion.

I came to know Andur in 1978, when we both were at Roskilde University, trying to persuade ourselves that we were Marxists of a kind. In 1981 we coauthored a fairly successful textbook in philosophy of science, still in use twenty-five years later. From time to time our ways have crossed, and we have cooperated in organizational matters, the arrangements of meetings etc. I have never understood why he finds philosophy, mathematics and logic to be more interesting than the physical sciences.

Krayer von Krauss, Martin Paul

Formerly at the Technical University of Denmark.

I graduated from the Royal Military College of Canada in 1996 with a bachelor's degree in Civil Engineering. Working as an Environmental Engineer, I found it difficult to find meaning in the compliance issues that dominated my day to day tasks. This disenchantment provided the motivation for me to pursue my studies in Environmental Management. Still in search for the reason for why we should protect the environment, I was drawn towards environmental ethics and policy. During the course of my M.Sc. at the Institute of Environment and Resources at the Technical University of Denmark (DTU), I met Prof. Poul Harremoës who introduced me to integrated environmental assessment and the debate on the Precautionary Principle. Together with him, I pursued a Ph.D. study focusing on the development of methods through which scientists and engineers participating in the regulatory process could communicate the uncertainty and ignorance that typically characterizes their assessments. My current research interests include uncertainty and quality assessment in policy relevant science, stakeholder involvement in the policy process, and emerging issues such as the risks of nanotechnology.

Together with Poul Harremoës, I taught the course Environmental Ethics and Management for engineering students at DTU. It was in this context that I first met Andur. I was soon inspired by Andur's humble attitude, his enthusiasm for ethics and the philosophy of science and the depth of his knowledge. The wide variety of examples he masters enable him to make abstract philosophical theories understandable to even the most pragmatic of engineering students. Unlike many I have met, he understands the importance of teaching ethics and philosophy to scientists and engineers.

Larsen, Jesper K.

Department of Mathematics and Physics, Roskilde University.

His research area include applied mathematical modelling, and he was one of the founders of the simulation project SIMA (SIMulation in Anesthesia).

Lützen, Jesper

Institute for Mathematical Sciences, University of Copenhagen.

My main interests are in mathematics and mechanics in the 19th and early 20th century. I am the author of *Joseph Liouville 1809–1882. Master of Pure and Applied Mathematics* (Springer 1990) and *Mechanistic Images in Geometric Form: Heinrich Hertz's Principles of Mechanics* (OUP 2005).

I have benefited very much from Andur's efficient fund raising and his enthusiastic work for Danish history and philosophy of mathematics.

Larsen, Heine

Department of Mathematics and Physics, Roskilde University.

He has had a leading role in the implementation of the SIMA simulator.

Madsen, Marlene Dyrløv

Systems Analysis Department, Risø National Laboratory, Roskilde and Section for Philosophy and Science Studies, Roskilde University.

Master in Philosophy and Communication from Roskilde University in Denmark in 2001. I work with patient safety, specifically issues regarding reporting and learning from human error and adverse events investigating how to build safe and just cultures within healthcare to improve patient safety, based on safety critical theory, human factors, theories of organizational learning and change and ethics. Current interests: ethics of patient safety and medical error, responsibility and accountability; truth telling, (open-disclosure) and apology in relation to patients, ethics of punishment and 'moral luck'.

During my graduate study at Roskilde University I had the honour of Andur's supervision on my master's thesis "Philosophy for Air Traffic Controllers: An investigation of the ethical issues to consider when handling human error in safety critical organizations." Andur has played a central role in influencing the course of my research. He insistently encouraged me to contact Henning Boje Andersen at Risø National Laboratory in the process of deciding on the topic for my master's thesis. A thesis which caught me on to interesting domains and fields of interdisciplinary research that I have been able to continue during my Ph.D., which Andur helped bring about.

Olsen, Jan-Kyrre Berg

Section for Philosophy and Science Studies, Roskilde University.

I became a Ph.D. in 2003 on a dissertation concerning the relation between temporal experience and time concepts in physics. My current research interests are within areas such as philosophy of technology and science, metaphysics, epistemology and philosophy of religion, more specifically the thematic is "pessimistic views on technological, economical and human progress". Presently I am editor-in-chief (with Stig Andur Pedersen and Vincent F. Hendricks) of Blackwell Publishing's *A Companion to Philosophy of Technology*. This is the first international reference work on technology. I am also editor-in-chief (together with Evan Selinger and Søren Riis) of Ashgate Publishing's *New Waves in Philosophy of Technology*.

Like so many others I have benefited largely from Andur's wisdom as supervisor. It was Andur that employed me in 1999 in a doctoral fellowship at Roskilde University, and that later on urged me to study philosophy of technology and thus got me interested in the idea of progress in our society. His genuine broadmindedness and friendliness has become an inspiration. And working together with Andur means that you have access to a large variety of information from sources like mathematics, philosophy of science, technology, physics, and philosophy of time, philosophy of language, history, formal logic, medicine, and to all sorts of thinkers in the international scientific community.

Olufsen, Mette S.

Department of Mathematics, North Carolina State University.

Her research is in the fields of biofluid dynamics, computational neuroscience, scientific computing, and interdisciplinary research in mathematical biology.

Ottesen, Johnny T.

Department of Mathematics and Physics, Roskilde University.

He is a leading researcher in the area modelling of physiological systems, especially in modelling the cardiovascular system and its control mechanisms.

Paris, Jeff

Department of Mathematics, University of Manchester.

I have been at Manchester University man and boy since I entered as a undergraduate in Chemistry in 1963. Within a year I had changed to Mathematics and have remained in that Department since, becoming a professor in 1984. Almost all my research has been in Mathematical Logic, though in more recent times sufficiently related to interests in Philosophy to get me elected to membership of the British Academy.

I feel I know Andur well though my main contact with him was over a period of just a few months in 1990 when I enjoyed a visit to Roskilde University. Maybe it's because he is such an affable fellow that that one quickly gets to thinking one has always known him.

Rosenberg, Raben

Center for Basic Psychiatric Research, Psychiatric University Hospital in Aarhus.

The activities of the Center comprise basic psychiatric research in biological psychiatry and psychiatric epidemiology as well as clinical services including a neuropsychiatric unit. Raben Rosenberg is author of several publications (books, articles) within biological psychiatry and psychopharmacology (anxiety, affective's disorders), biophysics (Dr.MSc. thesis on membrane transports), drug abuse and philosophy of science.

Henrik R. Wulff, Stig Andur Pedersen and Raben Rosenberg started their collaboration in 1977, teaching philosophy of science and ethics to medical students at the Faculty of Health Sciences, University of Copenhagen. One of the results of the collaboration was the textbook, "Philosophy of Medicine", since translated into several languages. The teaching and the textbook was very favourable received by the students due to a fruitful collaboration of a philosopher and mathematician (Stig Andur Pedersen), the gastroenterologist Henrik R. Wulff and the psychiatrist Raben Rosenberg. The scholarship of Stig Andur both as a philosopher and a scientist deeply integrated into the natural sciences were of utmost importance for the success of the teaching and the book.

Rump, Camilla

Centre for Science Education, University of Copenhagen.

Upon graduation, I became involved in a quality development project at the Technical University of Denmark, aiming at improving student learning outcomes through qualitative test of students' understanding, designed by the teachers. Since then I have been involved in research, teaching, and developing engineering and science teaching at university level. Current research interests include induction of physics students into the physicists' community.

I have been fortunate to have been involved in several projects with Andur over the years. So far, all have aimed at studying engineer's use of knowledge in practice in order to inform engineering education. Andur's participation was crucial to the insight into human thinking we got, because Andur's unique approach, which combines both cognitive and epistemic perspectives, shaped the projects. The supervision of Frederik V. Christiansen's Ph.D.-project with Andur has likewise provided opportunities to discuss the issue of learning and reasoning at the university level and beyond. You always get wiser from discussing with Andur.

Øhrstrøm, Peter

Department of communication, Aalborg University.

I have been interested in the history and philosophy of science since the 1970s. In particular, I have studied the logic and philosophy of time. I worked as a high school teacher from 1972 to 1982, in the period 1982–85 I was employed at the University of Aarhus, and from 1986 at Aalborg University.

I have known Andur for more than 25 years. He was the external examiner at my Ph.D. defense in 1980, and he was one of the opponents at my doctoral defense in 1988. Since then we have collaborated on a number of research projects. It has always been a pleasure to work together with Andur.

Index

.

www.ingramcontent.com/pod-product-compliance
Lightning Source LLC
Chambersburg PA
CBHW060954220326
41599CB00023B/3705